POLYMER SCIENCE AND TECHNOLOGY
Volume 5A

ADVANCES IN POLYMER FRICTION AND WEAR

POLYMER SCIENCE AND TECHNOLOGY

Volume 1 • STRUCTURE AND PROPERTIES OF POLYMER FILMS
  Edited by Robert W. Lenz and Richard S. Stein • 1972

Volume 2 • WATER-SOLUBLE POLYMERS
  Edited by N. M. Bikales • 1973

Volume 3 • POLYMERS AND ECOLOGICAL PROBLEMS
  Edited by James Guillet • 1973

Volume 4 • RECENT ADVANCES IN POLYMER BLENDS, GRAFTS, AND BLOCKS
  Edited by L. H. Sperling • 1974

Volume 5 • ADVANCES IN POLYMER FRICTION AND WEAR (Parts A and B)
  Edited by Lieng-Huang Lee • 1974

A Continuation Order Plan is available for this series. A continuation order will bring delivery of each new volume immediately upon publication. Volumes are billed only upon actual shipment. For further information please contact the publisher.

POLYMER SCIENCE AND TECHNOLOGY
Volume 5A

ADVANCES IN POLYMER FRICTION AND WEAR

Edited by

Lieng-Huang Lee

Xerox Corporation
Rochester, New York

PLENUM PRESS · NEW YORK AND LONDON

Library of Congress Cataloging in Publication Data

American Chemical Society International Symposium on Advances in Polymer
   Friction and Wear, Los Angeles, 1974.
   Advances in polymer friction and wear.

   (Polymer science and technology, v. 5)
   Includes bibliographical references.
   1. Polymers and polymerization—Congresses. I. Lee, Lieng-Huang, 1924-   ed.
II. Title.
QD380.A43   1974   547'.84   74-17059
ISBN 0-306-36491-3 (v. 5a)

First half of the Proceedings of the American Chemical Society
International Symposium on Advances in Polymer Friction and Wear,
held in Los Angeles, California, April, 1974

© 1974 Plenum Press, New York
A Division of Plenum Publishing Corporation
227 West 17th Street, New York, N.Y. 10011

United Kingdom edition published by Plenum Press, London
A Division of Plenum Publishing Company, Ltd.
4a Lower John Street, London W1R 3PD, England

Printed in the United States of America

Preface

Polymers and polymer composites have been increasingly used in place of metals for various industries; namely, aerospace, automotive, bio-medical, computer, electrophotography, fiber, and rubber tire. Thus, an understanding of the interactions between polymers and between a polymer and a rigid counterface can enhance the applications of polymers under various environments. In meeting this need, polymer tribology has evolved to deal with friction, lubrication and wear of polymeric materials and to answer some of the problems related to polymer-polymer interactions or polymer-rigid body interactions.

The purpose of this first International Symposium was to introduce advances in studies of polymer friction and wear, especially in Britain and the U.S.S.R. Most earlier studies of the Fifties were stimulated by the growth of rubber tire industries. Continuous research through the Sixties has broadened the base to include other polymers such as nylon, polyolefins, and polytetrafluoroethylene, or PTFE. However, much of this work was published in engineering or physics journals and rarely in chemistry journals; presumably, the latter have always considered the work to be too applied or too irrelevant. Not until recent years have chemists started to discover words such as tribo-chemistry or mechano-chemistry and gradually become aware of an indispensable role in this field of polymer tribology. Thus, we were hoping to bring the technology up to date during this Symposium, especially to the majority of participants, polymer chemists by training.

This Symposium was sponsored by the Division of Organic Coatings and Plastics Chemistry and co-sponsored by the Division of Cellulose, Wood and Fiber Chemistry and of Colloid and Surface Chemistry of the American Chemical Society. Many scientists from Japan, the Netherlands, the United Kingdom, and the U.S.S.R. kindly agreed to participate in the Symposium. The Conference later took place at the Los Angeles Convention Center between April 1 and 4, 1974.

Most of the original manuscripts appeared in the Coatings and Plastics Preprints (Vol. 34, No. 1, 1974). This Proceedings contains all revised or expanded papers, together with discussions, the Introductory Remarks by Session Chairmen, the Symposium Address, and a communication to the Editor. Following the sequence of sessions, this book consists of eight parts:

v

1.   Mechanisms of Polymer Friction and Wear
2.   Polymer Properties and Wear
3.   Characterization and Modification of Polymer Surfaces
4.   Polymer Surface Lubrication and Solid Lubricants
5.   Polymer Properties and Wear
6.   Friction and Wear of Polymer Composites
7.   Polymer Tribology Research in the U.S.S.R.
8.   Trends in Polymer Tribology Research

The first three parts are in Volume 1; the last five in Volume 2.

For the benefit of polymer chemists, in nearly every part
there is one review paper to cover the state of the art. Most
papers present recent research results in science or technology
pertaining to the subject matter. At the end of this book, a
general discussion was devoted to the future of polymer tribology
research. Thus, in this book, we attempt to cover the past, the
present, and the future of this important branch of interdisciplinary
science.

I am grateful to Professor David Tabor, Dr. J. J. Bikerman
and Dr. David Clark for the Plenary Lectures and Professor John D.
Ferry for the Symposium Address. I would like to thank Professors
D. Dowson, G. J. L. Griffin, K. C. Ludema, D. V. Keller, Drs.
D. H. Buckley and T. L. Thourson for chairing various sessions.
I am indebted to all authors for their fine contributions.

Acknowledgement is made to the donors of The Petroleum Research
Fund, administered by the American Chemical Society, for partial
support of this Symposium. The kind response by the U.S.S.R.
Academy of Sciences in sending Dr. A. I. Sviridyonok to present
two papers is accepted with gratitude. I sincerely appreciate
the assistance of the Xerox Corporation in inviting several
speakers to visit its research laboratories prior to the Symposium.
Finally, I would like to thank Ms. Judy Lewis for typing finished
manuscripts for this book.

Lieng-Huang Lee

August, 1974

Webster, New York, U.S.A.

Contents of Volume 5A

Contents of Volume 5B

PART EIGHT   Trends in Polymer Tribology Research

ADVANCES IN POLYMER FRICTION AND WEAR

Part A

PART ONE

Mechanisms of Polymer Friction and Wear

Introductory Remarks

Lieng-Huang Lee

Joseph C. Wilson Center for Technology, Xerox Corporation

Rochester, N. Y. 14644

This Session presents five papers dealing with mechanisms of polymer friction and wear. The Plenary Lecture will be delivered by Professor David Tabor of the Cambridge University. Dr. Tabor has spent over twenty years in studying various aspects of polymer tribology. His school has been the focal point to all those interested in this subject. His paper should bring all of us up-to-date on basic principles in the mechanistic study.

After Dr. Tabor's paper, I shall survey the effects of surface energetics on polymer friction and wear. I hope that my paper will provide some insight about various interactions at the interface. I shall constantly remind you that polymers, unlike metals, are highly deformable materials. Whatever we propose, we should consider relaxation properties of polymers.

Following my presentation, Dr. Savkoor will discuss the adhesion mechanism, specifically for rubbers. Rubbers, unlike rigid polymers, have high deformation loss; therefore, the loss tangent enters both the adhesion and the deformation components of the friction force. Since rubber was the first material to attract the attention of tribologists, we seem to know a lot more about rubber friction than about polymer friction.

Mechanistic studies require the magic touch of the organic chemist. Dr. Richardson will introduce his model compound study to learn the interactions between PTFE and clean iron film. Though there is no perfect model compound, this type of study can provide new clues to the interactions at the molecular level.

The last paper of this session will be presented by Dr. Southern.

His paper deals with fracture mechanic aspects of polymer wear. In
the past, we have been interested in how fast a polymer wears.
Drs. Southern and Thomas will show us why a rubber wears. The
fracture mechanics approach should be an interesting new development
in the study of polymer wear.

Friction, Adhesion and Boundary Lubrication of Polymers

David Tabor

Physics and Chemistry of Solids, Cavendish Laboratory

University of Cambridge, Cambridge, England

This review discusses the mechanism of polymer friction in terms of adhesion and deformation processes. It suggests that the deformation term is the only certain factor that can be directly related to the bulk viscoelastic properties of the polymer. The mechanism of adhesion is then discussed in terms of electric surface charges and van der Waals forces and it is suggested that a simple monistic view is not valid. The adhesion component of friction depends on speed, temperature and contact pressure in a manner implying a close correlation with bulk viscoelastic properties. This appears to be reasonably valid for rubbers but less so for thermoplastics. Reasons for this are discussed and some of the inadequacies of current theories of polymer friction analysed. In particular it is suggested that too much emphasis has been placed on "getting the right answer" and too little on determining what actually takes place during sliding itself. Certain aspects of fluid and solid lubrication are reviewed.

In fluid lubrication the importance of wettability and the role of electrical double layers are discussed. Solid lubricants are more effective as protective films and as an example the use of monolayers of

amphipathic molecules is described.
Reference is also made to chemical
modifications of the surface and to
the behavior of polymers containing
small amounts of "dissolved" additives.
Although certain aspects of polymer
lubrication are fairly well understood,
we require a more comprehensive picture
of the physical and chemical processes
involved.

INTRODUCTION

In a survey paper prepared for a Conference where only the titles of the papers are available it would, I think, be a mistake to attempt a comprehensive review of the whole field. It seems more sensible to provide a synoptic, if somewhat eclectic description of the major ideas that have emerged during the last twenty years and to subject them to critical - but kindly - scrutiny. The approach I shall adopt is based on the simple view that, although many mechanisms are involved in the frictional process there are two which are generally of most widespread importance: - first, the ploughing of asperities of the harder solid through the surface of the softer, and secondly the shearing of adhesive bonds formed at the regions of real contact.

THE FIRST BASIC IDEA

Deformation, Ploughing, Grooving, Mechanical Work

Most workers in the field accept the view that if a polymer slides over a rough surface some contribution to the friction will arise from the deformation of the polymer by the hard asperities. There are numerous terms for this in the literature but they are all concerned with a dissipative process by deformation. Except in the case where the asperities are sharp and actually tear the surface of the polymer the strains involved are small. Thus with rubber-like materials the ploughing process produces evanescent grooves and the energy loss is directly related to the hysteresial properties of the rubber[1]. If the adhesion (see below) between the asperity and the rubber is small these losses can be calculated in terms of the viscoelastic properties of the rubber and there is good agreement between theory and experiment (Fig. 1). If there is tearing, of course, the losses will be greatly increased.

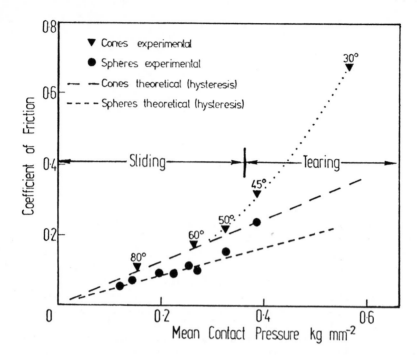

Figure 1.   Coefficient of friction as function of contact
            pressure in lubricated sliding on rubber.   Theory
            calculated on basis of deformation losses.

With thermoplastic polymers in the glassy condition such a
correlation raises an issue that will arise in later sections of
this paper.   Viscoelastic studies of polymers are usually carried
out at very small strains in the hope that the behavior will be
linear and the principle of superposition roughly valid.   If the
grooving process also involves relatively small strains one can expect
fair correlation.   Figure 2 shows an example where this occurs[2].
This does not refer to a direct grooving experiment produced by
sliding, but grooving produced by rolling a sphere over a P.T.F.E.
surface.

The calculation of the friction component due to deformation
covers a whole series of analyses ranging from the beautifully simple
picture of Dupuit[3] in 1839 to the modern sophisticated models of the
applied mathematicians (e.g. Hunter[4], in 1961).   There are more
equations than authors but in principle they all reflect a simple
physical idea: that energy is fed into the polymer ahead of the

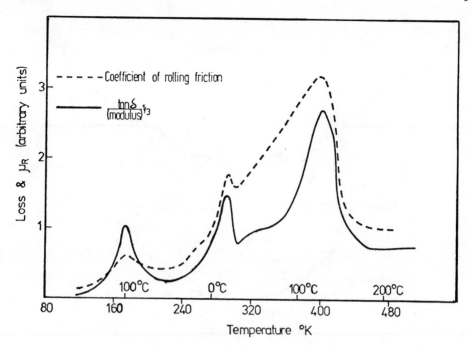

Figure 2. Comparison of grooving friction (in rolling of a
           sphere) with hysteresis loss property of P.T.F.E.

asperity, some of this is restored at the rear of the asperity and
urges it forward. The net loss of energy is related to the input
energy and the loss properties of the polymer at the particular
temperature, contact pressure, and rate of deformation of the process
(Fig. 3).

The agreement with calculation shows that the deformation term
involves the bulk properties of the polymer rather than some special
surface condition. In some situations this part of the frictional
process is the major one and in that case the maximum energy dissipation does not occur at the sliding interface but at a small depth
below the surface when the maximum shear stresses occur. This can
lead to subsurface failure[5] in P.M.M.A., to degradation of lignin
in the fibrilisation of wood pulp[6], to blistering of automobile
tires and to other types of subsurface damage.

There are two missing features in all this work. First there
is little satisfactory work on the detailed way in which material

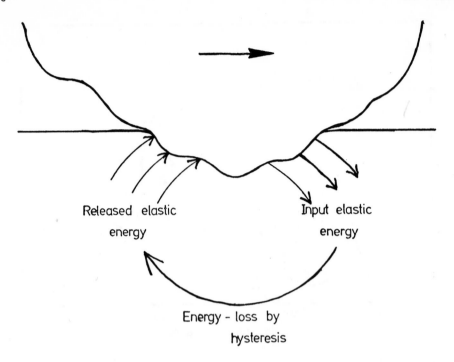

Figure 3.   Basic physical model of loss mechanism in deformation
            of polymer by a hard slider.

is deformed and piled up ahead of the asperity. This is related to
the interaction of the purely grooving process with any adhesion
that may occur at the interface. The idea originally proposed by
Bowden and Tabor[7],[8] that the grooving and adhesion term could be
simply added together remains simple minded. The physical picture
could do with a more searching analysis similar to that adopted in
slip-line-field theory of plasticity. A second factor is that at
the regions of real contact the polymer is subjected to local pressure
that, for glassy polymers, may well exceed a thousand atmospheres.
This in turn affects the viscoelastic properties of the polymer.
For example even if the strains are small a pressure of 1,000
atmospheres may raise the glass transition temperature by 10 - 20°C.
This produces corresponding changes in the modulus and loss-tangent
[9,10,11].

     A further factor is that at low temperatures (or high speeds)
the ploughing process may well lead to cracking. This can be greatly
accentuated if some adhesion occurs at the interface.

THE SECOND BASIC IDEA: ADHESION

The Forces of Adhesion

When bodies are brought into contact they will experience some
type of adhesion. Basically there is no mystery here since adhesion
at the regions of atomic contact is as natural as cohesion within
the bodies themselves and arises, broadly speaking, from the same
types of forces. Indeed the more important mystery is why, in some
circumstances, there appears to be no adhesion between contacting
surfaces.

With polymers the adhesion forces are generally regarded as
arising from two sources. One is electrostatic; if the materials
in contact have different electronic-band structures there may be a
displacement or even a flow of charge at the interface leading to
the formation of an electrically charged double-layer. Some recent
very elegant experiments by Schnabel[12], Krupp[13], Derjaguin et al.[14]
on the contact between polymeric particles and certain semiconducting
materials have shown that irradiation by uv light can change the
density of charge at the interface and hence greatly influence the
strength of adhesion. This work, greatly stimulated by the technology
of photocopying, shows that electrostatic forces are probably the
major factor in these systems. Derjaguin et al.[15] have gone further
than this. In the stripping of polymers from a rigid substrate (as
in the technological peel-test) they have observed appreciable
electrostatic charge and from calculations based on the work of
peeling have again attributed the major part of the adhesion to this
process. Since the peel-test involves appreciable loss of energy by
viscous deformation of the polymer (see the recent papers by Andrews
and Kinloch[16]) such calculations provide a weak basis for Derjaguin's
conclusions, and they have been much disputed by Skinner et al.[17],
Huntsberger[18] and others. There is however little doubt that electrostatic charges can play a part, of greatly varying importance, in
adhesion. For example von Harrach[19] has observed and measured
appreciable charge-separation in the stripping of metal films from
an insulator and Weaver[20] the effect of an electrostatic field on
polymer adhesion. More recently Davies[21] has observed charge
separation after loading and unloading a metal sphere on a rubber
surface. On the other hand, as we shall see in what follows the
adhesion of rubber to rubber can be quite strong although it is
hard to see how an electric double layer can be formed in such a
case. Derjaguin's emphasis on electrical charges constitutes a
monistic view that can be misleading.

The other source of adhesion with polymers arises from van der
Waals forces and, if there are certain polar atoms present, from
hydrogen bonding. We shall deal in what follows only with the former.
Consider for example the contact between a smooth rubber sphere and
a hard smooth flat surface. If the spheres and flat are brought

into contact with zero load and if there are no attractive forces
between them the area of contact, according to the classical laws
of elasticity, will be zero (Figure 4a).   However because the

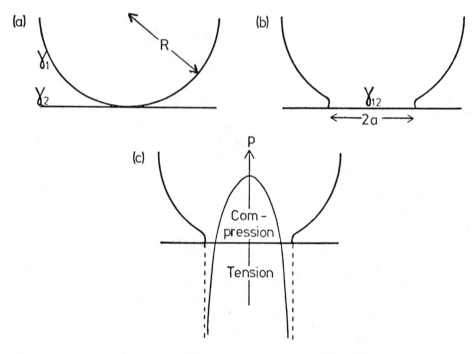

Figure 4.   Adhesion between elastic sphere and hard flat
surface as result of surface forces.

bodies attract one another the surfaces are drawn together as shown
in Figure 4b to make contact over a circle of radius a.   In principle
the area of contact could be calculated from the van der Waals forces
across the interface.   It is however much simpler to carry out the
calculation in terms of energies.   If the surface energy of the
rubber sphere is γ_1, of the other surface γ_2 and of the interface
γ_{12}, the area of contact πa^2 implies a destruction of surface
energy of amount πa^2 $(\gamma_1 + \gamma_2 - \gamma_{12})$ and this must be equal to the
stored elastic energy in the system.   The pressure distribution is
shown in Figure 4c in which it is seen that the central zone is
under compression, the edges under tension.   Since the applied force
is zero the integrated compressive force must equal the integrated
tensile force.   If now we wish to pull the spheres apart the smallest
force will begin to produce separation at the periphery of the contact

region (where the forces are already tensile): the separating force
will rapidly increase until a critical value Z is reached at which
the rate of release of stored elastic energy just exceeds the rate
of increase of surface energy arising from creation of free surface
at the interface. The surfaces will then pull apart. In a very
elegant analysis Johnson, Kendall and Roberts[22] have shown that for
the geometry of two spheres of radius R_1 and R_2 the elastic moduli
disappear from the final expression and Z has the value.

$$Z = \frac{3\pi R_1 R_2}{2(R_1 + R_2)} (\gamma_1 + \gamma_2 - \gamma_{12}) \qquad (1)$$

We may consider the results they obtained for experiments carried
out with a sphere of rubber (R = 2.2 cm) on a flat surface of identi-
cal rubber ($R_2 = \infty$). We assume that $\gamma_1 = \gamma_2 = \gamma$ and that the interface has virtually zero surface energy ($\gamma_{12} = 0$). Then

$$Z = 3\pi R \gamma \qquad (2)$$

The pull-off force was found to be 750 dynes which gives a value
of γ = 34 erg cm^{-2} (34 mJm^{-2}). This is a very reasonable value for
rubber. A curve showing the way in which the contact area varies as
a normal load is first applied and then reduced is shown in Figure 5.
The area of true contact is always increased by the van der Waals
forces around the contacting regions: the effect is small when the
normal load is large, but can be relatively large as the normal load
is reduced.

These ideas may be applied to extended surfaces. Consider for
simplicity a smooth flat rubber surface containing a few asperities
placed gently in contact with a smooth hard surface (Fig 6a). Contact
first occurs at the tips of the asperities: these are then drawn
down to give a finite contact area as indicated above (Fig. 6b).
If the surfaces between the asperity contacts are separated by a gap
of order 50 - 100Å, the van der Waals force across the gap will
exert an attractive force equivalent to a pressure of about 10^6 dynes
cm^{-2} (i.e. one atmosphere). The rubber will deflect most at the
centre of the gap and may make contact with the counterface. If
this occurs the surfaces will pull together and practically the
whole of the interface will come into molecular contact (Fig. 6c).

This is probably the reason for the strong adhesion observed
with relatively smooth soft rubber-like materials. A similar mechanism also probably applies to the adhesion between thin films of
polymer such as "cling film wrap". On the other hand if the materials
are hard and/or rough the area over which molecular contact occurs
may be very greatly reduced and the adhesion correspondingly
diminished.

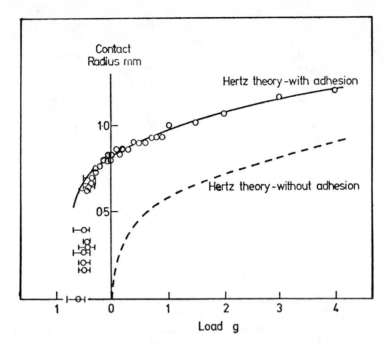

Figure 5.   Contact between elastic sphere and flat. As a result
            of surface forces the contact is larger than the
            Hertzian value and is finite for negative loads.

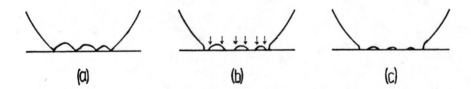

Figure 6.   Schematic diagram indicating how the contact area
            between rubber and a hard surface can increase to
            a large fraction of the geometric area.

    These concepts refer to an ideal elastic solid. Since rubbers
and polymers are viscoelastic materials, their deformation properties
are not reversible. In pulling surfaces apart some deformation
energy will also be expended and the observed work of separation

will be greater than that arising from the interfacial forces
themselves. The adhesion will, in fact, appear to increase with
the hysteresis-loss property of the polymer as the recent work of
Andrews and Kinloch[16] has shown.

Adhesion in the Friction of Rubber-Like Polymers

We turn to the part that adhesion plays in the friction of
polymers. It now becomes desirable to treat rubber-like materials
under one heading and polymers below their glass transition temperature under another.

If rubber adheres to another surface, work must be expended in
order to produce sliding since the adhesions at the interface must
be broken. The formation and breaking of adhesive bonds is not, of
course, a conservative energy system: between the making and breaking of bonds energy is dissipated mainly as vibrations within the
solids, that is as heat. The detailed mechanism of the energy
dissipation process is, indeed, a major problem in all theories of
rubber friction.

The most striking experimental results are those due to Grosch[23].
He measured the friction of rubber sliding over a glass surface
(prossessing a slightly irregular topography) over a wide range of
speeds and temperature. He did not, however, exceed a sliding speed
of about 1 cm sec^{-1} so as to avoid complications arising from
frictional heating. He then found that each friction-velocity curve
could be displaced on the velocity axis by an amount determined by
the temperature and the known viscoelastic properties of the rubber
using the WLF transform: the result gave a single master-curve as
shown in Figure 7.

Similar results, perhaps not so crisp, have been obtained by
other workers (e.g. Ludema and Tabor[2]). The maximum friction occurred,
at room temperature at a sliding speed of approximately 1 cm. sec.$^{-1}$.
In separate oscillatory tension-compression experiments Grosch found
that, for the same rubber, the viscoelastic loss reached a maximum
at room temperature at a frequency of about 10^6 Hertz. If the
friction and viscoelastic processes are due to the same mechanism a
sliding speed V of 1 cm sec^{-1} must correspond to a frequency ν of
10^6 Hertz. The connecting parameter is a length λ where $\lambda\nu = V$.
The value of λ is thus about 10^{-6} cm or 100Å. This corresponds to the
length of the segments involved in the deformation of rubber. A
segmental length of the same order is obtained in dielectric-loss
measurements on rubber[24]. The results thus suggest that there is
a dissipative process involving the attachment, straining and
detachment of rubber segments about 100Å long at the sliding interface.

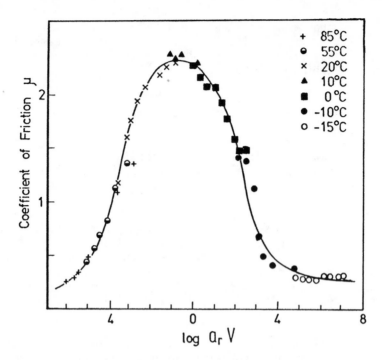

Figure 7.   Data from Grosch showing that the friction of rubber
            at various speeds and temperatures can be treated by
            the WLF transform to lie on a single "master" curve.

Theories of Rubber Friction

There are two very interesting and critical reviews of rubber
friction. The first by Schallamach[25] is brief, closely argued and
aimed at the specialist: the second by Bartenev[26] (in Russian) is
of a more elementary nature. There is also a useful book in English
by Moore[27a] but this is in the nature of an expository student text-
book rather than a critical review. A somewhat more critical version
is by Moore and Geyer[27b] which attempts to show that all theories
include a tan δ term.

The adhesion theories of rubber friction fall into two main
groups - molecular and macroscopic. The molecular theories are
associated with the names of Schallamach[28], Bartenev[29,30], and
Rieger[31], and have much in common. The basic idea is that bonds are
formed at the interface, strained and then broken, and one form or
another of Eyring's rate-process theory is applied. In effect one
finds that the theories end with two main factors multiplied together,

one of which decreases with frequency, the other of which increases
with frequency. The product gives a maximum at some specified
frequency (or sliding speed). With Schallamach's theory the bond
strength increases with the speed but the number of bonds formed
decreases with sliding speed. Bartenev[29] treated the rubber rather
like a viscous liquid so that the friction would increase monotonically
with sliding speed and later introduced the idea of a decrease in
the area of true contact with increasing speed (or decreasing
temperature). This again gives a maximum.

    The macroscopic theories of friction are associated with the
names of Bulgin et al.[32] Ludema and Tabor[2] and in a slightly different
way by Savkoor[33] while Kummer[34] has attempted to combine both
molecular and macroscopic approaches into a unified theory. In common
with Schallamach[25] I find it difficult to grasp the details of his
model. Bulgin et al. assume that rubber elements adhere to the
counter surface and are elongated during sliding until a critical
force is reached: at this stage the bond breaks. The element then
snaps back and returns part of the stored energy to the specimen,
the recovered energy being proportional to the resilience of the
rubber. The energy lost is proportional to the loss tangent of the
rubber and inversely proportional to the "hardness" of the rubber.
The results are in good agreement with observation but the model
does not at all make clear how any energy can be restored to the
rubber when the element snaps back. An even simpler analysis is
that due to Ludema and Tabor[2]. They assume that the adhesion is
always stronger than the rubber and that tearing takes place in a
layer about 100Å from the interface. They use Smith's[35] classical
measurements of the tear-strength of rubber as a function of temperature and tear-rate to deduce the tear strength of the interface:
they use the variation of modulus with rate of deformation to deduce
the area of contact. The product gives the shear strength of the
interface as a function of speed which is in reasonable agreement
with observation. Since they use Smith's tear-strength data they
are not required to answer the problem of energy dissipation.

    These theories of rubber friction have at least two major
defects in common. First they ignore the nature of the counterface:
the friction is apparently determined solely by the properties of
the rubber. But is the friction of rubber on glass the same as
rubber on P.T.F.E.? Secondly they do not consider whether sliding
takes place truly at the interface or within the rubber itself.
There is some evidence that in Grosch's experiments the low speed
sliding experiments gave no detectible transfer of rubber on glass
(i.e. sliding occurred at the interface): whereas at the higher
speeds transfer occurred implying that the rubber itself was being
sheared. Nevertheless the results lie on a single curve as though
only a single process was involved. Even within the confines of a
classical molecular model one may pose two questions not often raised
by its protagonists. First, if large strains occur in the surface

layers ought we to expect the viscoelastic properties of the rubber
to be similar to those deduced from measurements carried out at
relatively small strains. The answer appears to be (Thirion, private
communication) that the viscoelastic properties of rubber remain
essentially unchanged for strains of over 50%. What happens however
if the strains are appreciably higher than this? Under these
conditions the molecules very close to the interface may become very
strongly oriented parallel to the direction of sliding. Such semi-
crystalline material will have visco-elastic properties very different
from those of the bulk rubber. Bartenev and Schallamach have clearly
been aware of this complication.

Perhaps the major weakness of all these theories is that they
are too much concerned with "getting the right answer" and too little
concerned with what actually happens at the interface. This is
brought out most clearly by Schallamach's recent paper[36] in which
he described an optical study of the contact between a rubber hemispherical slider and a transparent perspex flat. He shows that
adhesion between the surfaces causes buckling of the rubber and
generates "waves of detachment" which traverse the contact area at
a high speed from front to rear. The energy dissipation process
must be very different from that proposed in most of the past theories.
These waves of detachment are observed with hard, as well as with
soft rubbers (Courtel[37]), but whether they always occur and under
what conditions is not yet established. Evidently we need a new
look, literally, at the problem.

In the more practical range of high speeds and loads other
factors become of increasing importance, particularly high temperatures and the possibility of thermal and mechanical degradation of
the rubber.

Adhesion in the Friction of Polymers

In the sliding of a polymer hemisphere over a clean hard smooth
surface there is often transfer of the polymer to the counter-surface.
It is natural therefore to attribute the friction to the shear-
strength of the polymer itself. In that case we might expect the
friction to depend on speed and temperature in a way that reflects
the strain-rate and temperature dependence of the strength properties
of the polymer. Broadly speaking such a correlation exists. Polymers
which show little dependence of strength properties on temperature
and deformation rate (e.g. cross-linked polymers like Bakelite) show
little dependence of friction on temperature and sliding speed. By
contrast thermoplastics show a much more marked dependence of friction
on sliding conditions. The detailed correlation is not however very
satisfactory.

By analogy with the behavior of rubber some workers have suggested

that the friction-velocity-temperature curves should be transformable
(by a suitable series of shifts) into a single master curve. In the
glassy range, of course, the shift would not be that given by the
W.L.F. transform but by an appropriate Arrhenius factor. This factor
may be determined from viscoelastic measurements on the polymer and
over a very limited range a master-curve can be constructed using
reasonable values of the Arrhenius shift. The good agreement is,
however, illusory. Frictional measurements carried out over an
appreciable range of temperatures and speeds cannot be fitted into
a single master curve based on viscoelastic measurements[2]. There
are at least two reasons for this.

First the individual regions of contact are under contact
pressures of the order of 1 Kbar: this will raise the temperature
of most relaxation processes by temperatures of the order of 20°C.
A recent paper by Vinogradov et al.[38] discusses this, though in a
rather indirect way. Secondly the viscoelastic loss measurements
are generally carried out at very small strains so as to lie in the
"linear-viscosity" range. By contrast the material at the interface
is sheared to failure. Quite different strength properties are
therefore to be expected. Some attempts to correlate friction with
shear strength properties measured under comparable conditions have
been described by Bahadur and Ludema[39a] but further work in this area
is required.

The Effect of Pressures

We have already seen that the effect of pressure is to raise
the temperature of various visco-elastic transitions in the polymer.
It is also clear that in many cases the polymer at the sliding interface is subjected to enormous shear strains so that, in effect, the
frictional process involves the shear strength of the polymer rather
than some "small-strain" viscoelastic property. We may, therefore,
ask how far the shear strength of a polymer is determined by contact
pressure and other variables. Bahadur[39b] has discussed some of the
data which he has obtained on the effect of strain-rate and
temperature. Here we describe the effect of pressure. Thin films of
polymer were deposited on a smooth glass surface and a hemi-spherical
glass slider slid over it[40]. The glass surfaces were "fired" to be
as smooth as possible. These surfaces were carefully examined after
sliding and no damage was observed. Consequently it may be assumed
that sliding took place within the film material or at the interfaces
but that no glass contact had occurred. The contact pressure could
be deduced from Hertz's equations assuming elastic deformation of
the glass. By using sliders covering a wide range of curvatures and
applying a wide range of loads the contact pressure could be varied
from about 10^7 Nm^{-2} to 3.10^{10} Nm^{-2} (1to 300 Kg mm^{-2}). Some typical
results are shown in Figure 8 for a number of polymers (similar

results have been obtained by Towle[41] and Bowers[42]). It is seen
that although the details vary from one material to another, for a
wide variety of materials the shear strength s increases with contact
pressure p in a manner that approximates to

$$s = s_o + \alpha p \qquad\qquad (3)$$

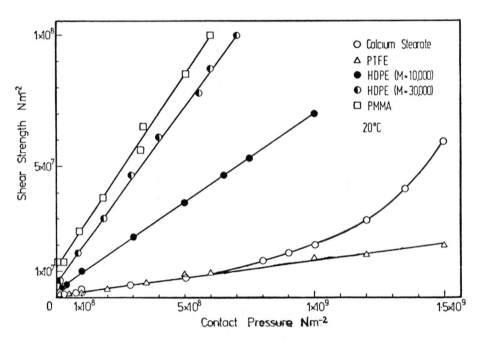

Figure 8.   Shear strength of a number of organic materials
as a function of mean contact pressure.

The shear strength does not change rapidly with sliding speed
but it is markedly temperature decendent. Further as is apparent
in Figure 9 the temperature dependence varies according to the
chemical nature of the polymer. Can these results be applied to
the friction of polymers? if we have a well-defined geometry, say
a smooth polymer hemisphere sliding over a smooth flat surface, we
may determine the geometric area of contact and assume it to be
approximately the same as the molecular area of contact. We may
then deduce the mean contact pressure. From the frictional force F
we may deduce the mean interfacial shear strength s. Further as
Adams[43] showed many years ago we may increase the contact pressure
between the polymer and the flat simply by increasing the load;
we then find that s increases with contact pressure though Adams
preferred a law of the type $s = s_o^{\alpha p}$. How does this compare with

the shear strength of a thin film of polymer trapped between hard
surfaces? It agrees reasonably well in magnitude. There are however
certain difficulties. Recently we have measured the friction between
a hard slider and a polymer flat in a chamber which could be subjected
to a high hydrostatic pressure using nitrogen gas as the pressure
medium[44]. The results show that a pressure of 1600 atmospheres
(16 Kg mm^{-2}) generally produces a drop in friction by a factor of
about 2. The behavior is far too large to be attributed to a
decrease in the ploughing term. The behavior appears to correlate
with the dependence of loss factor (tan δ) on pressure. But in view
of the strictures levelled in this review at other workers this
cannot be pressed any further. More work is required to clarify
the problem.

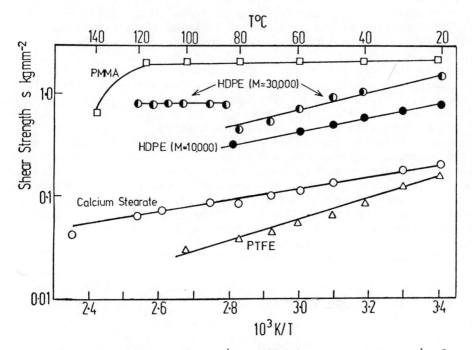

Figure 9. Shear strength (at a fixed contact pressure) of a
number of organic materials as a function of
temperature.

Incidentally the presence of a shear strength $s = s_o + \alpha p$
suggests that the frictional force may finally appear in the form

$$F = \beta + \mu W \qquad (4)$$

This resembles the binomial relation suggested by Coulomb and much favored by some workers. It has been used in many misleading connotations. In a detailed discussion of this it has been suggested (Tabor[45]) that if multinomial equations are to be favored in friction there are much better reasons for using a quinque-nomial (penta-nomial, 5-term) equation rather than a binomial.

Effect of Speed and Temperature

If sliding speeds are limited to an upper value of perhaps 1 cm s^{-1} the effect of speed on the frictional behavior should merely reflect the effect of rate of deformation on the properties of the polymer-counterface combination. If the interfacial adhesion is strong, the dependence refers to the properties of the polymer itself. If the adhesion is weak the dependence refers to the adhesion-dehesion processes at the interface. In principle there is no difficulty here. The same applies to the effect of bulk heating provided the temperature is not sufficient to produce irreversible changes in the materials. For higher speeds, however, and most practical systems will generally operate in such a range, the effect of speed will necessarily be very complicated. The adhesion component will produce heating at, or very close to the interface: the deform-ation component will produce subsurface heating. The temperature-rise due to both processes will modify the rheological properties of the polymer. In addition very high speeds may involve high shear-strains in the surface layers and possibly drastic modifica-tions in the structure. Thus the overall interaction of high speeds and frictional heating will affect the frictional behavior in a manner that will be extremely difficult to calculate. The situation will be further complicated if the temperature rise produces irreversible chemical changes.

We have already mentioned the role of sub-surface heating in well-lubricated systems where the major source of friction is the deformation term. Clearly speed, in such cases, is a very important factor.

The Interfacial Layer

The material at the interface may be in a very different condi-tion from the polymer in bulk even if no major chemical changes occur. For example with rubber, the surface layers may be so highly strained as to be virtually crystalline. With long chain polymers such as PTFE or polythene the surface layers may consist of molecules drawn out in the direction of motion so that sliding actually occurs over very highly oriented material[45]. With other polymers the material will be torn, sheared and heavily distorted. In some cases chain

rupture may occur. In addition, particularly if appreciable temperatures are generated, there may be oxidation, exudation of low
molecular weight homologues even carbonisation. All these physical
and chemical changes are produced by the sliding process itself.

Kragelskii, with his passion for classificatory terminology has
referred to this as the "third body" in the frictional process[47].
With this term he wishes to emphasize that the material involved in
the frictional process may have physical and chemical properties
very different from those of the parent material[47]. The point is
well made. It also implies that simple minded theories of polymer
friction may end up by describing processes occurring at the interface in a material which no longer exists there.

BOUNDARY LUBRICATION

Liquid Films and the Electrically Charged Double-Layer

The purpose of a lubricant is to provide a film that will prevent
solid-solid contact and that is, itself, easily sheared. The most
desirable lubricant is clearly a fluid since, unlike a solid, it can
undergo repeated shear indefinitely. To maintain a fluid film the
geometry of the surfaces, the speed and the viscosity of the fluid
must be such that a hydrodynamic pressure is developed in the fluid
sufficient to keep the surfaces apart. This is termed hydrodynamic
lubrication and in principle there is no contact between and no
wear of the moving surfaces. In practice the surfaces may not have
the required geometry, the speeds and/or viscosity may be too low
so that the pressure developed in the fluid is not sufficient to
keep the surfaces apart. Some contact will occur, at first on the
most pronounced surface asperities.

A possible way of achieving lubrication under these conditions
is to provide a pressure within the fluid by a completely different
mechanism. One method used in large machinery is by externally
pressurizing the system. Another is by forming an electrically
charged double-layer of similar sign on the surfaces. This is the
classical stablising force in colloidal systems. Such films, of
the order of 100Å thick, can successfully support a pressure of the
order of about 1 atmosphere. This is far less than the pressure
developed between metal surfaces at the tips of contacting asperities
but it is of the order of magnitude of the pressure developed between
soft rubbers (Roberts and Tabor[48]). Thus electrically charged double-
layers can in principle, be used to lubricate soft rubbers and
polymers. Roberts[49] even suggests that such a mechanism may operate
in synovial fluid in the lubrication of human joints. Although there
is direct evidence for the formation of such stable layers under
static conditions or during sliding at relatively low shear rates,
we do not know if under more practical conditions the charges are

disturbed and the load carrying capacity lost.

Amphipathic Organic Molecules

The next stage, by analogy with metals, is the lubrication achieved by thin films of amphipathic molecules such as stearic acid, stearamide, oleamide, etc. These materials may be adsorbed on the surfaces from solution or deposited directly as dry films[50]. They are not as effective on polymers as they are on metals partly because they do not adsorb so well[51]. Recently Senior and West[52] have shown that by suitably treating the polymer surface it can be made to provide more adsorption sites and in that case fairly effective boundary lubrication can be achieved. Of course if the chemically modified surface is worn away the effect is lost. The shear properties of the boundary lubricant are also important. For example in the experiments carried out on glass surfaces (see above[40]) the shear strength of oleamide was found to be about one half of that of stearamide at the same contact pressure. A similar difference is found in the lubrication of a material such as polythene; the friction is considerably lower with oleamide than with stearamide.

These surface films, even if effective, are gradually worn away by repeated traversals. One way of overcoming this is to incorporate the amide, as a fine dispersion, in the polymer itself. The amide diffuses to the surface and may be replenished as it is worn away. This phenomenon applies to a wide range of low molecular-weight additives (Bowers, Jarvis and Zisman[53], Cohen and Tabor[54], Briscoe, Mustafaev and Tabor[55]). Thin films of polymer containing such additives can be deposited on metal surfaces from an organic solvent. Under suitable conditions, such films can then provide effective lubrication of the metal in certain metal-working operations.

Molecular Interactions

In some cases very small molecules can act as lubricants. For example water is absorbed by nylon and plasticises the surface layers. So long as penetration of water is limited, the soft surface layer supported by the harder unplasticised substrate provides a low-friction system[53]. This type of molecular interaction is related to two other aspects of polymer lubrication. First, at the molecular level of boundary lubrication wettability of the polymer by the lubricant may be of considerable importance. For example water does not wet rubber and does not provide effective boundary lubrication. However a water-alcohol mixture with a surface tension of 33 mJm^{-2} wets the rubber and, under boundary conditions, produces an appreciable reduction in friction. Secondly, in the choice of polymer lubricants there is the general problem of selecting fluids and additives that will not accelerate stress cracking of the polymer, cause excessive softening or produce chemical degradation.

Under practical conditions of speed and load chemical reactions may readily occur, partly as a result of the presence of highly deformed (hence more reactive) material, partly as a result of frictional heating. With metals, for example, hydrocarbon lubricants are cracked and converted into organic peroxides and the subsequent behavior is governed by interaction between the metal and the hydrocarbon oxidation products (Vinogradov et al.[56]). In general hydrocarbon oils are not used with polymers but recently Koutkov and Tabor[57] have shown that similar reactions may occur between silicones and metal-polymer sliding combinations.

The Skid Resistance of Automobile Tyres

Finally we may revert to an early piece of work (Greenwood and Tabor[1]) describing the lubrication of rubber with films of soap (sodium and calcium stearate). If a hard sphere is slid over such a surface the shear resistance of the lubricant film may be so small that the friction is dominated by the deformation losses in the rubber. In that case as is shown in Figure 1, the sliding lubricated friction reflects the hysteretic or viscoelastic properties of the rubber. Since the strains are relatively small and the interfacial adhesion very weak, the simple quasi-static model of hysteresis losses applies well. Over an appreciable range of conditions the lubricated friction can be expressed in terms of the rate of deformation (or sliding speed), the temperature and the bulk hysteresis properties of the rubber. This has led to the use of high-loss rubbers on the tread of automobile tyres as a means of increasing their skid-resistance on wet or greasy road surfaces.

CONCLUSIONS

In this review I have discussed the general mechanisms of polymer friction primarily in terms of adhesion and deformation mechanisms. The deformation term is first discussed and the evidence suggests that this is the only certain factor that can be directly related to the bulk viscoelastic properties of the polymer. The mechanism of adhesion is then discussed in terms of electric surface charges and van der Waals forces and it is suggested that a simple monistic view is not valid.

With unlubricated surfaces there is general agreement with the view that the adhesion component is responsible for the major part of the friction of polymers. The adhesion component is not, of course, a constant but depends on speed, temperature and contact pressure in a manner suggesting a close correlation with bulk viscoelastic properties. This appears to be reasonably valid for rubbers but not so for thermoplastics. Reasons for this have been discussed and some of the inadequacies of current theories of polymer friction analysed. In particular it is suggested that too much emphasis has been placed on "getting the right answer" and too little on determining what actually takes place at the sliding interface. More attention needs to be paid to waves of detachment in the sliding of rubber, to the effects of contact pressure and large interfacial shear strains on the mechanical properties of the polymer: in addition a better understanding is required of the mechanical and chemical breakdown of the polymer produced by the sliding process itself.

The review also deals with boundary lubrication in terms of decreasing lubricant film thickness from the electrically charged double layer (ca 100Å thick), the monomolecular layer of amphipathic molecules (20-40Å thick) to the very thin adsorbed layers of small molecules that may be only 5-10Å thick. Reference is also made to chemical modifications of the surface and to the behavior of polymers containing a small amount of dissolved additives. It is evident that, in the boundary lubrication of rubbers and polymers, certain aspects are fairly well understood. On the other hand our knowledge is rather fragmentary. We need a more comprehensive picture of the physical and chemical reactions of lubricants with polymer surfaces, the properties of the surface films formed, and their mechanical and thermal stability as a function of speed, load and temperature.

ACKNOWLEDGEMENTS

I wish to express my thanks to Dr. B. J. Briscoe for critical and constructive comments.

REFERENCES

1. J. A. Greenwood and D. Tabor, Proc. Phys. Soc. 71, 989 (1958).
2. K. C. Ludema and D. Tabor, Wear, 9, 329 (1966).
3. A. J. E. J. Dupuit, Comptes Rendus, 9, 698: 775 (1839).
4. S. C. Hunter, J. Appl. Mech., 28, 611 (1961).
5. J. F. Archard, private communication.
6. D. Atack and W. D. May, Pulp and Paper Magazine of Canada, Conf. Issue (1958).
7. F. P. Bowden, A. J. W. Moore and D. Tabor, J. Appl. Phys., 14, 80 (1943).
8. F. P. Bowden and D. Tabor, The Friction and Lubrication of Solids, Clarendon Press Oxford, Vol. 1, 1950, Vol II, 1964.
9. A. Zosel, Kolloid Z. 199, 113 (1964).
10. P. R. Billinghurst and D. Tabor, Polymer, 12, 101 (1971).
11. E. Jones-Parry and D. Tabor, Polymer, 14, 617 (1973).
12. E. Schnabel, Lichtmodulierte Electrostatische Doppelschichthaftung, Dissertation, Karlsvuhe (1969).

13. H. Krupp, J. Adhesion, $\underline{4}$, 83 (1972).
14. B. V. Derjaguin, J. P. Toporov, I. N. Aleinikova and L. N.
     Burta-Gaponovitch, J. Adhesion, $\underline{4}$, 65 (1972).
15. B. V. Derjaguin, Research (London), $\underline{8}$, 70 (1955).
16. E. H. Andrews and A. J. Kinloch, Proc. Roy. Soc. (London)
     A $\underline{332}$, 385: ibid. 401, (1973).
17. S. M. Skinner, R. L. Savage and J. E. Rutzler, J. Appl. Phys.
     $\underline{24}$, 438 (1953).
18. J. R. Huntsberger, Treatise on Adhesion and Adhesives, Ed.
     R. Patrick, Vol. 1, p 119, Marcel Dekker Inc., New York (1967).
19. H. von Harrach and B. N. Chapman, Proc. Int. Conf. Thin Films,
     Venice, Vol. II, 157, (1972).
20. C. Weaver, Faraday Soc. Special Discussion, Adhesion of Metals
     to Polymers, Paper 2, Nottingham, U.K., (1972).
21. D. K. Davies, J. Phys. D. (Appl. Phys) $\underline{6}$, 1017, (1973).
22. K. L. Johnson, K. Kendall and A. D. Roberts, Proc. Roy. Soc.
     (London) A $\underline{324}$, 301 (1971).
23. K. A. Grosch, Proc. Roy. Soc. (London) A $\underline{274}$, 21 (1963).
24. A. Schallamach, private communication.
25. A. Schallamach, Rubber Chem. Techn., $\underline{41}$, 209 (1968).
26. G. M. Bartenev, Treniye i Iznashivaniye Polimerov, (1973).
27a. D. F. Moore, The Friction and Lubrication of Elastomers,
     Pergamon Press, Oxford (1972).
27b. D. F. Moore and W. Geyer, Wear, $\underline{22}$, 113 (1973).
28. A. Schallamach, Wear, $\underline{6}$, 375 (1963).
29. G. M. Bartenev, Doklad Akad. Nauk SSSR, $\underline{96}$, 1161 (1954).
30. G. M. Bartenev and A. I. El'kin, Wear, $\underline{8}$, 8 (1965).
31. H. Rieger, Kaut. Gummi Kunstst. $\underline{20}$, 293, (1967).
32. D. Bulgin, G. D. Hubbard and Walters, M. H. Proc. Rubber Tech.
     Conf. 4th, Inst. Rubber Industry, London, p 173 (1962).
33. A. R. Savkoor, Wear, $\underline{8}$, 222 (1965).
34. H. W. Kummer, Unified Theory of Rubber and Tire Friction,
     Pennsylvania State University Engineering Res. Bulletin B-94
     (1966).
35. T. L. Smith J. Polymer Sci., $\underline{32}$, 99 (1958).
36. A. Schallamach, Wear, $\underline{17}$ 301 (1971).
37. R. Courtel, Comptes Rendus, Acad. Sci., $\underline{277}$, A479 (1973).
38. G. V. Vinogradov, A. I. Yel'kin, G. M. Bartenev and S. Z. Bubman,
     Wear, $\underline{23}$, 33 (1973).
39a. S. Bahadur and K. C. Ludema, Wear, $\underline{18}$, 109 (1971).
39b. S. Bahadur and K. C. Ludema, J. Appl. Polymer Sci., $\underline{16}$, 361 (1972).
40. B. J. Briscoe, B. Scruton and F. R. Willis, Proc. Roy. Soc.
     (London) A $\underline{333}$, 99 (1973).
41. L. C. Towle, J. Appl. Phys., $\underline{42}$, 2348 (1971).
42. R. C. Bowers, J. Appl. Phys., $\underline{42}$, 4961 (1971).
43. N. Adams, J. Appl. Polymer Sci., $\underline{7}$, 2075 (1963).
44. B. Briscoe, E. Jones-Parry and D. Tabor. In Press (1973).
45. D. Tabor, Chapter in book on the Physics of Surfaces, edited
     by J. Blakely, in press.

46. C. M. Pooley and D. Tabor, Proc. Roy. Soc. (London) A 329, 251 (1972).
47. I. V. Kragelskii, The Nature of the Friction of Solids Conf., Minsk, 1970. p. 262, Nauka i Technika, Minsk (1971). In Russian.
48. A. D. Roberts and D. Tabor, Proc. Roy. Soc. (London) A 325, 323 (1971).
49. A. D. Roberts, Nature, 231, 434 (1971).
50. T. Fort Jr., J. Phys. Chem., 66, 1136 (1962).
51. R. C. Bowers, W. C. Clinton and J. A. Zisman, Ind. Eng. Chem. 46, 2416 (1954).
52. J. M. Senior and G. W. West, Wear, 18, 311 (1971).
53. R. C. Bowers, M. L. Jarvis and W. A. Zisman, Ind. Eng. Chem. (Res. and Development) 4, 86 (1965).
54. S. C. Cohen and D. Tabor, Proc. Roy. Soc. (London) A 291, 186 (1966).
55. B. J. Briscoe, V. Mustafaev and D. Tabor, Wear, 19, 389 (1972).
56. G. V. Vinogradov, N. S. Nametkin and M. I. Nossov, Wear, 8. 93, (1965).
57. A. A. Koutkov and D. Tabor, Tribology, 3, 163 (1970).

DISCUSSION OF PAPER BY D. TABOR

L. H. Lee (Xerox Corporation): As the Chairman of the Symposium, I would like to thank Dr. Tabor for this excellent lecture on polymer tribology. During the past twenty years, Drs. Tabor and Bowden, from the Cambridge University, have published important papers in elucidating mechanisms of polymer friction and wear. This plenary lecture, in part, is the manifestation of the balance between theoretical and experimental approaches taken by Dr. Tabor and his colleagues.

R. R. Myers (Kent State University): In dealing with lubricating type of experiments, aren't you always going to be confronted with the wedge of liquid which gets in between the sliding surface and the substrate? In that case, do you not have hydrodynamic lubrication when sliding a stylus over a lubricated surface? If so, the viscoelastic properties of the lubricant are important considerations.

D. Tabor: Yes! Sliding over a lubricant film will always involve viscous or visco-elastic or even plastic work. This must be added to the deformation losses produced by the stylus in the substrate. In reference 1 of the review paper it is shown that at very small loads the losses in the lubricant may become appreciable compared with those in the substrate. At higher loads the position is reversed.

K. L. Mittal (I.B.M. Corp.): Would you comment on a very basic question: What is the minimum thickness of adsorbed films required to mask the effect of the substrates?

D. Tabor: At its simplest we may answer this question as follows. A monolayer of adsorbed molecules is sufficient to mask the surface forces emanating from contacting substrates. Direct measurement of van der Waals forces (1), adhesion measurements in high vacuum (2) and contact angle measurements (3) illustrate this point. A much thicker layer is required to prevent mechanical interaction between the contacting surfaces. In the case of a sphere on a flat the film thickness must be rather greater than the diameter of the circular contact region (4) in order to supress the interaction of the substrates.

1. J. Israelachvilli and D. Tabor, Proc. Roy. Soc. London A331, p. 19, (1972).
2. See, for example, B. J. Briscoe and D. Tabor, Faraday Spec. Disc. of Chem. Soc. 2, p. 7, (1972).
3. Contact Angle Wettability and Adhesion, Advances in Chemistry Series, 43, American Chemical Society, (1964).

4. G. M. Hamilton, L. E. Goodman, Transactions of ASME, Journal
    of Applied Mechanics, **p.** 371, June (1966).

K. Tanaka (Kanazawa University, Japan):   What is the relationship
between the shearing strength of a lubricant film and the viscous
resistance of the film?   I have the impression that physicists do
not use the term "shearing strength" when referring to a viscous
liquid.

D. Tabor: When we talk of the shearing strength of a lubricant
film, we are thinking of the film as a solid.   The shearing strength
is thus a strength property resembling, say, the critical shear
stress of a metal.   It may depend on shear rate, but over a fairly
wide range of conditions it is a material constant.   If, however,
we think of the lubricant film as being a liquid, the stress required
to produce shear depends critically on the shear rate.   Over a wide
range of conditions, the shear stress will indeed be proportional
to the shear rate.   Consequently, these two terms are not really
interchangeable.

Effect of Surface Energetics on Polymer Friction and Wear

Lieng-Huang Lee

Joseph C. Wilson Center for Technology, Xerox Corporation

800 Phillips Road, Webster, New York 14580

Friction and wear involve solid-to-solid contacts governed by Van der Waals and electrostatic interactions on the surface of a friction pair. Thus, surface energetics could play a major role in determining polymer friction, if deformation of bulk were not accompanying the contact phenomena. This paper discusses both the relevance and irrelevance of surface energetics to polymer friction.

Polymer wear is not well understood and may be divided into abrasion, fatigue and roll formation. The effects of fracture surface energy and tearing energy instead of surface free energy on polymer wear are discussed. Surface free energy appears to influence only the polymer transfer.

I.   INTRODUCTION

This paper will discuss the relevance and the irrelevance of surface energetics to polymer friction and wear. This survey covers most of the important works published in the past and during this recent Symposium[1]. Since basic principles related to polymer friction have been reviewed by Tabor[2], and by Savkoor[3], the scope of this discussion will be limited to pertinent friction mechanisms with emphasis on surface interactions. First will be a brief discussion of the following friction processes: polymer sliding, elastomer sliding, lubricated polymer sliding, polymer rolling.

Then the main part of this paper deals with the effect of surface energetics on sliding friction, followed by examining multiple adhesion components of friction force - backgrounds and actual examples. Thus, it is necessary to distinguish van der Waals and electrostatic interactions. For Van der Waals interactions, we describe basic components of each type of force for both microscopic and macroscopic bodies.

In the last section of this paper, several types of polymer wear are briefly mentioned. Since wear involves cutting and tearing, included is a summary of some fundamentals on fracture of polymer for abrasive wear, fatigue wear and roll formation. Surface fracture energy and tearing energy, instead of surface free energy, are discussed. Surface free energy, though not affecting the above three wear mechanisms directly, appears to influence polymer transfer due to adhesive wear.

Throughout this survey we carefully show the effect of deform- ation on the adhesion mechanism of friction. We hope that with a better understanding of surface interactions and viscoelastic properties of polymers, we can appreciate the significance of the adhesion mechanism.

II.   MECHANISMS OF FRICTION

A.   Polymer Sliding Friction

Polymer sliding friction force may consist of two components[4]: F_a, the adhesion component and F_d, the deformation component

$$F_{sliding} = F_a + F_d.$$

The relative importance of these two components is strongly influenced by such factors as the nature of counterfaces, the load, the sliding speed, the temperature, the environments, the thickness of the polymer, and the lubricant. In the literature, several extreme cases have been reported. For example, Tanaka[5] concluded that the friction of a

polymer is due to the tangential resistance produced by the deformation of its surface neighboring the contact region rather than the
shearing of the functions formed at the contact region as described
by the adhesion theory. On the other hand, Bahadur and Ludema[6]
found that the major contributing factor was the adhesion component
for polymer sliding friction. As will be seen, the basic problem
is the difficulty in distinguishing or separating one component from
the other. This may have been the source of confusion regarding
conflicting theories of polymer friction.

1. <u>Adhesion Component</u>. Theoretically, the adhesion component
deals with the external friction which is affected by surface energetics of counterfaces. In metal friction[4], this takes the form of
welded surface asperities which result in junction growth. In
polymer friction, there is generally no junction growth, the adhesion force is presumably induced by molecular interactions between
the counterfaces. In metal friction, the adhesion term F_a results
primarily from the force required to shear the junctions and is thus
proportional to the true contact area A and the shear strength S

$$F = A S. \tag{1}$$

If the asperities deform plastically, the real contact area will be
equal to the quotient of the normal load W and the mean yield pressure
of the softer material

$$A = \frac{W}{P}. \tag{2}$$

Thus, the coefficient of friction (adhesion) μ_a will equal the ratio
of the interfacial shear strength to the yield pressure.

$$\mu_a = \frac{F}{W} = \frac{AS}{AP} = \frac{S}{P}. \tag{3}$$

The same reasoning has been applied to polymer adhesion.
Shooter and Tabor[7] found that the friction of clean metals sliding
on polymers is about the same as that of polymers sliding on themselves. Over a moderate load-range the friction was roughly constant
and appeared to be due primarily to the shearing of the interface.
The shear strength[7] calculated from the friction test was found approximately equal to the bulk shear strength. However, in those
cases, the shearing takes place within the bulk of the polymer
rather than at the interface because the whole of the polymer
material around the contact area deforms as demonstrated by Tanaka[5].
The shearing resulting from mere application of pressure affects
the deformation component of the friction as well as the adhesion
component.

If both interfacial shear strength and the true contact area
are measurable, then both quantities should be affected by surface

and bulk properties of a polymer. We examine possible factors affecting these two parameters in the following list (after Kummer's arrangements for rubber friction)[8].

Interfacial Shear Strength

    Rigid Counterbody
        Structure: morphology and polarity
        Thermal Properties: specific heat, thermal conductivity

    Polymer
        Additives: antioxidant, mold release agent, plasticizer
        Adsorbent: chemically or physically attached
        Electrical Properties: dielectric constant, electrical
                    conductivity
        Rheological Properties: tan δ, viscosity
        Structure at Interface: crosslink density, crystal-
                    linity, morphology
        Thermal Properties: glass temperature, specific heat,
                    thermal conductivity, thermal
                    history
    Lubricant
        Film Thickness
        Rheological Properties: shear strength, viscosity
        Structure: polarity
        Thermal Properties

    Operating Factors
        Contact Time
        Environments: gases or vacuum
        Pressure
        Sliding Velocity
        Temperature

Contact Area

    Rigid Counterbody
        Microroughness and macroroughness

    Polymer
        Microroughness and macroroughness
        Rheological properties

    Operating Factors
        Contact Time
        Environments: gases or vacuum
        Load
        Sliding Velocity
        Temperature

From the above list, it is evident that there are some factors directly related to surface energetics. For example, rigid counterbody and polymer structures at the interface, adsorbents, and lubricant can affect surface energetics. However, it is also clear that besides surface energetics many other factors can affect either shear strength or the real contact area, and both of these components can determine the friction force.

The major problem in understanding the mechanism of friction is the lack of information regarding structural changes at the interface. Therefore, the application of new characterization techniques to the study of polymer friction should lead to a new interpretation of friction results. The uses of ESCA, as described by Clark[9]; of field ion micrograph (FIM) and auger spectroscopy, by Buckley[10]; of esr, by Bely and Sviridyonok[11]; and of SEM by others have already given new insight to the subtle structural changes at contact. Furthermore, the close examination of polymer melting, by Tanaka[12], has shown that sometimes the interfacial layer has a different superstructure if frictional heating is substantial.

Besides the structure at the interface, the true contact area is also very difficult to measure. The contact area is influenced primarily by the normal load. Both acoustic and optical methods have recently been discussed by Kragelsky et al.[13]; the acoustic method appeared to offer many advantages. Before the true contact area can be clearly determined, we can not accurately measure the friction force due to the adhesion component.

2.   Deformation Component.   Since the adhesion component can not be readily separated from the bulk properties of polymer, it might not be easy establishing an independent deformation component when F = AS relationship is used. However, in the case of rolling friction[14,15,16] and lubricated sliding friction[17], the major component involved is the deformation of a polymer. The deformation losses can be accounted for by the energy of the ploughing of a polymer by asperities on the counterface. Thus, the deformation component involves the loss angle of a polymer and is affected by the external load

$$F_d = K_d (W)^n \tan \delta, \tag{4}$$

where $\tan \delta = E''/E'$,
and E'' = loss modulus,
    E' = storage modulus.
    K_d = deformation constant
    W = load
    n = constant

The loss tangent is related to the internal friction between polymeric chains[16] according to the following equations:

$$\tan \delta = \frac{\psi}{2\pi} , \qquad\qquad (5)$$

$$\text{and} \quad \psi = \frac{\Delta(E)}{(E)} , \qquad\qquad (6)$$

where ψ = internal friction,
          or specific damping capacity,
$\Delta(E)$ = energy dissipated per cycle,
(E) = vibration energy.

Equation (5) indicates that the internal friction within the bulk of a polymer is an integral part of the deformation component and the total friction force.

Deformation loss is also a collective term, including the special cases of hysteresis losses, grooving losses, and viscous losses. Hysteresis losses occur, for example, in viscoelastic materials, where deformation is not lasting and part of the deformation energy is recovered -- partial energy feedback. Grooving losses are observed with plastics and elastic materials stressed beyond their yield point, and they imply permanent deformation -- no energy feedback. (The term "scratch hardness" is used in connection with a grooving test for minerals that indicates primarily cohesion losses.) By separating the deformation losses within a solid into hysteresis and grooving losses, we can avoid the sometimes used term "elastic hysteresis", which is incorrect because "elastic" precludes hysteresis.

It may be argued that viscous losses properly belong under the heading of interface losses. From the macroscopic viewpoint, and considering boundary and hydrodynamic friction, this may be correct. On a molecular scale, however, the energy is dissipated within the sheared fluid and must therefore be regarded as a bulk loss.

Examining briefly the following factors affecting the deformation component -

          Rigid Counterbody
          Asperity Density
          Microroughness and macroroughness

Polymer
    Rheological Properties: tan δ, hysteresis
    Structure: crosslink density, crystallinity, morphology
    Thermal Properties: thermal conductivity, specific
                 heat

Lubricant
    Rheological properties

Operating Conditions
    Load
    Sliding Velocity
    Temperature

It is apparent that surface energetics of the polymer do not directly influence the deformation component. However, it is important to bear in mind that the deformation component can be indirectly affected by the adhesion component because high interfacial adhesion can reduce the damping of the polymer.

    3. <u>Relative Importance of Two Components</u>. The co-existence of two components does not necessarily imply that there are two independent mechanisms. Sliding friction (unlubricated) may well be controlled by the adhesion mechanism; but for polymer friction, deformation is an intrinsic property influencing adhesion. We have seen different results in the literature regarding the relative importance of these two components as determined by different methods. These results may reflect the methods by which the friction processes were determined and should not be used to establish any friction mechanism.

    For surfaces with thick asperities, or under a heavy load, the deformation component (or the ploughing term) can indeed be important. A method used by Istomin and Kuritsyna[18] could separate the two components. They used a small chromium-hardened steel ball (dia. 6mm and hardness ∇ 10) sliding on polymers. The actual contact area was calculated from $S = b^3/12R$, where b is the track width and R the radius. The deformation component was determined in the presence of a lubricant and is proportional to b^3 because the bulk of sample transferred is proportional to b^3 (Fig. 1a). The adhesion component F_a can be determined according to the difference $F - F_d$ and is proportional to b^2 (Fig. 1b).

    Another set of results on the ratios of the two components (Table 1) were obtained by Kuritsyna and Meisner[19].

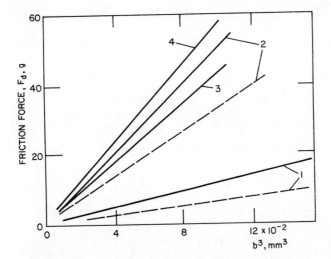

Fig. 1a. Relation of the Deformation Component of the Friction
Force to the Width of Friction Path.
1. PTFE; 2. Polycaprolactam; 3. Polyethylene;
4. Polyamide-b-68 - at 20°C, --- at 70°C.

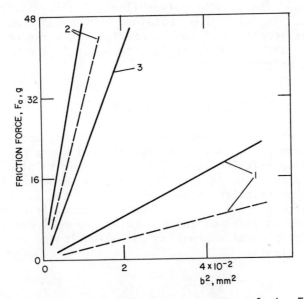

Fig. 1b. Relation of the Adhesion Component of the Friction
Force to the square of path width.
1. PTFE; 2. Polycaprolactam; 3. Polyethylene HD
- at 20°C; --- at 70°C.
(Data of N. P. Istomin and A. D. Kuritsyna (Ref. 18))

TABLE 1

ADHESION AND DEFORMATION COMPONENTS OF FRICTION FORCE FOR POLYMERS
(Data of A. D. Kuritsyna and P. G. Meisner (Ref. 19))

Polymers	T °C	F def.	Hardness Kg/mm^2 Passage Under Load for 5 Min.	F adh.	Contact Angle 0
PTFE	20	4.0	3.1	0.1	105
	70	2.2	1.5	0.05	-
Polyethylene, HD	20	15.6	3.5	0.51	86
Nylon-6	20	17.0	7.7	1.02	65
	70	12.0	4.2	0.79	-
Polyamide P-68	20	20.2	8.5	-	-

They found F_a/F_d for certain polymers to be: PTFE, 0.025; polycaprolactam resin and fiber, 0.06; polyethylene, 0.032. Their results
agree, in principle, with those obtained by Tanaka[5]. However, without knowing the roughness of the counterfaces, it can not be concluded
from their data that the deformation component is the major factor
in determining polymer friction.

B. Elastomer Sliding Friction

    This subject has been well reviewed by Conant and Liska[20],
Kummer[8], Schallamach[21], Moore and Geyer[22], and Moore[23]. In principle,
elastomer sliding friction is similar to polymer sliding friction.
The major characteristic is that the adhesion component F_a for
elastomer sliding friction is also affected by tan δ. Bulgin,
Hubbard and Walters[24] first showed that the total friction force
can be expressed in terms of tan δ. Later, Kummer[8] derived the
following semi-empirical expression for unlubricated elastomer sliding
friction coefficients:

$$\mu = \mu_a + \mu_h ,$$

$$= \left[k_a \frac{E'}{P^r} + k_h \left(\frac{P}{E'}\right)^n \right] \tan \delta \qquad (7)$$

where μ_a and μ_h are adhesion and hysteresis components; k_a, k_h adhesional and hysteretic constants for the friction pair; E' modulus, P nominal pressure, r and n are constants.

Yandell[25] applied the mechano-lattice analysis to describe the sliding friction of elastomers. The coefficient of sliding friction was shown to depend solely on the damping properties of rubber. He suggested that the load dependence of the coefficient of friction, usually observed on the smoother surfaces and attributed to adhesion, could be due to hysteresis or deformation. This idea could have led Bikerman[26] to conclude that the adhesion theory is incorrect in explaining polymer friction phenomena. Practically all other factors affecting elastomer sliding friction[8] were similar to those for polymer sliding friction. Surface energetics which enter the constant k_a are chiefly controlled by polymer structure and adsorbents at the interface. Thus, adhesion is still playing an important role in elastomer friction, although the adhesion term is indeed affected by the hysteresis properties of a rubber.

C. Lubricated Sliding Friction

The deformation component can be determined by the lubricated sliding friction. In the presence of a lubricant, the shearing of the interfacial junctions or contacts can be prevented; therefore, the total frictional force essentially equals the deformation component[17]. Under the lubricated conditions, surface energetics and shear strength of the lubricant control the friction instead (see Section III. C.).

The mechanism of boundary lubrication for polymers[27] is generally similar to that of metals. Lubricants are known to be ineffective for polymeric surfaces, partly because of plasticization, and plasticization could weaken the interfacial forces and reduce both the shear strength and the yield pressure of the polymer. This weakening effect can be illustrated with the following equations[27]

$$F = A \left[\propto S + (1 - \propto) S_u \right] \tag{8}$$

and

$$\mu = \frac{\propto S}{P} + \frac{(1 - \propto) S_u}{P} \tag{9}$$

where $\propto$ is the fraction of unlubricated area; S strength of the soft polymer, and S_u shear strength of the lubricant film. If plasticization takes place, the first term remains unchanged but the second term will increase owing to the lowering of yield pressure.

D. Polymer Rolling Friction

Similar to the lubricated sliding friction, rubber rolling friction generally involves only the deformation component[14,15]. Flom[16] later found that the rolling friction theory can be extended to thermoplastics. Flom[16] showed that the coefficient of rolling friction λ can be expressed:

$$\lambda = \lambda_{(BULK)} + \lambda_{(SURF)} ; \tag{10}$$

$$= \lambda_{def} + \lambda_{adh} ,$$

and
$$\lambda = K_b \tan \delta \left(\frac{W}{Ga^2}\right)^{1/3} + C_s \left(\frac{a^2}{WG^2}\right)^{2/3} , \tag{11}$$

where λ = rolling friction coefficient,
a = radius of the ball,
G = shear modulus,
K_b = bulk constant,
C_s = $K_s \pi [\frac{3}{8}[(1-\sigma^2)/(1 + \sigma)]^{2/3}$,
σ = Poisson's ratio,
K_s = surface constant.

For hydrogen-bonding polymer, e.g. nylon, a small surface effect was detectable[16]. Thus, the coefficient should be described according to Equation 11 to include the second term. Besides this special case, the surface effect is generally negligible, and the coefficient of friction can be simplified to be a function of $\tan \delta$.

III.   SURFACE ENERGETICS AND FRICTION

A.   Surface Energetics and Unlubricated Sliding Friction

If adhesion mechanism is indeed the major mechanism for polymer unlubricated sliding friction, we should be able to detect various effects of surface energetics upon the friction process. On the contrary, it has been very difficult to find any clear-cut correlation. One of the reasons is that the adhesion component in itself contains more complicated viscoelastic elements of a polymer than those involved in the deformation component, as discussed in the preceding section. Another reason is that the total friction force sometimes contains a measurable deformation component especially when the counterface is rough.

In light of the similarity between the relaxation theory for friction[28,29] and that for adhesion[30,31] we may use the analogy of an activated complex to describe the bond-breaking and the bond-forming processes as suggested by Hatfield and Rathmann[32]. In the free-energy profile (Fig. 2) the activation energy for bond-forming is the sum of the free energy of activation for the deformation $F^{\ddagger}$ and the work of adhesion W_a

$$E = F^{\ddagger} + W_a , \qquad\qquad (12)$$

where $W_a = F_A + F_B - F_{AB}$,

F_A, F_B = surface free energy of counterfaces,
F_{AB} = interfacial free energy.

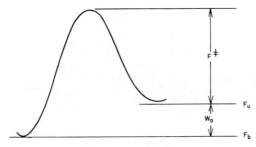

Fig. 2. Free energy profile-adhesion system: F_U, energy min.-
unbounded molecules; F_B, energy min.-bonded molecules;
W_a, energy of adhesion bond; $F^{\ddagger}$, free energy of
activation.

If the polymer friction at the rubbery state indeed involves the adhesion mechanism, the activation energy already contains a deformation energy component $F^{\ddagger}$ which is generally larger than the thermodynamic work of adhesion, W_a. The presence of $F^{\ddagger}$ prevents a clear comparison of W_a's among polymers, and also introduces unnecessary confusion about the validity of the adhesion mechanism. Because of the stronger influence of the deformation energy than the thermodynamic work of adhesion, one may argue that the adhesion mechanism is groundless.

Any argument of this subject generally originates from the lack of understanding of modern polymer adhesion theories. Polymer adhesion largely involves the deformation of the polymer[33-35], and to a lesser extent the thermodynamic work of adhesion as depicted by the absolute rate theory. Only for a very thin film[36] does the thermodynamic work of adhesion determine the total work of adhesion. The thicker the polymer, the stronger the influence of the deformation properties[37].

From the above discussion, it is easier to expect the lack of general correlation between frictional force and surface energetics even though adhesion is the basic cause of frictional resistance for polymers. The best correlation between frictional force and surface energetics, in terms of γ_c, was first noted by Bower et al.[38,39] for

halogenated ethylene polymers. However, the correlation becomes
less evident for other polymers[27] as shown in Figure 3.

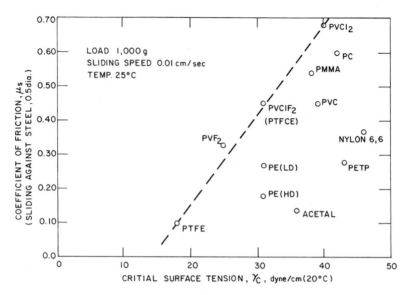

Fig. 3. Effect of Critial Surface Tension on Static Friction
         Coefficient of Polymers Against Steel
         (This data from NRL (Ref. 27))

Later, an empirical relation was suggested by West and Senior[40] to
account for the effect of surface energy on friction force

$$F = A.S \left(\frac{\gamma_c - 15}{11} \right) \qquad (13)$$

The authors claimed that with the above correction for F = A.S they
were able to obtain a better correlation between S/p and μ for
polymers studied by King and Tabor[41].

    Recently, Savkoor[3] found a thermodynamic relationship which
could support the adhesion theory of friction:

$$W_a = \gamma_L (1 + \cos \theta), \qquad (14)$$

where W_a is the work of adhesion; γ_L surface tension of a liquid;
and θ the contact angle. Two linear plots between cos θ and μ of a
rubber compound on various track surfaces were obtained for water
and mercury respectively. Their results suggest that a direct
correlation between friction force and the work of adhesion W_a may
indeed exist.

    To demonstrate the importance of separating out the deformation component prior to the comparison of surface energetics, we chose the polymers of the same modulus range from the work of Tanaka[5] and compared the friction coefficients with critical surface tensions γ_c[42]. Since modulus affects the true contact area, A, for polymers with the same modulus range, at least we can keep one of the variables in the F = A.S relationship fairly comparable.

Without knowing the extent of the deformation component, if any, in Tanaka's total friction force data[5], we find the correlation in Fig. 4 to be rather encouraging for polymers within a similar range of modulus. In other words, the coefficient of friction μ appears to increase with the increase in surface energy or critical surface tension. This is especially interesting because the same data led Tanaka[5] to believe that deformation is the friction mechanism, especially in the case of steel slider.

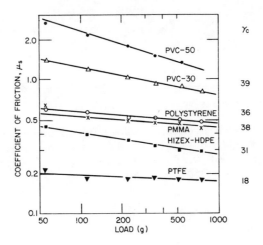

Fig. 4.   Load dependence of the coefficient of static friction μ_s observed between the steel hemisphere, 0.24 cm in radius and the plate of polymers.
(Data of K. Tanaka (Ref. 5))

Ideally, the comparison between friction force and surface energetics should be carried out for the adhesion component F_a alone because the deformation component F_d is not directly affected by surface energetics. In reality, there is not much data available which can be clearly identified for each component. For what was available, we obtained a linear correlation between F_a and γ_c in Fig. 5.

Since the data[19] involved has only three points, we can not conclude whether this trend could apply to all other polymers. However, the above two examples can serve to demonstrate the hidden role of surface energetics in polymer friction, especially the adhesion component. In the coming sections, we shall illustrate the effect of other specific surface interactions on unlubricated polymer sliding friction.

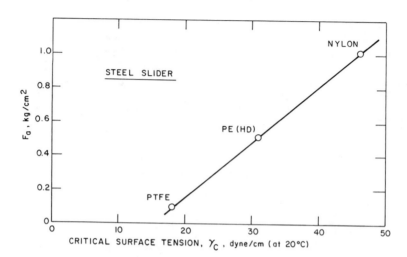

Fig. 5.   Relationship between adhesion component of friction
          force and critical surface tension.
          (F_a data of A. D. Kuritsyna and P. G. Meisner (Ref. 19)

Though the above examples are unrelated, they all tend to suggest that there is some correlation between surface energetics and the adhesion component. However, as discussed in the beginning of this section, adhesion is also a kinetic process, and the activation energy consists of the deformation energy and the work of adhesion. Unless all deformation energies for different polymers are nearly identical at a certain temperature, it is not easy to compare adhesion or friction directly with surface energetics.

B.   Multiple Adhesion Components

The adhesion component of the unlubricated sliding friction for an elastomer may be expressed in terms of tan δ, the loss angle[23]

$$F_a = K_a \, \sigma_o \left(\frac{W}{H} \right) \tan \delta \qquad (15)$$

where K_a is the adhesion constant; σ_o, the maximum stress on a unit area; W, load, and H, hardness. This equation can be extended to the friction of other polymers in their rubbery state.

According to Ludema and Tabor[43], the effect of tan δ on F_a of unlubricated sliding for glassy polymers is less evident than those for rubbery polymers. Thus, surface effect (van der Waals or electro-statics[44]) could be more readily detectable for glassy polymers than for rubbery polymers. Since van der Waals forces[45] consist of dispersion forces, orientation forces, induction forces, and hydrogen bonding, we may then write F_a

$$F_a = F_a{}^d + F_a{}^o + F_a{}^i + F_a{}^h + F_a{}^e \qquad (16)$$

where superscripts d, o, i, h, e denote dispersion, orientation, induction, hydrogen bonding and electrostatics.

In Equation 16, we also note that the surface effect determines only the constant K_a. Then specifically, we can write

$$K_a = K_a{}^d + K_a{}^o + K_a{}^i + K_a{}^h + K_a{}^e. \qquad (17)$$

In the following sections, we shall briefly describe various forces and their relations to polymer friction.

1.   Van der Waals Forces Between Microscopic Bodies
   1.1.   Dispersion Forces.   One of the basic molecular forces acting at the interface is the dispersion force involving the simul-taneous excitation of both molecules. The perturbation responsible for dispersion forces is the instantaneous coulomb interaction between the electrons and nuclei in the two molecules. London[46] derived an approximate formula for the dispersion energy E_d for two spherical molecules in terms of their polarizabilities α and ionization potential I:

$$E_{disp} = -\frac{3}{2}\left(\frac{I_A I_B}{I_A + I_B}\right)\left(\frac{\alpha_A \alpha_B}{r^6}\right), \qquad (18)$$

Where r is the distance between the two molecules. It is important to realize that the London interaction is present in all pairs of atoms or molecules. The London dispersion forces, thus, form the basis of adhesion between solid bodies.

Later, Casimir and Polder[47] showed that the London theory must be modified to account for the interaction between molecules at large distances, e.g. greater than 100Å. The retardation energy may be expressed

$$E = -\frac{23}{8\pi^2}\left(\frac{\alpha_1 \alpha_2}{r^6}\right) \cdot \frac{hc}{r}, \qquad (19)$$

where h is Plank constant, and c the velocity of light. The dependence of the energy on distance is as r^{-7}, i.e. there is even more
rapid dropoff of energy with distance when the "retardation" effect
in the London force comes into play.

The presence of the retardation effect magnifests the importance
of the contact distance between counterbodies in friction. If the
distance is larger than 100Å, the adhesion force due to van der Waals
interactions could be very small.

1.2. Orientation Forces. Besides the most basic non-polar
interaction, dispersion forces, there are polar interactions between
molecules of counterbodies, e.g. the dipole-dipole interaction
(Keesom), the dipole-induced dipole interaction (Debye) and hydrogen
bonding. The Keesom interaction (orientation)[48] is temperature
dependent and the energy is expressed as

$$E_{orien.} = -\frac{2\mu_1^2\mu_2^2}{3kTr^6},$$ (20)

where μ is the permanent dipole moment; k the Boltzman constant and
T the absolute temperature. The best example for this type of interaction is between polyvinyl chloride and acrylonitrile copolymer[49].
This interaction is strong enough to cause an increase in the adhesion
component of the friction force, F_a,[49] as shown in Fig. 6.

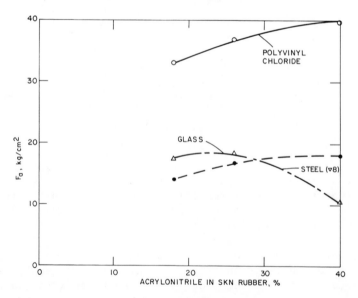

Fig. 6.   Effect of Polarity of Counter Body on Friction Force
          of Butadiene-Acrylonitrile Rubber
          (Data of A. Konstantinova (Ref. 49))

These data[49] were originally presented in such a way without realizing the specific interaction between these two polymers. We plot their data in Fig. 6 and find that the increase of F_a with the nitrile content is the best evidence for the Keesom type dipole-dipole interaction.

An interesting example is the interaction between polychloroprene and ice recently presented by Southern and Walker[50]. Due to the polar interaction, the friction coefficient at maximum of μ-v curves for polychloropropene was higher than those for styrene-butadiene rubber and natural rubber.

1.3. Induction Forces. Another polar interaction is the induction between a permanent dipole and an induced dipole[51]. The energy for this type of polar interaction (Debye) between molecules is expressed

$$E_{ind} = - \frac{\alpha_1 \mu_2^2 + \alpha_2 \mu_1^2}{r^6} , \qquad (21)$$

where α is the polarizability. For example, an HCl molecule will polarize a He atom, and an attractive force results. The essential difference between polarization forces and dispersion forces is that whereas the former arise from virtual transitions involving the passive partner, the virtual transitions responsible for the latter involve simultaneous excitation of both molecules.

An interesting study carried out by Dukhovskoy, et al.[52] could demonstrate the importance of surface orientation effect. By irradiating PTFE with fast helium atom in vacuum (10^{-5} torr), the friction coefficient against steel increased, but the coefficient returned to the original level after the exposure ended (Figure 7). However, the same treatment reduced the friction coefficient of poly-ethylene and polypropylene nearly to zero (Fig. 8). Both of these two effects were temporary. The authors suggested that the formation of specific structure was probably due to molecular orientation in the surface layer during the irradiation process. This type of induced orientation, though not well understood, could be caused by any one of three van der Waals interactions discussed in this section.

1.4. Hydrogen Bonding. Lastly, hydrogen bonds arise also from dipole-dipole interactions[45]. The strengths of hydrogen bonds vary between 1 to 10 K cal/mole. The polymers with amide or ester groups tend to hydrogen-bond to other substrates of hydrogen-bonding tendency. An excellent example is the strong friction force between nylon and nylon (Fig. 9)[53].

Recently, Tanaka[12] observed the high friction coefficient of Nylon 6 on glass, and the values were higher than 1.3 at all speeds and reached the highest value of 2.1 at 10 cm/sec. The high friction is presumably caused by hydrogen bonding between Nylon 6 and glass.

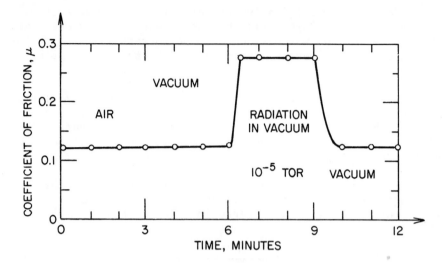

Fig. 7.   Effect of the Irradiation by Fast Helium Atom on
           Friction of PTFE.

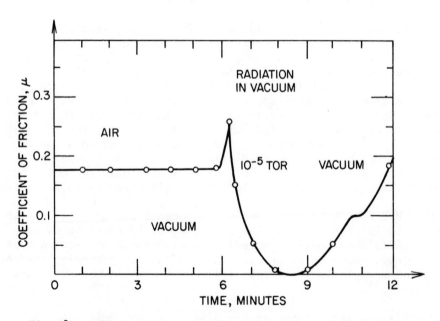

Fig. 8.   Effect of the Irradiation by Fast Helium Atom on
           Friction of Polyethylene.
           (Data of E. A. Dukhovsky et al. (Ref. 52))

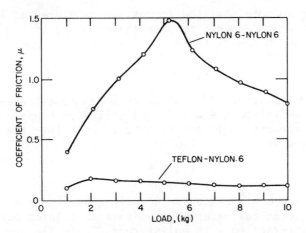

Fig. 9. Friction of Nylon 6 on Teflon and Nylon 6 itself.

V = 5 cm/sec.
(Data of M. Watanabe (Ref. 53))

2. Van der Waals Interactions Between Macroscopic Bodies and Adhesion Forces

2.1. Hamaker Constants. Most of the above discussions on van der Waals forces are for the microscopic bodies. In reality, the contacts are made between macroscopic bodies[54], especially in the case of friction and wear. We shall briefly discuss the factors involved in calculating van der Waals forces in macroscopic bodies and the effect of adsorbed layer of gases on adhesion.

For macroscopic bodies in contact, the energy of interaction is, in the first approximation, made up of the sum of all interaction energies between all the atoms in counterbodies. If the distance Z_O between the two macroscopic bodies is very small in comparison to other dimensions of the system, the summation procedure can yield results as shown in Table 2.

TABLE 2. Van der Waals Interactions at Small Separations

	Force	Energy
2 Thick Plates	$\dfrac{A}{6\pi Z_O^3}$	$\dfrac{A}{12\pi Z_O^2}$
Plate-Sphere	$\dfrac{AR}{6Z_O^2}$	$\dfrac{AR}{6Z_O}$
2 Spheres	$\dfrac{AR}{12Z_O^2}$	$\dfrac{AR}{12Z_O}$

It is important to note that the forces can be expressed in two factors, one of which depends only on the geometry of the system (R, the radius and h the distance between the two counterbodies), whereas the other factor, the Hamaker constant (A), is determined by the nature of the interacting materials.

The Hamaker constant[55,56] can be obtained by different methods. In this study, we follow the work of Lifshitz[57], Krupp[44], in using the Lifshitz - van der Waals constant $\hbar\bar{\omega}$ to calculate the Hamaker constant, arrived at

$$A = \frac{3\hbar\bar{\omega}}{4\pi} \qquad (22)$$

where $\hbar = h/2\pi$ and $\bar{\omega}$ = averaged angular frequency. From the $\hbar\bar{\omega}$ data of Krupp[58] for identical materials, we have calculated both $\hbar\bar{\omega}$ and A for materials interacting with polystyrene (Table 3).

TABLE 3.   Calculated Lifshitz - van der Waals Constants and Hamaker Constants for Polystyrene Interacting With Other Materials.

	$\hbar\bar{\omega}$ (eV)[12]	$A_{12} \times 10^{20}$ (J)
PS - Graphite	4.59	17.0
PS - Au	4.52	17.0
PS - Ag	4.23	16.3
PS - Ge	3.67	13.8
PS - Cu	3.57	13.8
PS - Diamond	3.57	13.8
PS - Si	3.34	13.2
PS - Si + Oxide (130Å)	3.01	11.3
PS - Al_2O_3	2.64	10.1
PS - CdS	2.62	10.1
PS - KCl	1.66	6.3
PS - H_2O	1.40	5.3
PS - SiO_2 (Glass)	1.24	5.0

2.2.   Adhesion Forces Derived From van der Waals Interactions. The data in Table 3 may be used to calculate adhesion force derived from van der Waals interactions. Thus, the force between a sphere and a plate is expressed

$$F = \frac{AR}{6Z_o^2} = \frac{\hbar\bar{\omega}}{8\pi Z_o^2} \qquad (23)$$

If the substrate undergoes an elastic deformation[54], Equation (23) should be modified as

$$F = \frac{AR}{6Z_o^2} \left(1 + \frac{A}{6\pi Z_o^3 H} \right) \tag{24}$$

where H is the modulus (or micro-hardness) of the substrate.

If the distance is larger than 100Å, the retardation force[59] should be calculated for a sphere and a plate

$$F' = \frac{2\pi B}{3} \frac{R}{Z_o^3} \tag{25}$$

where B is a material constant different from the Hamaker constant.

Buckley and Brainard[10] recently showed chemical bonding between PTFE and tungsten under high vacuum. This may be an interesting example for the adhesion originated from van der Waals interactions. However, good adhesion is not limited to the experiments carried out in vacuum. The adhesion between two mica sheets were actually measured in air by Tabor et al.[59].

2.3. Effect of Adsorbed Gas Layer on Adhesion. The roles of an adsorbed gas or a gaseous medium on adhesion have been studied by Krupp[60]. The adhesion between gold particles and silicon was affected by oxygen monolayer, and an oxide layer of 130Å did lower both the Hamaker constant and the adhesion forces (Fig. 10). However, there is still measurable adhesion in the presence of oxygen. Thus, in the presence of gaseous media, adhesion force could still be an important resistance force to friction.

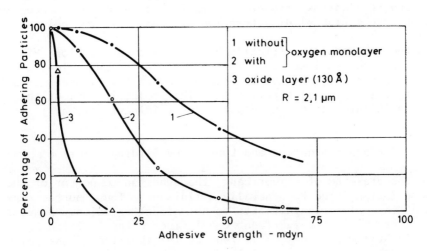

Fig. 10. Effect of Adsorbed Gas on Adhesion of Gold Particles
          on Silicon.
          (Data of H. Krupp (Ref. 60))

3. Electrostatic Forces. Electrostatic forces have been shown
to be one of many driving forces for adhesion by Derjaguin[61] and by
Skinner and co-workers[62]. In recent years, Krupp[60] and his school
also published interesting results in electroadhesion. Despite
these studies, there have been serious doubts about the importance
of electrostatic contribution on adhesion, especially to insulative
polymers.

The role of electrostatic interaction on polymer friction is
also not clear[4], though frictional charging is a familiar phenomena.
According to the adhesion theory of friction, the process involves
primarily the shearing of junctions at the contact regions. It has
been shown by Shooter and Tabor[6] that the shearing process is similar
to the shearing of the bulk polymer. For ductile polymers, the major
part of fracture energy is the deformation energy; the surface effect,
e.g. electrostatics, plays only a minor part. For example, Derjaguin
and Toporov[1] recently found that there was some electrostatic contribution to rolling friction force which is generally smaller in magnitude than the sliding friction force.

The weak electrostatic contribution is illustrated with one of
the typical examples of a triboelectric pair. Polymers as well as
other materials can be classified into a triboelectric series, i.e.
a list in which the sign of the charge developed on a given surface
by contact with another can be predicted from the relative positions
of the two materials in the series (Fig. 11)[63]. If we pick a triboelectric pair from the two extreme positions of the series, we should
expect a strong electrostatic attraction between the pair. For
example, nylon and teflon are the two extremes on the triboelectric
series, and if electrostatic attraction is the major resistance, we
should observe a high sliding friction between these two polymers.

On the contrary, the friction between nylon and teflon (Fig. 9)
is very low and much lower than that between nylon and nylon, as
shown by the results obtained by Watanabe et al.[53]. Indeed, this is
an excellent example illustrating the minor role, if any, of the
electrostatic interaction in unlubricated sliding friction.

C. Surface Energetics and Lubricated Sliding Friction

In the presence of a lubricant, the adhesion component of the
sliding friction force F_a generally diminishes. Thus, surface energetics of the polymeric substrate do not affect friction. However,
surface energetics of a lubricant can influence the friction.
Presumably, the weakly held film reduces the shearing of the polymer
surface and hence the interfacial adhesion. Owens[64] found a linear
correlation between critical surface tension γ_c of lubricants incorporated in vinylidene chloride-acrylonitrile copolymer and static
friction of cellulose films (Fig. 12).

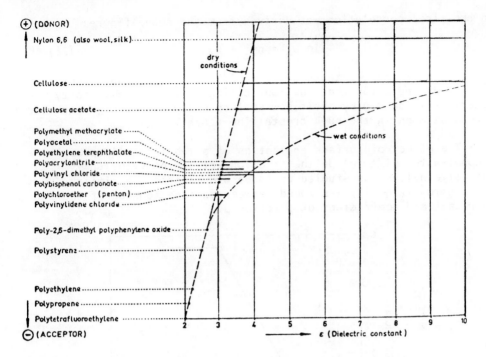

Fig. 11.   Triboelectric Series of Polymers
(Ref. 63)

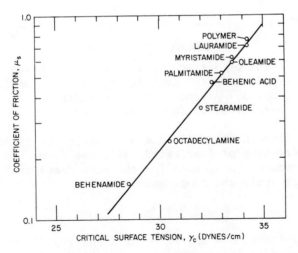

Fig. 12.   Static Friction vs. Critical Surface Tension for
Lubricated Films.   (Data of D. K. Owens (Ref. 64))

An empirical correlation was derived from the above figure:

$$\ln \mu_s = n\gamma_c + k, \qquad\qquad (26)$$

where n and k are constants equal to 0.28 and 9.9, respectively.
This study indicates that the lubricating action of those fatty
materials is caused by their low surface energies and high structural
strengths when in the bulk crystalline state.

The effect of surface energetics of a lubricant (isopropyl
alcohol-water mixture) on the friction between polyethylene
and polyethylene was studied by Cohen and Tabor[65]. A similar
correlation (Fig. 13) was found between surface tension of the
lubricant and coefficient of friction μ.

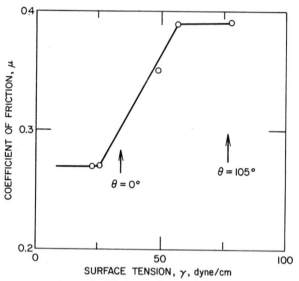

Fig. 13.   Effect of Surface Tension of the Lubricant on Poly-
           ethylene on the Coefficient of Friction (Lubricant-
           Isopropyl Alcohol - Water Mixtures).
           (Data of S. C. Cohen and D. Tabor (Ref. 65))

Hilyard[66] studied the influence of elastic compliance and surface
energy on the friction of lubricated cellulose film. An empirical
relation was proposed to account for the dependence of frictional
force on contact angle and compliance J:

$$F = KJ^n \, f \, (\cos \theta), \qquad\qquad (27)$$

where K is a constant; n = 2/3 if Hertz's equation is obeyed. The
friction was found to increase with J and cos θ. However, the exact
correlation between F and f (cos θ) was not established.

Senior and West[67] measured the friction between chemically modified polyethylene sheets and a polyethylene hemisphere under boundary lubricating conditions. They found that for the modified surfaces the coefficient of friction generally decreased with decreasing surface tension of the lubricating liquid.

IV. SURFACE ENERGETICS AND POLYMER WEAR

The basic factors which influence polymer wear[29,68] are pressure, coefficient of friction, surface texture, sliding velocity, elastic modulus, strength and fatigue resistance. Surface energetics may affect polymer wear through the adhesional friction. Thus, a direct correlation between surface energetics and polymer wear is not apparent.

Three major mechanisms of polymer wear are abrasion[69], fatigue[70], and roll formation[71]. Abrasive wear is caused by the sharp protrusion on the counterface; tearing, microcutting and scratching can result on the polymer surfaces (Table 4). Fatigue wear is induced by the blunt projections on the counterface. As a result of fatigue wear, the polymer undergoes cyclic deformation. On a smooth surface, a viscoelastic material can fail by roll formation at the sliding interface. Thus, wear by roll formation is common among polymers when bulk deformation exists[71].

TABLE 4

CLASSIFICATION OF TYPES OF WEAR

(Adapted from Ratner and Farberova (Ref. 69))

Material	Examples	Wear on Abrasive Paper		Wear on Metal Gauze	
		Contact Conditions	Type of Wear	Contact Conditions	Type of Wear
Highly elastic	Rubbers	Elastic-partly plastic	Tearing	Elastic	Tearing and fatigue
Elastic	Filled rubbers Polyolefins Polyamides	Plastic-partly elastic	Cutting and partly tearing or fatigue	Elastic-partly plastic	Tearing and fatigue and partly cutting
Plastic-elastic	Plasticized P.V.C.,P.T.F.E.	Plastic	Cutting	Plastic-elastic	Cutting tearing, fatigue
Rigid	Polystyrene P.M.M.A. Thermosetting resins (epoxy, polyester phenolics)	Plastic	Cutting	Plastic-partly elastic	Cutting, and partly fatigue

A. Abrasive Wear

Giltrow[72] showed that wear rates can be correlated with cohesive energies of polymers. Polymers with high cohesive energies have low wear rates. Cohesive energy which is a measure of intermolecular bond strength can also affect surface energetics of polymers[73,74].

All homogeneous isotropic polymeric materials involving rupture and tearing can be divided into three categories: glassy, visco-elastic and rubbery polymers. Glassy polymers have high moduli which are not time-dependent. The materials are generally hard and have tensile strengths. Thus, the Griffith criterion[75] can be applied to determine the critical stress of the crack growth.

$$\sigma_B = \left(\frac{2\gamma_s E}{\pi C}\right)^{1/2} = \left(\frac{\mathscr{G}_c E}{\pi C}\right)^{1/2} \tag{28}$$

where σ_B is the tensile strength of the material; $2C$, crack length; E, modulus; γ_s, surface free energy; $\mathscr{G}$ the strain energy release rate defined by Irwin[76]; $\mathscr{G}_c$ is the critical value at which the fracture occurs. Orowan[77] proposed that surface free energy γ may also include a term which arises from the plastic work done in deforming the material near the fracture surface as the crack propagates. Thus, $\mathscr{G}_c$ should be the sum of the following two terms:

$$\mathscr{G} = 2\gamma_s + \Psi, \tag{29}$$

where Ψ the irreversible work dissipated in plastic deformation.

Berry[78] defined the above surface energy as fracture surface energy. The values for the fracture surface energies of glassy polymers (Table 5) are generally 1000 times larger than the surface free energies γ_s or γ_c critical surface tensions. Thus, the work of plastic deformation Ψ is far greater than γ_s of a polymer. The role of surface energetics, as discussed for polymer friction, is only a minimal part in the brittle fracture of glassy polymers during the abrasive wear process, and yet it is still an indispensable part.

TABLE 5. Fracture Surface Energies (10^5 erg cm^{-2})
(Ref. 78)

Method	Polymer	
	Polymethylmethacrylate	Polystyrene
Cleavage (Benbow)	4.9 + 0.5	25.5 ± 3
Cleavage (Svensson)	4.5	9.0
Cleavage (Berry)	1.4 ± 0.07	7.13 ± 0.36
Tensile (Berry)	2.1 ± 0.5	17 ± 6

Davidge and Tappin[79] showed that the fracture energy in the Griffith-Irwin equation changes after the crack propagation starts. The initial value for glassy polymers (e.g. PMMA) is generally larger than the value of the overall fracture process, whereas the former for graphite is smaller than the latter. For PMMA and glass, there is a multiplicity of crack sources during fracture initiation, but for graphite subsidiary, cracking takes place as fracture proceeds.

In the case of viscoelastic material, the concept of tensile strength has lost its traditional meaning. The tear follows a time dependent process. Thus, the Griffith theory is not applicable to the tearing process; surface energy plays no apparent role.

Rubbery materials have low moduli which are generally not time-dependent. We can say that true tensile strengths exist if the materials undergo high degree of deformation. Under this condition, tearing[80] can take place along the entire deformed path with a high propagation rate.

The tearing energy of a highly elastic material can be calculated from the applied deformations[81]. For example, for a test piece containing an edge crack of length c deformed in simple extension, a generalization of finite strains of Griffith's formulation gives

$$T = 2KWC \tag{30}$$

where W is the strain energy density in the bulk of the rubber and K is a slowly varying function of the strain determined empirically.

An intrinsic material parameter T_O, critical tearing energy, has been found to be 100 times greater than surface free energy of a rubber[82]. Lake and Lindley[83] determined T_O and tensile strength for several rubbers (Table 6) but found no correlation between the two parameters. The value of T_O is in the region of 5×10^4 erg/cm^2 for all rubbers examined, whereas the values of large-scale catastrophic tearing are between $10^6 - 10^8$ erg/cm^2, and within the range of fracture surface energies for glassy polymers.

Champ, Southern and Thomas[84] recently applied the tearing criterion to the study of rubber abrasion. Their results were expressed as the crack growth per cycle r as a function of maximum T, tearing energy value, attained during each cycle (approximately $10^5 < T < 10^7$ erg/cm^2)

$$r = BT^{\alpha} \tag{31}$$

The exponent α varies from about 2 for natural rubber to 4 or more for non-crystallizable unfilled rubbers, and B is a constant.

TABLE 6.   Experimental Values of the Crack Growth Limit T_o and the Tensile Strength for Various Rubbers.

Rubber	$T_o (10^4$ erg/cm^2)		Tensile Strength $(10^6$ dyn/cm$^2)$
	At Atm. Pressure	In Vacuo $(10^{-6}$ torr)	
natural rubber (cispolyisoprene, crystallizing)			
mainly polysulphide crosslinks	2.5	6.5	265
mainly polysulphide crosslinks*	4	-	275
mainly monosulphide crosslinks	1.5	4	230
carbon-carbon crosslinks	2	-	200
isomerized natural rubber (60% trans-polyisoprene, non-crystallizing)	7	-	15
synthetic cispolyisoprene* (crystallizing)	7	-	230
butadiene-styrene copolymer* (non-crystallizing)	6	-	27
cis-polychloroprene* (crystallizing)	7	-	130
butyl rubber (weakly crystallizing)	4	4	18

These data are taken from Lake & Lindley (Ref. 83), where details of the vulcanizates are given. Rubbers marked with an asterisk contained added antioxidant, which can increase the T_O value at atmospheric pressure as shown by the results for natural rubber. The tensile strength was determined in accordance with B.S.903.

B.   Fatigue Wear

Fatigue wear is difficult to be separated from other wear processes, e.g. abrasive wear. For example, the predominant wear process for a polymer sliding on a gauge is fatigue. A convenient way in characterizing, approximately, the fatigue properties of a polymer[70] is by the use of the following expression:

$$\eta = \left(\sigma_0 / \sigma \right)^b , \qquad (32)$$

where η is the number of cycles to failure, σ_0 the static strength corresponding to a single application of a stress, and σ the applied stress. The exponent b is a material property and is > 1. It is important to note that the fatigue properties of a material related to wear are not those of the bulk, but of the interfacial layer.

Fatigue is usually considered as a mechanical process of failure, involving crack initiation and propagation. Thus, at the molecular level, the failure can be considered to be a thermal activated process.

According to Zurkov[85], the time-to-break under a constant external stress σ takes the form of an Arrhenius relation:

$$\tau = \tau_o \exp\left(\frac{U_o - \beta\sigma_B}{kT}\right),\qquad(33)$$

where τ_o, U_o and B are material constants; k Boltzmann's constant. τ_o and U_o are close to the natural period of lattice oscillations and interatomic binding energy (or cohesive energy). The constant β is a measure of the effectiveness of the stress in overcoming the bond activation barrier U_o.

Free radical formation was identified with e.s.r. during the fracture process by Zurkov[85], Peterlin[86] and DeVries[87]. Thus, polymer wear due to the fatigue as well as the abrasive processes can lead to molecular degradation or thermal degradation of polymers. The degradation should predominantly take place on the polymer surface as shown by Bely and Sviridyonok[11].

C. Roll Formation

Originally, roll-formation was proposed by Reznikovskii and Brodskii[88] to be limited to high elastic materials. New evidence[71] shows that sliding friction of polymers with varying degrees of ductivity can also undergo roll formation. Furthermore, polymers usually considered to be brittle at room temperature can be ductile under load (20 to 30 Kg/cm^2) about 80° under their T_g's.

Interestingly, the transition from brittle to ductile behavior[89] follows a similar activation process as discussed for fracture:

$$\tau_r = \tau_{or} \exp\left(\frac{\Delta H_a^* - \Psi\sigma}{RT}\right),\qquad(34)$$

where τ_r is time-to-yield; τ_{or} a constant $\sim 10^{-12} - 10^{-13}$ sec; ΔH_a^* the activation energy for the glass transition; Ψ is a stress-concentration coefficient similar to a in Zurkov's equation, and R is the gas constant.

On the basis of Eyring's rate theory, an expression similar to Zurkov's equation can be derived as

$$\tau = \tau_o \exp\left(\frac{U_o - \alpha\sigma}{RT}\right),\qquad(35)$$

where τ is the time to fracture; τ_o, material constant similar to τ_{or}; U_o, the activation energy for the brittle fracture; α, a stress concentration factor as β in Zurkov's equation.

From equations (35) and (36), Aharoni[71,89] concluded that when

$$\tau > \tau_r,$$

the material will yield in a ductile manner, and that when

$$\tau < \tau_r,$$

the polymer will undergo a brittle fracture. Amorphous polymers whose ΔH_a^* is much larger than their relative U_0 generally fail in a brittle manner. These criteria were applied to determine the mechanism of roll formation.

D. Polymer Transfer

In view of the indirect roles of surface energetics in numerous forms of polymer wear, we can not establish a unique correlation between surface energetics and each of the above discussed mechanisms. The transfer of polymer to the counterface does not follow a certain set of rules or regularities. Pooley and Tabor[90] concluded that the low friction and the light transfer of PTFE and polyethylenes are not affected by surface energy, e.g. γ_c, or crystal texture of the polymer, or the crystallite size. The transfer is essentially due to smooth molecular profiles. On the other hand, Tanaka[91] observed the formation of a long band on PTFE and determined the wear rate to be a function of the width of the bands rather than the crystal- linity. This wear mechanism bears a similarity to that of delamin- ation[92].

The transfer of metal to the polymer surface was first observed by Rabinowicz and Shooter[93]. This type of transfer was assumed to be caused by the strong local adhesion. On the other hand, the transfer of polymer to metal has been frequently reported[94]. Owing to the polymer transfer, the kinetic friction generally reduces.

Sviridyonok et al.[95] studied the transfer between polymer and polymer. Several polymers with known critical surface tension γ_c were used for the study: PTFE, (18.5 dyne/cm), polyethylene (31 dyne/cm); PMMA - (39 dyne/cm); polyethylene terephthalate (PETP) - (43 dyne/cm); polycaprolamide (PCA) - (46 dyne/cm). The results indicate that a polymer generally transfers from less to more polar polymer surfaces. For highly polar polymers, the transfer mechanism appears to be accompanied by formation and interaction of free radicals. This interesting work shows that finally surface energetics appear to be involved directly in polymer transfer.

CONCLUSIONS

This survey has shown basic roles of surface interactions during the polymer sliding friction process. Surface energetics influence only the adhesion component of the friction force. Even for the adhesion component, deformation loss can greatly control the friction, especially for rubbery polymers. For example, the loss angle appears to be a common denominator for both adhesion and deformation components. Thus, the internal friction as well as the external friction are operating simultaneously under load for most polymers at their rubbery state.

The adhesion component, in fact, consists of various interaction components. We may express the friction force as an adduct of different surface forces:

$$F_a = F_a^d + F_a^o + F_a^i + F_a^h + F_a^e,$$

where superscripts d, o, i, h, e denote dispersion, orientation, induction, hydrogen bonding and electrostatic interactions. In general, van der Waals interactions are much more important than the electrostatic interactions.

Surface energetics do not affect polymer wear as directly as polymer friction. The commonly used parameters, e.g. fracture surface energy or tearing energy, contain only a small portion of surface free energy. The deformation energy predominates the total energy loss. However, surface energetics are still fundamentally important.

Most published evidence on polymer friction led one to believe that the adhesion mechanism can explain many unrelated phenomena provided that the relaxation aspect is taken into account. It appears that as long as polymers are concerned, relaxation controls both adhesion and friction. Thus, a clear demarcation between adhesion and deformation components is unattainable, and many unnecessary arguments about mechanisms can be avoided if relaxation[96] is considered to be the intrinsic property of polymer.

REFERENCES

1. L. H. Lee, (Ed.), Coatings and Plastics Preprints, 34, No. 1 (1974), (Also this volume, Advances in Polymer Friction and Wear, Plenum Press, New York (1974)).
2. D. Tabor, Ref. 1, p. 203 (1974).
3. A. R. Savkoor, Ref. 1, p. 220 (1974).
4. F. P. Bowden and D. Tabor, The Friction and Lubrication of Solids, Oxford at the Clarendon Press, Vol. I (1950) and Vol. II (1964).
5. K. Tanaka, J. Phys. Soc. (Japan), 16, 2003 (1961).
6. S. Bahadur and K. C. Ludema, Wear, 18, 109 (1971).
7. K. V. Shooter and D. Tabor, Proc. Phys. Soc., B65, 661 (1952).
8. H. W. Kummer, Eng. Res. Bull. B-94, Pennsylvania State University, July (1966).
9. D. T. Clark, Ref. 1, p. 290 (1974).
10. D. H. Buckley and W. A. Brainard, Ref. 1, p. 298, (1974), Also Wear, 26, 75 (1973).

11. V. A. Bely and A. I. Sviridyonok, Ref. 1, p. 416 (1974).

12. K. Tanaka and Y. Uchiyama, Ref. 1, p. 346, (1974).

13. I. V. Kragelsky, V. A. Bely and A. I. Sviridyonok, Ref. 1, p. 416 (1974).

14. D. Tabor, Phil. Mag., 43, 1055 (1952).

15. D. Tabor, Brit. J. Appl. Phys., 6, 79 (1955), and Proc. Roy. Soc. (London) A 229, 198 (1955).

16. D. G. Flom, J. Appl. Phys., 31, No. 2, 306 (1960), 32, No. 8, 1426 (1960).

17. J. A. Greenwood and D. Tabor, Proc. Phys. Soc., 71, 989 (1958).

18. N. P. Istomin and A. D. Kuritsyna, Mashinovedeniye, No. 1, 104 (1965).

19. A. D. Kuritsyna and P. G. Meisner, "Plastics as Anti-friction Materials", Izd. AN SSSR (1961).

20. F. S. Conant and J. W. Liska, Rubber Chem. and Technol. 33, No. 5, 1218 (1960).

21. A. Schallamach, Rubber Chem. and Technol., 41, 209 (1968).

22. D. F. Moore and W. Geyer, Wear, 22, 113 (1972).

23. D. F. Moore, The Friction and Lubrication of Elastomers, Permagon, Oxford (1972).

24. D. Bulgin, G. D. Hubbard and M. H. Walters, Proc. 4th Rubber Tech. Conf. (London), 173, May (1962).

25. W. O. Yandell, Wear, 17, 229 (1971).

26. J. J. Bikerman, Ref. 1, p. 244 (1974).

27. R. C. Bower and C. M. Murphy, "Status of Research on Lubricants, Friction, and Wear", NRL Report 6466, January (1967).

28. A. Schallamach, Proc. Phys. Soc., 66B, 386 (1953).

29. G. M. Bartenev and V. V. Lavrentev, Polymer Friction and Wear, (Russ), "Chemistry", Publisher, Leningrad, (1972), Translated by D. B. Payne into English, Editor L. H. Lee, to be published (1974).

30. K. Kanamaru, Kolloid Z. U. Z. Polymere, 192, 51 (1963).

31. T. Hata, Recent Advances in Adhesion, Ed. L. H. Lee, p. 269, Gordon and Breach, N. Y., London, and Paris (1973).

32. M. R. Hatfield and E. B. Ratemann, J. Phys. Chem., 60, 957 (1956).

33. A. N. Gent and J. Schulz, Recent Advances in Adhesion, Ed. L. H. Lee, p. 253, Gordon and Breach, N. Y., London, and Paris (1973).

34. J. R. Huntsberger, Treatise on Adhesion and Adhesives, Ed. R. L. Patrick, Vol. 1, p. 119, Marcel Dekker, New York (1967).

35. E. H. Andrews and A. J. Kinloch, Proc. Roy. Soc. (London), A332, 385 and 401 (1973).

36. K. Kendall, J. Phys. D. Appl. Phys. 4, 1186 (1971).

37. D. H. Kaelble, Treatise on Adhesion and Adhesives, Ed. R. L. Patrick, Vol. 1, p. 169, Marcel Dekker, New York (1967).

38. R. C. Bowers, W. C. Clinton and W. A. Zisman, Mod. Plastics, 31, 131, February (1954).

39. R. C. Bowers and W. A. Zisman, Mod. Plastics, 40, 139, December (1963).

40. G. H. West and J. M. Senior, Wear, 19, 37 (1971).

41. R. F. King and D. Tabor, Proc. Phys. Soc., B66, 728 (1953).
42. W. A. Zisman, "Contact Angle, Wettability and Adhesion",
     Advances in Chemistry Series No. 43, p. 1, American Chemical
     Society (1964).
43. K. C. Ludema and D. Tabor, Wear, 9, 329 (1966).
44. H. Krupp, Advan, in Colloid and Interface Sci., 1, 111 (1967).
45. R. J. Good, Treatise on Adhesion and Adhesives, Ed. R. L.
     Patrick, Vol. 1, p. 119, Marcel Dekker, New York (1967).
46. F. London, Z. Physik, 63, 245 (1930).
47. H. B. G. Casimir and D. Polder, Phys. Rev. 73, 360 (1948).
48. W. H. Keesom, Phys. Z., 22, 126 (1921), 23, 225 (1922).
49. N. A. Konstantinova, Thesis, MGPI, (1967).
50. E. Southern and R. W. Walker, Ref. 1, p. 282 (1974).
51. P. J. W. Debye, Phys. Z., 21, 178 (1920), 22, 302 (1921).
52. E. A. Dukhovskoy, A. M. Ponomaryev, A. A. Silin and V. L. Talrose,
     Intern. J. of Polymeric Mater., 1, 203 (1972).
53. M. Watanabe, M. Karasawa and K. Matsubara, Wear, 12, 185 (1968).
54. M. Van Den Temple, Advan. Colloid Interface Sci., 3, 137 (1972).
55. H. C. Hamaker, Physica, 4, 1058 (1937).
56. J. Gregory, Advan. Colloid and Interface Sci., 2, 396 (1969).
57. E. M. Lifshitz, Sov. Phys. JETP (Engl. Transl.), 2, 73 (1956).
58. H. Krupp, J. Colloid and Interface Sci., 39, No. 2, 421 (1972).
59. D. Tabor and R. H. S. Winterton, Nature, 219, No. 5159, 1120
     (1968).
60. K. Krupp, Recent Advances in Adhesion, Ed. L. H. Lee, p. 123,
     Gordon and Breach, New York, London, and Paris (1973).
61. B. V. Derjaguin and N. A. Krotova, Adheziya, Aca. Sci. Publ.,
     Moscow (1949).
62. S. M. Skinner, R. L. Savage and J. E. Rutzler, Jr., J. Appl.
     Phys., 24, 438 (1953).
63. D. W. Van Krevelen: Properties of Polymers, Correlations With
     Chemical Structure, p. 218, Elsevier, Amsterdam (1972).
64. D. K. Owens, J. Appl. Polymer Sci., 8, 1465 (1964).
65. S. C. Cohen and D. Tabor, Proc. Roy. Soc. (London), A291, 186
     (1966).
66. N. C. Hilyard, Brit. J. Appl. Phys., 17, 927 (1966).
67. J. M. Senior and G. H. West, Wear, 18, 311 (1971).
68. R. P. Steijn, Metals Eng. Quart. 9, May (1967).
69. S. B. Ratner and I. I. Farberova, Soviet Phys. 8, (9), 51 (1960).
     Also J. K. Lancaster, Wear, 14, 223 (1969).
70. M. M. Reznikovskii, Soviet Rubber Technol., 19 (9), 32 (1960).
71. S. M. Aharoni, Wear, 25, 309 (1973).
72. J. P. Giltrow, Wear, 15, 71 (1970).
73. J. L. Gardon, J. Phys. Chem., 67, 1935 (1963).
74. L. H. Lee, Interaction of Liquids at Solid Substrates, Advan.
     Chem. Series, No. 87, p. 106, American Chemical Society (1968).
75. A. A. Griffith, Phl. Trans. Roy. Soc., 221, 163 (1921).
76. G. R. Irwin, J. Appl. Mech., 24, 361 (1957).
77. E. Orowan, Rept. Prog. Phys., 12, 185 (1949).
78. J. P. Berry, J. Appl. Phys., 34, 62 (1963).

79. R. W. Davidge and G. Tappin, J. Mat. Sci., 3, 165 (1968).
80. N. A. Brunt, Kolloid Z. u. Z. Polymere, 239, No. 1, 36 (1970).
81. G. J. Lake and A. G. Thomas, Proc. Roy. Soc. (London), A300, 108 (1967).
82. L. H. Lee, J. Polym. Sci., A-2, 5, 1103 (1967).
83. G. J. Lake and P. B. Lindley, J. Appl. Polym. Sci., 9, 1233 (1965).
84. D. H. Champ, E. Southern, A. G. Thomas, Ref. 1, p. 237 (1974).
85. S. N. Zhurkov and E. E. Tomashevsky, Proceedings of the Conference on Physical Basis of Yield and Fracture, p. 200, Oxford (1968).
86. A. Peterlin, Poly. Eng. Sci., 9, 172 (1969).
87. K. L. DeVries, D. K. Roylance and M. L. Williams, Intern. J. Fract. Mech., 1, No. 2, 197 (1971).
88. M. M. Reznikovskii and G. I. Brodskii, Sov. Rubber Technd., 20, (7), 13 (1961).
89. S. M. Aharoni, J. Appl. Polym. Sci., 16, 3275 (1972).
90. C. M. Pooley and D. Tabor, Nature, Phys. Sci., 237, 88 (1972).
91. K. Tanaka, Y. Uchiyama and S. Toyooka, Wear, 23, 153 (1972).
92. S. Jahanmir, N. P. Suh and E. P. Abrahamson, Wear, 28, 235 (1974).
93. E. Rabinowicz and K. V. Shooter, Proc. Phys. Soc., B65, 671 (1952).
94. D. Tabor and R. F. Willis, Wear, 13, 413 (1969).
95. A. I. Sviridyonok, V. A. Bely, V. A. Smurugov and V. G. Savkin, Wear, 25, 301 (1973).
96. V. V. Lavrentev and V. L. Vakula, Ref. 1, p. 417, (1974).

DISCUSSION OF PAPER BY L. H. LEE

S. Bahadur (Iowa State University): How would you explain physically the relationship between adhesion and viscoelastic losses? It so seems to me that the tan δ term in the adhesion friction force relationship is merely to fill in as a constant to provide the equality relationship (if any)?

L. H. Lee: The loss tangent term is not merely a constant. It is real as demonstrated by so many authors dealing with elastomer adhesion.

For elastomers and polymers, adhesion can not be achieved without deformation. Therefore, in some work like Dr. Tanaka's paper, J. Phys. Soc., Japan, 16, 2003 (1961) there appeared to be little "adhesion" contribution but all "deformation" contribution to the friction force. In fact, both contribute to the friction force, but the importance is that the adhesion component also involves the deformation of polymer bulk. Therefore, some time the argument about friction theory becomes a matter of semantic.

J. J. Bikerman (Case-Western Reserve University): Perhaps I can answer your question. Those who believe in the adhesion theory of friction assume that frictional force is more or less proportional to the molecular area of contact. To obtain this area, the polymer must be deformed. This deformation is supposed to depend on the viscoelastic properties of the material; thus, a relation between these properties and the frictional force is achieved.

R. R. Myers (Kent State University): I would think that if we were to accept Dr. Bikerman's explanation, then we would have an inverse dependence of friction upon the moduli. You would get a larger area of contact for a small modulus than you would for a large modulus. In other words, you would find that it is compliance that would be more appropriate for the explanation.

L. H. Lee: I have pointed out that the viscoelastic properties of polymer is not only affecting the contact area but also the shear strength. The work on the separation of A and S was carefully done by Drs. Ludema and Bahadur.

S. Bahadur: In emphasizing the contribution to sliding friction from polymer deformation, the role of adhesion should not be overshadowed. As described in your paper, the total friction is comprised of the deformation and adhesion components. The former accounts for the energy loss due to damping because of the loading and unloading cycle that the polymeric material in the substrate undergoes in a sliding situation. The theoretical calculations as well as the experimental results show that this contribution is very small in the case of thermoplastics. A proportionately much larger contribution is due to adhesion at the interface because of which there is additional deformation at the sliding interface which has been envisioned as shear occurring at a considerably high strain rate. In the case of lubricated sliding which eliminates the bulk of adhesion the shear deformation at the interface is practically absent. The adhesion component of friction has been correlated with limited success by Ludema and myself with the contact area which is governed by the elastic modulus of the polymeric material and the interface shear strength which is practically equal to the bulk shear strength of the polymeric material. We should therefore be cautious here not to loose sight of the difference between the viscoelastic losses which arise due to damping in the material and the shear deformation at the interface which is due to adhesion.

K. L. Mittal (IBM Corporation): You have only three points in the plot of Fa vs γ_c. In the light of various γ_c's (γ_c^A, γ_c^B, γ_c^C, c.f. Hata (1972)), do not you think that a plot of F_a vs γ_s using a variety of polymers will be more meaningful. Your limited linear plot can be quite misleading.

L. H. Lee: Indeed, cricital surface tension γ_c is only an approximate solid surface tension γ_s. However, for polymers, some γ_c values are rather close to γ_s values. Of course, an ideal plot should be based on γ_s and perhaps with more data points. The plot in the text is an illustration for a better fit of few data points after F_a values were used instead of the total friction force, F, or the coefficient of (total) friction, μ.

For solid surface tension data of polymers, you may refer to Dr. Wu's paper in Recent Advances in Adhesion, (Ed. L. H. Lee, Gordon and Breach, New York, London and Paris, 1973).

Adhesion and Deformation Friction of Polymer on Hard Solids

Arvin R. Savkoor

Laboratorium voor Voertuigtechniek, Technische Hogeschool

Delft, The Netherlands

The main objective of this paper is
to present a detailed view of various aspects
of adhesion and deformation friction based
mainly upon the research findings of this
laboratory. An attempt is also made to
analyse the large number of results reported
elsewhere in order to develop a coherent
presentation. Certain fundamental aspects
of friction are discussed at first from a
phenomenological point of view. A discussion
of the law of friction relating to motion
follows with a study of the stick-slip
phenomena. The next part deals with the
isothermal adhesive friction of elastomers
on nominally flat and smooth surfaces.
Various factors such as physico-chemical
nature, sliding speed, temperature, pressure
and viscoelastic properties are studied.
A quantitative theory to predict the frictional
behaviour from physical data specifying the
above factors is given along with experimental
results obtained in support of the theory.
For a given friction pair and normal load,
dependence upon the speed and temperature
may be calculated from the viscoelastic
functions. The magnitude of the friction
coefficient in a given viscoelastic state
depends upon surface adhesion, which
in turn is related to contact angle data,
and upon the real area of contact which

may be evaluated from elastic and surface
statistical data. Certain other factors
influencing friction such as the presence
of water film and an applied electrical
potential are briefly considered. Finally
the deformation friction arising from
inelastic behaviour of materials is considered
along with a discussion of existing theories.

INTRODUCTION

PRELIMINARY CONSIDERATIONS

SURFACE PROPERTIES RELATED TO FRICTION

QUALITATIVE DESCRIPTION OF THE MECHANISM

FUNDAMENTAL ASPECTS

FACTORS AFFECTING FRICTION

RELATIVE MOTION AND TEMPERATURE

DISPLACEMENT DEPENDENCE IN INITIAL FRICTION

NORMAL LOAD, NOMINAL AREA, PRESSURE AND SURFACE ROUGHNESS

MATERIALS FOR FRICTION PAIR

    Elastomer Properties

    Counterfaces of Hard Solids

EVALUATION OF THEORETICAL MODELS

A QUASI-QUANTITATIVE THEORY

    Friction of Asperities

    Details of Calculations

    Other Factors

DEFORMATION FRICTION

CONCLUSIONS

INTRODUCTION

Friction implies a resistance to relative motion of bodies in contact where no obvious impediment to such motion seems to exist. Some of the early investigators recognized the futility of looking at the surface roughness in the hope of locating a source of purely mechanical obstruction. In search of a raison d'être, they undertook to study the interaction taking place between surface molecules during the motion. In view of the scale and the complexity of the problem, anything more than a qualitative indication can hardly be expected from such an approach. Moreover as more evidence becomes available, it is being appreciated that friction is not merely a surface effect; deformation of bodies also plays an important part. This inference has been emphatically brought out in numerous papers dealing with the friction of polymers. Knowledge of the details involved is not yet complete. It is felt however, that the field is at an advanced state of development.

A detailed study of polymer friction and wear is rewarding, not only for its own sake, but also for the sake of understanding the mechanism of friction in general. The performance of theoretical models of the mechanism may be assessed with a high degree of confidence. The reason being that, the dependence of friction upon various phenomenological quantities is so pronounced that the interpretation of results from experiments is not obscured by the difficulties of obtaining reproducible results.

Relative motion of contacting bodies usually refers to what may be described as "sliding". The relative motion occurs between points on the surface of contact. On the other hand, relative motion of bodies which occurs during free rolling does not involve any relative motion of points on the contact surface. Whereas the work done by the frictional forces in sliding is easy to perceive, it is not evident at first sight how the resistance to rolling comes into being. In recent years,it has been generally accepted that the inelasticity of materials is responsible for the rolling friction of polymeric solids. Since the mechanism of rolling friction has been well understood, recent efforts have been directed to the solution of well defined problems in continuum mechanics. A brief review of the progress made in this field will be described later.

On the other hand, understanding of the mechanism of sliding friction has been sketchy. This is partly owing to the fundamental nature of the difficulties which arise, in accounting for the complex factors governing friction. The other reason is that a consistent interpretation of experimental results is not generally easy. It is intended here to present an elaborate account of the mechanism of sliding friction with emphasis placed on the part dealing with adhesive friction of elastomers. An attempt has been made to develop a quantitative theory which admits the use of tangible properties. While considering the various aspects of sliding friction, it seems appropriate to examine and clarify some of the vague concepts revealed in the literature. Certain topics which have attained definite forms are stated briefly. There is no aspiration of embarking on a comprehensive survey of the literature.

PRELIMINARY CONSIDERATIONS

Sliding friction manifests itself in various forms depending upon the type of surface and the materials of the bodies. Obviously no single descriptive model of the mechanism could be applied to the entire range of the phenomenon. It is convenient, at the outset, to distinguish between surfaces which are dry from those which are wet. When the surfaces are wetted by a film of liquid, it is known that the coefficient of friction is generally reduced. At sufficiently high speeds of sliding, the liquid film provides for a partial or a complete load support by a mechanism of hydrodynamic or elasto-hydrodynamic lubrication. These aspects deal primarily with the mechanism of lubrication and fall outside the scope of this paper. When the sliding speed is very low, the hydrodynamic mechanism may be absent and a condition of boundary lubrication is obtained. The coefficient of friction is usually lower than that on a dry surface. It may be expected, however, that some relation exists between the two coefficients. It appears that the presence of a boundary lubricant modifies the mechanism of dry friction. In this context, the first step to be taken is that leading to the understanding of dry sliding friction. The main body of the paper is devoted to that purpose.

Sliding friction in its most mundane form arises on account of a mechanism involving adhesion of surfaces, deformation of bodies and a process leading to a local failure. Before attempting to be more specific, it is useful to consider certain features of surface interactions which are intimately related to the various details of the mechanism. Since the nature of surfaces plays a dominant role in friction, it is worthwhile to point out the surface properties which are relevant to the study of sliding friction.

SURFACE PROPERTIES RELATED TO FRICTION

Significant progress has been made in respect of the geometrical and physical properties of surfaces of hard solids. Relatively little is known about the surfaces of polymers. It is generally accepted that solid surfaces are covered by arrays of asperities and that their contact occurs over discrete "real areas of contact". Amongst the various descriptions of surface topography, that regarding the microtexture is of some consequence. It is convenient to describe the microtexture as "nominally smooth", if the slopes at various points on the surface are continuous. If, on the other hand, a surface possesses points where the slopes are discontinuous, it may be described as a sharp textured surface. Unfortunately, this aspect of surface roughness has received less attention than it deserves.

It is a common experience that substantial wear is associated with friction of sharp textured surfaces. In contrast to this, wear resulting from friction of nominally smooth surfaces is usually insignificant. Schallamach[1] observed that practically no wear occurred when a block of rubber was made to slide repeatedly on nominally smooth surfaces of hard solids. An instructive contribution of Archard[2] deals with the various forms of wear which are related to the number of encounters between asperities. It is interesting to note that the coefficient of friction on a nominally smooth surface is higher than that on a sharp textured surface. The different relations between friction and wear justify the distinction made between the two textures. It may be seen in the following section how the characterization of texture is related to the details involved in the mechanism of sliding friction.

The view that sliding friction on dry surfaces is mainly caused by adhesion between surfaces has been more or less established. The qualification "more or less" is used because of a lack of direct convincing proof. Objections have been raised by some workers to throw doubts on the adhesion theory. On closer examination it appears that most of the objections result from misunderstanding of loosely defined concepts. Recent work on this point will be described in the section dealing with the fundamental aspects. It is seen there that a surface topographical description alone is not sufficient for understanding the mechanism; surface chemical effects are of great significance to friction and wear.

The chance of finding a chemically clean surface is probably as remote as that of finding a perfectly flat surface. Generally, the surfaces of solids are contaminated by thin solid films consisting of adsorbed material. The thin boundary layer may display physical and chemical properties which are different from those of the bulk material.

The layer adheres to the bulk material so strongly that it is

not generally easy to remove the layer completely. On the other hand,
it is likely that the boundary layer has many faults within it.
Although some advance has been made in defining the properties of
boundary layers, the information is not adequate for the purpose of
estimating the strength of the layers. In order to make progress
it has been necessary to follow an approximate procedure. In particular, the property of surface energy is taken to serve the purpose
of the strength. Bowden and Tabor have described an alternate
approach based upon the concept of critical shear stress.

QUALITATIVE DESCRIPTION OF THE MECHANISM

A faint picture of the mechanism begins to emerge, when in
addition to the foregoing considerations, Green's inference[3],regarding
the kinematic conditions of contact,is taken into account. It may
be noted that the asperities of surfaces come into contact and adhere
over their sloping sides, adjacent to the summits. During the
relative motion of bodies, deformation and stress concentration may
occur within the real area of contact where the surfaces adhere.
It is conceivable that the process would terminate at some stage of
deformation, on account of the finite strength of adhesion. The
limit set to the process by the occurrence of failure depends upon
the surface strength. On nominally smooth surfaces, such a limit
may be attained naturally and without any loss of material in the
process. On the other hand, a premature termination of the process
may occur, if, during the tangential deformation of the bodies, an
asperity on a surface is forced into being cut by an edge of the
sharp textured countersurface (Figure 1). Although it is possible
to think of a few variants regarding the details of the cutting
process, the ultimate result is that the material is cut off either
in a single encounter or within a few such encounters. A part of the
asperity material which is left behind on the countersurface may
eventually be freed by a mechanism such as the one described by
Rabinowictz. Subsequent to the failure, dissipation of a small
amount of residual elastic energy may occur through damped vibrations
of the intact portion of the material.

The primary cause of friction on both the types of surfaces is
adhesion. In both the cases, deformation of material takes place.
The difference between the details of the mechanism in the two cases
arises from the different manner of failure. The reason of the
inverse relation observed between friction and wear is now clear.
The magnitude of the frictional resistance depends upon the condition
of failure. On a sharp textured surface, the point of failure is
determined by the intensity of stresses caused by a sharp edge and
the strength of the material. At the present time a suitable technique of measuring the sharpness is not available.

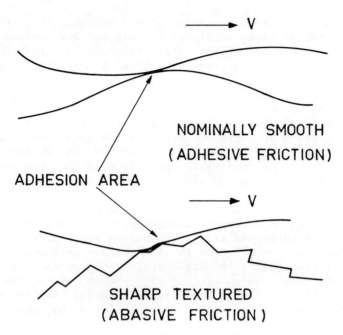

Fig. 1

Considerable progress has been made in describing the strength properties of polymers. It is known that the strength properties exhibit a viscoelastic behaviour, characteristic of polymers.

Considering the foregoing remarks, it is convenient to distinguish between the friction on the two types of surfaces by a suitable choice of names. Without going into the semantics, and in view of the fact that failure occurs at the surface of a nominally smooth solid, it seems appropriate to use "adhesive friction" to denote the friction of nominally smooth surfaces. Friction of sharp textured surfaces may be described as abrasive friction. The scope of the detailed discussion hereafter will be limited to the topic of adhesive friction.

FUNDAMENTAL ASPECTS

One of the most intriguing and at the same time controversial point of inquiry concerns with the primary cause of friction. The question is not merely academic, as some may like to believe; many a pragmatic attempt to understand and control friction of practical contrivances has not been successful. Moreover, some reflection upon this question is an essential prelude to the development of a quantitative theory.

Bowden and Tabor[4] have presented strong evidence in order to
establish that the adhesive origin of friction on dry surfaces
accounts for a major share of the total resistance offered to
sliding. Despite this, misgivings have often been spelt, since no
so-called "force of adhesion" is noticeable during normal separation
of bodies in contact under ordinary conditions. Regarding the
implication of our understanding of matter, it would seem only
natural that the molecules within the real area of contact adhere.
With reason, it would be proper to describe the behaviour in normal
separation as anomalous. It is to be remembered that an explanation
of the anomaly should be compatible not only with the observations
regarding normal separation but also with those regarding friction.

In that context, the explanation that the presence of contaminants and asperities on surfaces is responsible for the negligible
adhesion under ordinary conditions is not adequate. Under similar
conditions, the force of friction is of the order of the normal
load. Nor is it reassuring to note that a strong adhesion and high
friction coexist in an environment of high vacuum. Some other workers
introduced a notion of elastic recovery of material, but the notion
is not precise enough to be effective in resolving the misgivings.

Significant progress has been made very recently, with the help
of the concepts of continuum mechanics and fracture mechanics. The
approach is based upon the following interpretation of adhesion: In
phenomenological terms, adhesion prescribes a constraint of "no
relative motion" for surface points within the real area of contact.
There are two common modes of overcoming adhesion. In the so-called
"adhesion test", normal separation of bodies is caused by pulling
them apart. The other mode occurs during frictional sliding. In
this case, the separation occurs tangentially, i.e. along the plane
of contact.

The above approach offers two significant advantages. Since
phenomenological quantities are involved, it is possible to set up
critical experiments in order that the theory may be verified.
Secondly, by virtue of the well defined nature of the formulation,
the approach indicates where to look for the source of the anomaly.
It is therefore worth describing the essential features of the recent
investigations without going into details of the analysis.

The problem of normal separation of elastic spheres making
contact has been studied by Johnson et al.[5] by making use of an
energy approach. They found that the contact area under a given
normal load was larger than that expected using the classical Hertz
(Figure 2a) theory. The increase in the area of contact is caused
by the presence of surface energy. As the normal load is reduced,
the contact area decreases in a stable manner. At zero load, a
small but finite area is obtained. Further reduction of normal load
amounts to pulling apart of the solids. It was found that the contact

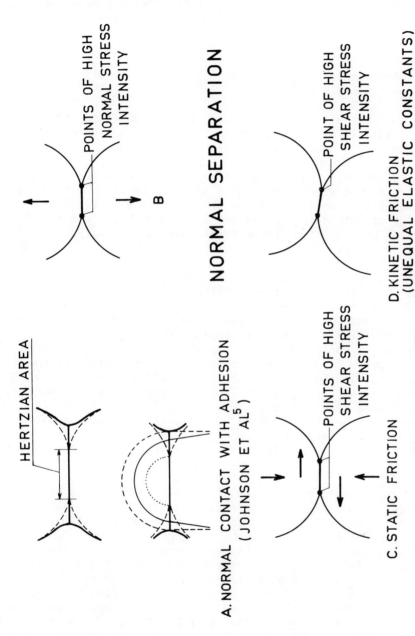

Fig. 2

area can withstand a small tensile load before a process of unstable
separation is initiated. The present author[6] investigated the same
problem by looking more closely at the stress intensity near the
circular boundary of the contact area (Figure 2B). It was shown
there that the results obtained by limiting the local stress intensity
were equivalent to those obtained from the energy approach. In effect,
the essential condition dictated by energy requirements,and the
sufficiency condition,provided by the high stress intensity,are
satisfied simultaneously.

The author[6] has also studied the problem of static friction.
Starting from an initial state in which the spheres are normally
loaded, a tangential force is applied and slowly allowed to increase
(Figure 2C). It is shown that the contact area, which was initially
larger than its Hertzian value, begins to decrease with increasing
tangential force by a process of stable normal separation. At a
certain value of the tangential force, the contact area attains its
Hertzian value corresponding to the normal load. The intensity of
normal stresses at this stage is so small that further normal
separation is unlikely.

The area remains constant. As the tangential force is increased
further, the intensity of the shear stresses at the boundary of the
contact goes on increasing. At a certain value of the tangential
force, a sliding mode may be initiated. Considerations of stability
indicate that the sliding mode is unstable. Shearing starts from
the boundary of the contact area and rapidly progresses towards the
centre. The process finally leads to tangential separation of the
bodies. The conditions of initiation were obtained by making use
of both the energy and stress intensity approaches. Although both
the approaches essentially give equivalent results, the former is
related to physical quantities which may be estimated independently.
The stress intensity approach does not yield an absolute limit; it
is nevertheless useful in predicting the initiation of failure, on
a comparative basis.

In a recent study[6], the problem of kinetic friction has been
formulated in accordance with the kinematic conditions of Green[2].
The problem has been analysed for two cases of elastic constants.
In the first, the elastic constants of the two bodies are equal.
The analysis has many features which are common to those of the
problem of static friction. In the second case, one of the bodies
is taken to be rigid and the other is regarded as being elastic
(Figure 2D). The geometry of deformations, in this case, is strongly
asymmetric. The energy approach is not convenient, since it requires
a prior knowledge of the shape of the region of failure. The solution
has been obtained from the considerations of stress intensity. The
process of failure initiates at the foremost point on the leading
edge of the contact area. It may be expected that the sliding mode
of failure is unstable.

These studies are based upon the notions of fracture mechanics. Hitherto, the process of initiation and the question of stability has been studied. In view of the potentiality of the methods, it appears that a comprehensive study of many other forms of asperity interaction may be contemplated. For example, the asperity interaction which arises when the normal force is not constant, but oscillates, which may lead to a failure involving combined modes. This may influence the value of the coefficient of friction.

It remains now to examine the theoretical results described above. Carefully performed experiments of Johnson et al.[5] clearly indicate that the process of normal separation is accurately described by the theory. Similar experiments regarding the tangential separation have yet to be performed. In both the experiments it would be desirable to know the precise location of the plane of failure. It is conceivable that failure occurs in normal separation by a mere detachment of the adhesive bonds. It may not involve any disruption of the boundary layers. On the other hand, the failure by tangential separation may be effected by shearing the boundary layer off. The difference between the values of strength for the two modes may be the obvious cause of the anomaly. Courtney-Pratt and Eisner[7] have studied the details of breakdown of surface layers of contacting metallic bodies. They observed that the normal forces cause very little breakdown of surface films; marked breakdown occurs only when sufficiently large tangential forces are applied.

Even in a hypothetical situation where separation by either modes occurs within a boundary layer, it is unlikely that two values of strength are equal. Since the boundary layer may hardly be regarded as isotropic, it would not be surprising to observe different values of strength in adhesion and friction.

It has been tacitly suggested that the tangential separation involves shearing through the boundary layer. Furthermore, it has been assumed that the failure is initiated locally. On the other hand, if the boundary layer consists of a ductile material, failure may take place by plastic flow. In that case, a criterion of failure may be given as a critical value of the average shear stress. It remains to be seen how far these details represent the actual process.

Hitherto, the discussion referred to the details of the mechanism of friction of elastic bodies. In friction of polymers, it is necessary to consider the viscoelastic properties of the materials. In first thoretical treatment, it would be convenient to undertake a linear viscoelastic stress analysis. A detailed analysis of the mechanism of separation is not available at present. In the absence of such an analysis, it has been necessary to use an approximate heuristic approach based upon the consideration of energy.

    Irrespective of the details of the process of failure, it is
evident that a certain amount of energy is needed in order to effect
the breakdown of adhesion. Since the process involves the formation
of new area, the energy required depends upon the extent of the new
area. This is true, regardless of the manner of formation,
whether it occurs by detachment of adhesive bonds or by shearing
the boundary layer. It is therefore reasonable to represent the
strength property by a specific surface energy. The magnitude of
this quantity may, after due consideration of the details of failure
be suitably chosen.

    In the friction of polymers, the material surrounding the real
area of contact is inelastically deformed. In order to effect
separation, the work done by an external force has to provide for
any loss of energy which occurs during deformation, in addition to
the energy that is needed to overcome adhesion. Consequently, the
resistance offered to sliding in adhesive friction of polymers
includes a part arising from the inelastic deformation. It is worth
noting that the part owes its existence to adhesion. It depends,
therefore, upon the strength of adhesion. It has nothing to do with
the mechanism of deformation friction which occurs during free
rolling.

                        FACTORS AFFECTING FRICTION

    There are few circumstances more disconcerting to an engineer
than a lack of knowledge about the expected force of friction which
has to be reckoned in a particular application. It is not practicable
to apply the knowledge of the details of asperity interactions to
practical problems involving large surfaces. It is desirable to
obtain constitutive relations between the force of friction and other
phenomenological quantities of the sliding process. These relations
are meaningful, only when they hold for a reasonably small part of
the surface containing a group of asperities. This condition is
usually valid in ordinary friction, at least on a statistical basis.
Much of our present understanding of the constitutive relations has
been formulated in the form of the laws of friction.

    Classical approaches of Amonton and Coulomb are of great value
in formulating the laws of friction. It is indeed surprising, considering on one hand, the limited number of experiments and the crude
experimental facility available at that time, and on the other hand,
the inadequate knowledge of the mechanism, that the classical work
has yielded viable laws of friction.

    By its very notion, friction involves the concepts of contact,
relative motion, the bodies and their surfaces. It suggests itself
that the laws of friction have to describe the factors which are
related to the intensity of contact, the details of relative motion,

the material of the bodies, and the characteristics of the surfaces. The intensity of contact depends upon the normal load, the nominal area of contact, the deformation properties of materials and the topography of the surfaces. The details of the relative motion may be described firstly by a logical statement, depending upon whether relative motion between the surfaces occurs. Secondly, when such motion occurs, the relative displacement with its time derivatives may be considered. Although not explicitly expressed in the original notion of friction, time may be an important factor even when there is no relative motion. Finally, friction may be regarded as a characteristic property of the materials and their surfaces.

The deformation properties of viscoelastic materials are functions of both time and temperature. Obviously, in addition to the aforementioned factors, temperature has to be considered. Since some of the factors are interrelated as far as their influence upon friction is concerned, it is convenient to study them in groups.

RELATIVE MOTION AND TEMPERATURE

This is probably the most distinct feature of polymer friction. In order to examine the dependence of friction on relative motion, it is at first convenient to consider that the temperature is constant. The isothermal coefficient of friction μ may be related to the relative displacement x, measured at the surfaces and time t reckoned from some characteristic event. Under kinetic conditions

$$\mu = \mu(x, \dot{x}, \ddot{x}, \ldots, t) \quad \text{for } \dot{x} \neq 0 .$$

In this expression the dots denote the time derivatives. An inequality arises when no relative motion takes place.

$$\text{If } \dot{x} = 0, \mu(t) < \mu_o(t) ,$$

where $\mu_o(t)$ is a time dependent limiting value of the static coefficient of friction.

Of the considerable amount of work done on kinetic friction, that pertaining to the dependence of friction upon uniform sliding speed has been the most exhaustive. Known experimental evidence[8] suggests that the speed dependence of friction is by far the major factor during motion.

It is now generally recognized that in the region of low sliding speeds, the coefficient of friction increases with speed. At a certain speed, the coefficient attains its maximum value. On further increasing the speed, the coefficient decreases. The variation of the coefficient with the logarithm of speed has a characteristic bell-shaped form. This was first reported by Grosch[9]. Among others, Ludema and Tabor[10], and the author[8] have obtained similar results in their experiments. The qualitative feature of the variation has been confirmed. There are two regions of speed in which the previous work has not been satisfactory. These are the region of very low speeds and the region in which friction decreases with increasing speed. There is also some confusion in the interpretation of the experimental results obtained before a steady value of friction is attained. Most of the reported experimental studies have been performed with the help of tribometers employing a constant speed drive facility. In order to maintain isothermal conditions, the maximum drive speed is usually restricted to a few centimeters per second. At higher sliding speeds, a rise in temperature may occur during sliding on account of frictional heating.

Static Friction

Results obtained by Grosch[9], Ludema and Tabor[10] suggest that the coefficient of friction at very low speeds of sliding tends to approach a finite constant value. This important feature of the friction-speed relation has often been neglected; It is of great significance both in theoretical studies and practical problems. Theoretically speaking, the static coefficient may be considered as a direct manifestation of adhesion during tangential separation. In practical applications involving rolling contacts, the calculations involving regions of adhesion and slip are based upon the assumption and the magnitude of the static coefficient.

Direct determination of the magnitude of the limiting value of static coefficient of polymers is a difficult, if not an impossible, task. One of the reasons is that the variation of friction with speed has an asymptotic nature as the speed tends to zero. The other reason is that it is difficult to decide precisely when the relative motion commences; relative motion of points within the real area of contact has to be observed. Measuring displacements at other points, introduces additional difficulties since the time dependent creep properties of the material have to be accounted with. The limiting value of the coefficient of friction may be expressed more conveniently in an asymptotic form rather than in the original form.

$$\lim_{V \to o} \mu(V) \to \mu_o .$$

Our experimental results[8] show that the above representation is meaningful. In the region of very low speeds where the variation of kinetic friction is small, a good estimate of the limiting static coefficient is obtained from the value of the kinetic coefficient at any one speed. The estimation may be made more quickly if the

requirement of a very low speed is somewhat relaxed by conducting
the test at a sufficiently high temperature. This point becomes
clear, in a later section, where the interrelation between speed
and temperature effects is considered.

It also appears from our results that the limiting value of
the static coefficient does not show any specific dependence upon
the initial time of contact. Admittedly, this has been observed
over a limited range of contact times owing to obvious limitations.

DISPLACEMENT DEPENDENCE IN INITIAL FRICTION

It is usually observed that the force of friction in the constant
speed tests initially increases with the relative displacement before
attaining its steady value. Some authors[11],[12] have considered "an
initial force of friction" to describe the value of the tangential
force when the test specimen is about to slide. If a relation of
the form $\mu(x,\dot{x})$ is considered to describe the kinetic coefficient,
the initial coefficient μ_i may be expressed as

$$\mu_i(V) = \mu(\varepsilon,V),$$

where ε is a small noticeable displacement. Since the precise
instant of transition from the state of rest to that of relative
movement is not known, $\varepsilon > o$. The value of ε is arbitrarily chosen
and moreover from the following discussion it appears that the
assumption of $\dot{x}=V$ is not correct when ε is small. It is, therefore,
doubtful whether the initial friction has any distinct physical
significance. It is worth reminding that physically meaningful
results from experiments which intend to describe the transition
from rest to relative motion, can only be obtained from the relative
displacements of points on the real area of contact. This task, as
seen earlier in the previous section, is a formidable one.

Difficulties have also been experienced sometimes in interpreting
the experimental results because the dynamics of the measuring system
has not been taken into account. In order to clarify the observations
regarding the increase of friction with displacements, a simple
analysis of the dynamics was performed[13]. On the assumption that
the coefficient of friction is a function of speed alone, it was
found that the tangential force initially increases with the sliding
distance and finally attains its steady value. The result is not
unexpected, since the body starts from rest and therefore cannot
attain the imposed speed at once. Some displacement is needed before
the sliding speed equals the imposed speed. It appears that this
explanation accounts for a considerable part of the displacement
dependence of friction. This would indicate that a term involving
the displacement dependence although small, may exist in the constitutive relation. A physical reason based upon the orientation of

molecules during sliding has been offered by Schallamach[1]. It is
also possible to think of a simpler explanation such as an action
of wiping the surface clean during the motion. Further work on
this point is needed before any definite conclusions may be drawn.
A tentative conclusion is that the displacement dependence is so
small that it does not seem to be important enough to justify further
investigation; the modest aim of avoiding the possible confusion in
the interpretation of experimental data is served. On the other hand,
the possible effects of non-uniform motion cannot be ruled out.

Non-uniform Motion

In a general description of relative motion, it is also desirable
to consider the influence of higher derivatives of the relative
displacement. As yet, no such specific effort is known to the author.
Some experimental work has been done in which instead of prescribing
a constant drive speed, a constant tangential force is applied. The
ensuing motion then describes the relation between friction and speed.
It has been noted by the author[8] that the friction-speed relations
obtained from the constant speed and the constant force tests are
alike. A similar result has been found by the author using his
theoretical model. These results suggest that the coefficient of
friction is mainly dependent upon the speed. Moreover the coefficient
is more likely to be a point function rather than a path function of
speed. With the scant information available, the observation made
here may only be looked upon as tentative. In the friction of metals,
it has been stated by Rabinowicz[14] that the past history of motion
over a certain critical distance is an important factor. Such a
path dependence has not yet been studied in detail.

It appears that a detailed study of the stick-slip behaviour
may provide some clues to the behaviour of friction under a non-
uniform motion.

Negative Slope of Friction-Speed Curve and Stick-Slip

Although it is not always stated explicitly, much of the currently known experimental work has been unsatisfactory for speeds
exceeding the speed at which the maximum friction is obtained. The
variation of friction with speed in that region is such that friction
decreases with increasing speed. Despite the fact that the drive
mechanism imposes a constant speed, the sliding speed of the
surface of the specimen is not found to be uniform. The tangential
force varies periodically between its maximum and minimum values.
The results do not directly serve the purpose of obtaining the
friction-speed relation. Both the inertia forces, and the actual
sliding speeds are unknown.

Some of the authors[15] have taken the average value of the tangential force to represent the force of friction corresponding to a value of sliding speed which is equal to the imposed drive speed. Some others[9] consider the highest value of the periodic force for that purpose. Considering the aforementioned difficulties, the assumptions involved in these representations are questionable. In view of the sweeping nature of the assumptions, it would be necessary to reconsider the description of the friction-speed relation in this region.

There are two ways of obtaining the friction-speed relation. One of them, is to measure all the details of the motion which occur during the stick-slip vibrations. The force of friction corresponding to the various actual slip speeds may be deduced from the details. The other way is to alter the measuring system in such a way that no stick-slip occurs. The frictional force may be obtained simply and related directly to the uniform speed of motion. In our experiments, the second method has been followed. This has an additional advantage of a theoretical nature, over the first method. The motion is uniform; the complications which can arise from an unknown influence of higher derivatives in non-uniform motion do not apply.

It has been pointed out by Blok[16], that the negative slope of the friction-speed curve is one of the causes of stick-slip. In the equations of motion of the system, the negative slope introduces a term of negative damping. Even for moderate loads, the negative damping coefficient is very large for low sliding speeds. A non-linearity of the damping term makes the limit cycle behaviour possible. Stick-slip is a manifestation of the behaviour. A simple way to eliminate the stick-slip is to introduce a positive damping which is larger than that resulting from the negative slope. Conventional viscous dampers were found to be inadequate for that purpose.

An interesting solution resulted from these considerations. The positive damping could be obtained by making use of another friction pair, the friction-speed relation for which has a positive slope in the speed range of interest. By choosing a sufficiently large normal load, it is possible to obtain an adequate positive damping. This second pair may be mechanically coupled to the original pair. The tandem system so obtained, was found to be capable of executing uniform motion[17]. The friction-speed relations of some of the rubber compounds have been obtained in this manner. A typical friction-speed relation in the region of negative slopes, which has been obtained for a friction pair consisting of a block of SBR compound of the type used in tyre treads, and a stainless steel plate is shown in Figure 3. In the range of test speeds used, the friction-speed curves of many rubber compounds have positive slopes at room temperature and negative slopes at temperatures lower than 0°C. In order to determine the region of negative slopes, the tests have to

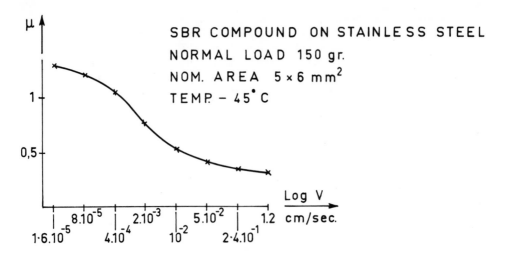

Fig. 3. Falling Friction-Speed-Curve Measured Without
         Stick-Slip

be carried out at low temperatures. The constant speed tribometer
is housed in a chamber with a temperature control facility. The
range of temperature control extends from -65°C to +130°C. The
speed range of the tribometer drive extends from 10^{-5} cms/sec to
1 cm/sec. The stabilizing friction pair consisting of a rubber
specimen, similar to the test specimen, and a hard solid, was mounted
outside the climate chamber. The two pairs were coupled mechanically
as shown in Figure 4. The test arrangement may also be seen in that
figure.

     For the same test pair, some results have been obtained to show
the variation of tangential forces which occurs without the use of a
stabilizing pair. A comparison of the forces recorded without and
with the modification are displayed in Figure 5.

     The saw-tooth form of variation obtained at the lower speeds
is clearly different from the quasi-harmonic form which occurs at
the higher drive speeds. It may be seen that the correct values of
the forces obtained without stick-slip are closer to the highest
values of the periodically varying forces at low speeds, and the
average values of the forces in the higher speed region. The ampli-
tude of the periodic forces, is large at the low speed end of the
range and is small at the higher speeds. Amongst the known experimental results in the literature, the results obtained by Grosch[9]
from the considerations of the highest force values, are less

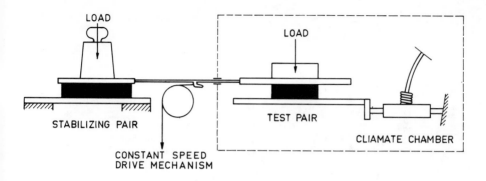

Fig. 4.   Schematic Representation of Tandem Measuring System

susceptible to inaccuracies. On the other hand, the friction-speed relations obtained from a consistent use of the average values of forces are seriously in error.

Having obtained the true friction-speed relation for uniform motion, it is interesting to analyze the detailed relation between the frictional force and the motion which occurs during stick-slip. It would then be possible to investigate the influence of the terms which involve the higher derivatives of displacement in the constitutive relation. A beginning in that direction has recently been made in our laboratory. In that context, it is worth noting that a phase plane analysis was undertaken by Hunt et al.[18] in order to investigate the stick-slip motion. They found that the analysis based upon speed dependent frictional properties could not account for their experimental results. They have suggested that friction depends, in some way, on the other details of motion.

Higher Speeds and the Interrelated Speed-Temperature Effect

Until now the discussion has dealt with the speed dependence which was restricted to speeds less than a few centimeters per second, in order to maintain the temperature constant. At higher sliding speeds, the temperature rises above the ambient temperature. This introduces complications in two ways. Firstly, it is not easy to

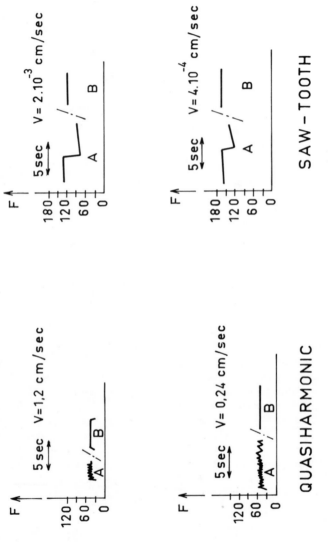

Fig. 5. Comparison of Force Traces (A) with and (B) Without Stick-Slip. SBR on Stainless Steel Temp. - 45°.

determine the average temperature of the interface. Furthermore, "flash temperatures" may occur during the sliding process. Secondly, the higher temperatures mainly occur in a layer beneath the surface. It is not clear which value of temperature should be considered effective in governing the friction properties. Before attempting to tackle this difficult problem, it is desirable to know the dependence of friction upon temperature under conditions of uniform temperature.

On examining together a number of isothermal friction-speed relations obtained at various uniform temperatures, it suggests itself that these relations are strongly correlated in some way.

Grosch[9] observed that a friction-speed curve obtained at one temperature could be matched with a different curve obtained at another temperature by shifting the curve sufficiently along the logarithmic speed axis. It is now generally recognized that the speed and temperature dependence of friction of polymers is similar to the well known time and temperature dependence of viscoelastic properties of polymers. The interrelation between the influences of speed and temperature in effecting a given change in the coefficient of friction, may be expressed empirically in the same way as that used to describe the rate and temperature effects in the viscoelasticity of thermorheologically simple materials. The interrelation is usually valid over a wide range of temperatures, extending from the glass transition temperature Tg of the polymer to a temperature of approximately Tg + 100°C. Furthermore, the interrelation may be approximately expressed using the W.L.F. equation. It applies to a wide range of polymers[9]. The horizontal shift log a_T which is needed for matching*the isothermal friction-speed relations obtained at various temperatures T, may be written as a function of T;

$$\log a_T = \frac{-8.86 \ (T-T_s)}{101.5 + T-T_s}$$

where T_s, characteristic temperature of the polymer is related approximately to the glass transition temperature T_g by

$$T_s \simeq T_g + 50°C.$$

Unfortunately, there are not many experimental results available to the purpose of confirming the aforementioned relations quantitatively. Besides the complications arising from stick-slip, a major difficulty which is typical of friction experiments is that the values of the coefficients are sensitive to surface conditions. In particular, the level of friction appears to be lower at temperatures below 0°C. It has been reported by Bartenev et al.[11] that condensation of atmospheric moisture is responsible for lowering the level of friction. The practical solution to this problem is to provide for special moisture traps or to carry out the measurements in vacuum.

In any case, it is but a conjecture, to presume that the surface
conditions during the tests at various temperatures are identical.
These factors have been mentioned here to emphasize the difficulties
involved and to remind that the relations discussed may be regarded
only as rough approximations. On the other hand, these are the
only available relations and are certainly adequate in describing
the interrelated effects of speed and temperature in a semiquan-
titative manner.

The significance of the interrelation is that the determination
of the friction-speed relation is possible, at least theoretically,
in the range of high speeds. It is possible to predict the friction
coefficient which may theoretically be obtained at higher speeds if
isothermal conditions prevailed. The information is obtained from
an actual isothermal test carried out at a temperature, which is
much lower than that at which the information is desired. It also
follows from the discussion that a single function of speed and
temperature may be used to express the dependence of friction upon
these two factors. A simple expression for the function may be used
viz, log $V.a_T$. It is convenient to regard the condition of isothermal
sliding at speeds V and temperatures T, such that $V a_T$ is a constant,
as a viscoelastic state of sliding.

The interrelation described here may also be studied experimentally, keeping the sliding speed constant for various values of the
ambient temperatures. Bartenev et al.[11] have reported results from
such experiments. No new information is generated from these experiments which is not contained in the aforementioned discussions.

The influence of non-uniform temperature distribution which
occurs when the actual sliding speeds are high, remains to be investigated. This problem is of importance in the theory as well as in the
practice. Some experimental efforts have been made by Rieger[19], but
the information available is not sufficient to warrant a detailed
analysis.

In some problems it may be of interest to know the friction
properties over a much wider range of temperatures than that considered so far. At temperatures largely exceeding $Tg + 100°C$ the properties of polymers may drastically change. The mechanism of friction
is strongly dependent upon the ultimate properties of the material.
It has so little resemblance to the one discussed in this paper,
that a separate study would be desirable. At the other end of the
temperature range, i.e. at temperatures below the glass transition
temperature, the frictional behaviour does not seem to be so strongly
related to the viscoelastic properties of polymers[10]. Since the
behaviour in this region of temperatures has not been deemed to be
important enough for practical purposes, there is a dearth of information on this topic.

NORMAL LOAD, NOMINAL AREA, PRESSURE AND SURFACE ROUGHNESS

This group of factors describes the physical nature of contact of a given friction pair, in a given viscoelastic state (V,T). Depending upon these factors, the intensity of contact which is intimately related to friction, is defined. It has been pointed out by Bowden and Tabor[4] that the area of real contact is usually much smaller than the nominal area on account of the presence of surface roughness and that the determination of the real area is of fundamental importance in tribology. This is undoubtedly justified, considering the fact that the failure of adhesion occurs at the real area of contact. The real area has been regarded as the key parameter in the description of the intensity of contact.

The most viable of the laws of friction is that describing the proportionality between the force of friction and the normal load regardless of the actual size of the nominal area of contact. It has appealed to many workers that the law could be explained, if both the frictional force and the normal load are proportionally related to the real area of contact. This line of thinking has stimulated many detailed investigations which describe the mechanism of proportionality between the normal load and the real area of contact. Although this is an important achievement by itself, it would be an overstatement to express complacent remarks that the law of friction is thereby explained. The other relation between the real area and the force of friction cannot be taken for granted, especially when the complex process of the failure of adhesion is considered. In order to obtain this relation, the details of the mechanism discussed previously, have to be quantified.

The relation between the normal load and the real area is much less dependent upon the complex process of friction. It may be obtained, as far as static contact is concerned, from a clearly defined problem in solid mechanics if the surface topography and the deformation properties of the materials are known. In the case of sliding contacts, it would be necessary to modify the analysis in order to make use of Green's[3] conditions, that the surfaces slide in parallel planes. It appears however, that the essential features of the analysis are not different from those in the case of static contact. The notions involved in the analysis of static contacts will be mentioned in the following.

Bowden and Tabor[4] originally gave an explanation of the linear relation between normal load and real area of contact, based upon plastic deformation which may occur because of the high local pressure at the tips of the asperities. For metallic bodies this would lead to a "cold welding" at the tips of the asperities. It may be noted that their explanation of the linear relation also provides for a mechanism of adhesion.

On the other hand, the linear relation may also be explained from considerations of elastic deformations, if the features of the topography are taken into account. There is lively activity in this field, following the works of Archard[2], Greenwood and Williamson[20]. It is worthwhile taking note of their important results, because they are relevant to the friction of elastomers. Moreover, there is no experimental work available on the details of surface topography of elastomers. Since elastomers deform even under very light loads, it is not easy to measure the topographical features.

For elastic contacts, it has been shown[20] that the real area increases proportionately with the normal load by a process involving contact of a larger number of asperities, in such a way that the mean area per contact is constant. The nominal pressure governs the penetration of surfaces and consequently the ratio of the real area to the nominal area of contact. As long as the normal load is kept constant, it follows that the real area is independent of the nominal area. A detailed analysis of the influence of the factors considered in this section upon the penetration of surfaces and the real area of contact is available in the aforementioned works.

It is worth reminding that the experimentally observed linear relation between the frictional force and the normal load is in general only approximate. Even if the relation between the frictional force and the real area is assumed to be linear, it is not necessary to expect an exact, linear relationship between the load and the real area.

Known experimental results regarding the friction of elastomers in a given viscoelastic state of sliding indicate that the friction coefficient μ decreases as the nominal pressure p increases. It is generally difficult to determine the relation accurately, owing to the spread in the experimental data. From our recent experimental work on elastomers[8], it is approximately found that over a range of speeds and temperatures the coefficient in a given viscoelastic state may be expressed as

$$\mu \, \alpha \, (p)^{-0.1 \text{ to } 0.2} .$$

Apart from this relation, no significant influence has been found if the nominal area and load are independently varied.

No simple formula is obtained for expressing the dependence of friction on surface roughness. It was observed that the dependence is weak, so far as it concerns nominally smooth surfaces.

The discussions in this section suggest, that the influence of the factors considered here is to scale the isothermal μ - V relation. Usually the scaling factor is small.

MATERIALS OF FRICTION PAIR

It was pointed out in an earlier section that the viscoelastic properties are strongly related to the frictional properties of the polymers. Ferry[21] has described some typical examples of viscoelastic responses of various polymers. The factors upon which the response depends are the molecular weight, the structure (amorphous or crystalline), the test temperature in relation to the glass transition temperature and the type and amount of foreign material which is usually added in commercial preparations for obtaining certain additional desirable properties.

The present state of knowledge of friction is hardly capable of handling such a wide variety of material factors. Most of the experimental studies that are available, deal with the friction of elastomers. The prime motivation has been to develop suitable rubber compounds for application to tyre treads, with a view to optimize their properties in relation to friction and wear. The study of friction of plastics has been of a cursory nature. Some systematic efforts have been made by Steijn[22] and recently by Ludema and Tabor[10].

It appears that the main aim of the experimental work, has until now been to show the dependence of polymer friction on speed and temperature and to establish the role of viscoelastic properties in that dependence. In that context, the influence of the test temperature in relation to the glass transition temperature has received much attention. The influence of some other factors has also been studied; since only a few results are available, the conclusions may be regarded only as tentative.

In order to minimize complications, most of the studies describe the friction of polymers on hard solids. The friction-speed relation then, displays the influence of the viscoelastic properties of the polymer. The hard solid countersurfaces have usually been steel and glass.

In our work, a few tests have been carried out in which elastomers slide upon rigid polymers. This has been done with a view to obtain more information on the role of the countersurface rather than to study the friction between two viscoelastic bodies. The test temperature chosen is such that the viscoelastic properties of the rigid polymer do not play any important role, whereas the elastomer displays strong viscoelastic effects. The rigid polymers provide low energy surfaces; the question, how the friction-speed relation of an elastomer is modified by the countersurface, may then be studied.

Elastomer Properties

It appears that the glass transition temperature of an amorphous
polymer is an important characteristic. Grosch[9] obtained the isothermal friction-speed curves of four different rubber compounds on
a glass countersurface. The rubber compounds used were styrene-
butadiene, acrylonitrile butadiene, butyl and isomerized natural
rubber. The curves were all reduced to 20°C, making use of the W.L.F.
transform. The peak values of friction were found to occur at
different speeds. In this series, acrylonitrile butadiene rubber
has the highest Tg. It attains its peak value of friction at the
lowest of the four speeds. At the other extreme, isomerized natural
rubber having a low Tg, attains its peak friction at the highest of
the four speeds. The other two compounds lie between the extremes
in the expected order.

In our experiments, it was felt that a better comparison of
properties would be achieved if the test compounds belong to a
particular series of copolymers. To that purpose, a series of
styrene-butadiene copolymers, in which the styrene content varied
from 0 to 30% were used. A compound of butadiene rubber with high
cis content, and a compound of natural rubber were also included in
the test programme. A specimen of natural rubber compound served
as a reference specimen; whether the condition of the countersurface
is maintained constant or not may be checked. Figure 6 shows the
results obtained at 20°C on a glass surface with the help of a constant speed tribometer.

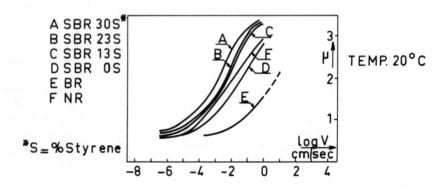

Fig. 6.   Friction of Various Compounds on Cleaned Dry Glass

As the styrene content increases, it is known that the glass transition temperature becomes correspondingly higher. From the results obtained, it is observed that the speed at which the coefficient is maximum, decreases with increasing styrene content. Walters[23] reached a similar conclusion from his experimental results. He obtained the friction-temperature relations at constant sliding speed. The temperature of maximum friction at a constant speed, increases linearly with the glass transition temperature. It appeared however, that in a graphical plot of the linear relation the point corresponding to the results obtained with butyl rubber falls distinctly outside the line. In our work with butyl rubber such an anomalous behaviour has been confirmed. Instead of relating the speed of maximum friction to Tg, Grosch[9] related it to the frequency at which the loss modulus is a maximum. That correlation has been found to be valid to several rubbers including butyl rubber. On the other hand, it is not surprising to expect that besides Tg, there are other parameters which have an important influence upon friction properties.

The data available regarding the influence of other characteristics of elastomers is scant. Grosch[9] reported a definite influence of crystallization on friction. A preferential orientation of the crystallites in the direction of sliding may cause the friction to be reduced. It has also been suggested there, that the addition of filler particles to a rubber compound does not substantially alter the speed at which friction is maximum. This may be expected, considering the fact that Tg does not alter substantially. Grosch[9] has reported that the magnitude of peak friction decreases with increasing filler content. In our experiments, the carbon black loading was varied from 10 parts to 50 parts, but no significant difference in magnitudes was observed. Bartenev[11] has pointed out that the magnitude of the peak slightly increases when the styrene content of a SBR copolymer is increased. He attributed it to an increase in polarity owing to the increased number of styrene groups. In view of the surface sensitivity of friction experiments, it is difficult to assess the significance of the small differences, in coefficients.

With a view to understand broadly, how the magnitude of friction and the shape of the friction-speed curve are affected by the material properties, more tests were carried out in our laboratory. Isothermal friction-speed curves have been obtained at various temperatures. A number of rubber compounds were used in the test. It is usually too difficult to compare the various test results since the reproducibility of experiments is not always good. At low temperatures, a general reduction in friction was observed at all speeds. The explanation offered by Bartenev has been considered in a former section. By reducing the moisture content, it was possible to obtain consistent results.

A compound of butadiene rubber, which was rated low in friction at room temperature, gave high friction coefficients at low temperatures. The result obtained for BR sliding on a glass plate are shown in the Figure 7. The difference between the magnitudes of the peak coefficients of different rubbers on a given countersurface is small (compare Figures 6,7).

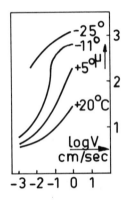

Fig. 7. Friction of Butadine-Rubber on Clean Dry Glass

Isothermal friction-speed curves obtained for various rubbers on a polyethylene countersurface are shown in Figure 8. Comparing this figure with Figure 6 obtained on a glass countersurface, it appears that the typical shape of the friction-speed curve for a given rubber compound on the two different countersurfaces are very similar. The level of friction over the range of sliding speeds on the glass surface is higher than that on the polyethylene surface.

It appears from our results that the peak value of friction depends mainly on the countersurface of the hard solid. The influence of the elastomer properties upon the peak value is small. On the other hand, the shape of the friction-speed curve depends to a great extent on the elastomer properties. The speed at which peak friction is obtained, depends on the shape of the curve. There is practically no change in the shape if a different countersurface is used.

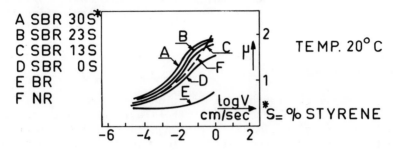

Fig. 8. Friction of Various Compounds on Polyethylene

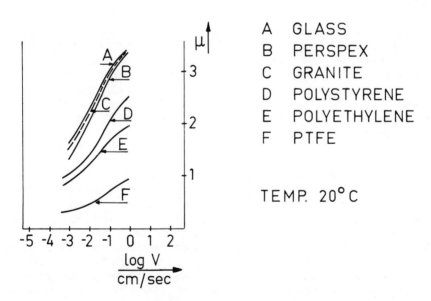

Fig. 9. Friction-Speed-Curves for a SBR-Compound on
         Various Tracks

Countersurfaces of Hard Solids

In order to examine the validity of the aforementioned inference
regarding the role of the countersurface in the determination of the
magnitude, more tests were carried out on various countersurfaces
of hard solids. Furthermore, the purpose of the tests was to study
the relevant properties of the countersurface, which are responsible
for causing the level of friction to be what it is.

Figure 9 shows the isothermal friction-speed curves of a SBR
compound on various countersurfaces. It is evident, that the form
of the various curves is similar. The magnitude of the friction
coefficient in a given viscoelastic state depends strongly on the
countersurface. It appears as though the friction-speed curves are
scaled proportionally by factors which depend upon the countersurfaces
of the harder solids.

The physical and chemical properties of the materials of these
solids are so diverse that there is hardly any correlation to be
expected between these properties and the levels of the friction-
speed curves. Materials such as glass, stone and PMMA in clean
state (in the engineering sense), give equally high values in this
series, whereas PTFE gives the lowest value as expected. There is
no correlation between the peak values and the surface roughness
(CLA values).

A few tests were conducted using smooth and rough countersurfaces
of PTFE. It was observed that the friction on the smooth surface was
slightly larger than that on the rough surface. The differences in
the observed levels of the friction-speed curves are too large to be
explicable by any arguments based upon differences in surface
roughness. The materials of the solids used as countersurfaces were
all considerably stiffer than the elastomers. In a given viscoelastic
state, the real area of contact on these various countersurfaces,
from Hertzian considerations, can not be appreciably different.

Being a surface sensitive property, it would be logical to
expect that the magnitude of friction in a given viscoelastic state
depends upon a surface physico-chemical parameter. It may be recalled
from the section on fundamental aspects, that ths most obvious parameter is the surface energy. It was found to be convenient to
quantify this aspect, using the data on contact angles of drops of a
liquid on the various materials. Contact angles on the various
materials have been usually measured directly, using a low power
microscope. In some cases the data was obtained from photographic
enlargements. The liquids used were distilled water and mercury.
Since it was difficult to measure the contact angles of mercury
directly, some of the results have been obtained by the sessile
drop method.

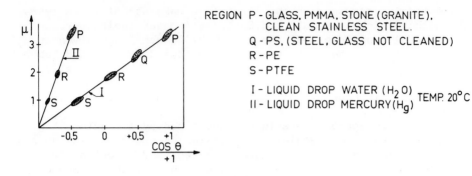

Fig. 10.   Relation of Contact Angles to Peak Friction

Figure 10 shows the relation between the peak values of friction coefficients of SBR elastomers on the various countersurfaces and the cosine of the contact angles measured or obtained from the reported data in the literature. Because of the spread in the data, the results have been presented by grouping them in various regions.

EVALUATION OF THEORETICAL MODELS

Experimental studies have hitherto led the theoretical work on friction. In particular, the experimental results obtained by Schallamach[1], Bartenev[11] and Grosch[9] have stimulated the recent progress in this area. Advances in the understanding of rubber friction have been reviewed by Schallamach[24] in 1968. In that review, he presented a detailed description and a discussion of the theoretical models from his point of view. It is not necessary, therefore, to give an account of these models. On the other hand, a brief consideration of certain important differences in the interpretation and assessment, from our point of view, does not appear to be redundant. Although none of these models provide detailed quantitative information, they may be regarded as exploration of certain approaches to the solution of the complex problem. With a view to develop a quantitative theory, it has been desirable to examine the potentiality of the different models.

The models described in the literature make use of three types of approaches. Schallamach[25] and Bartenev[11] based their approach upon a molecular picture. Of the two, Schallamach[25] has given an elaborate account. The works of the author[26] and, Ludema and Tabor[10] describe models based upon phenomenological approaches. Finally, a third type of approach has been described by Kummer[27] and Rieger[28]. They utilize their concepts of "electrical roughness of surface".

Although the approach previously described by the author[26] has been more successful in obtaining results which are close to those observed experimentally, it can hardly be regarded as definitive. It will be clear from the following discussion that this approach is physically not unreasonable. The essential features of the mechanism of friction described in a previous section are aptly described by the model.

It may be recalled from the introduction that an approach based purely upon molecular considerations may be useful to obtain a qualitatively meaningful picture. There are obvious limitations so far as the quantitative results are concerned. The first type of approach taken by Schallamach[25] has such limitations. Moreover, his theory failed, in spite of an assumption regarding the existence of adhesion, to predict a finite value of the static coefficient of friction. There are many basic inconsistencies in his model. The most serious objection is that his theory predicts finite friction over a wide range of speeds when no energy barrier of adhesion exists. In the absence of adhesion, his model does not contain any external agency which may account for the finite friction. At high sliding speeds, the model predicts vanishing friction even when strong adhesion is present. On closer examination, it appears that all these untenable results may be traced back to the oversimplified description of the rate process.

The third type of approach introduces a concept of electrical roughness which gives rise to a sinusoidally distributed force field. The force field excites the asperities of rubber during the relative motion. No static coefficient is predicted by the theories. Although Rieger has predicted that a maximum value of friction occurs at a definite speed,* there is no indication of the order of magnitude. It is difficult to conceive how the various details of the mechanism may be quantified. At present, the concepts implied in their models do not appear to be tangible.

The author[26] and, Ludema and Tabor[10] have taken approaches which are based upon a phenomenological description of the mechanism. The approaches are somewhat similar so far as the qualitative aspects of the relation between the viscoelastic properties and friction are concerned. In both the approaches, the relation is explained by theories which are based upon the strain-rate and temperature dependence of both the real area of contact and the strength of the

interface. The details of the explanation however, are different.
These details are in both the cases tangible.

        Ludema and Tabor[10] use an approach which is conceptually very
simple and clear. The friction force is given by the product of
real area and ultimate shear stress. In order to explain the
dependence of friction on viscoelastic properties, they consider
the relations between the latter and each of the two terms of the
product taken separately. The speed of sliding governs the rate
dependence of shear strain. There are two main difficulties in
their approach. Firstly, in order to introduce the variation of
shear strength with strain-rate (and therefore speed) and temperature
they assume that the adhesion between rubber and the harder countersurface is as strong as the rubber itself. This assumption leads to
an overestimation of the order of magnitude of the friction force.
Moreover, the deficiency in their model resulting from an inadequate
consideration of the constraint imposed by adhesion, leads to an
untenable conclusion that the magnitude of friction is not surface
sensitive as long as the real area of contact is maintained. It may
be seen,in figure 11,that the viscoelastic dependence of friction is
scarcely related to the absolute level of the friction-speed curves
on the various harder solids. Moreover, the mechanism described is
not consistent with the observations of inappreciable wear in the
friction of elastomers. The second difficulty arises in relating
the strain-rate to the speed of sliding. It is assumed that the
strain-rate which defines the viscoelastic properties involved in
the calculation of real area of contact is much lower than that
involved in the determination of the ultimate shear stress at a
given speed of sliding (and a given temperature). The assumption
is not an unreasonable one. Their argument, based upon a strain
rate which depends upon the diameter of a contact circle does not
appear to be strong. The results obtained, show a variation of
friction with speed, which is qualitatively similar to the variation
found in the experiments. There is a poor quantitative agreement,
though, with respect to the extent to which the magnitude of the
friction coefficient varies with the sliding speed.

        The assumption that the strength of the interface equals the
strength of rubber is an unnecessary restriction; the viscoelastic
dependence of failure can be explained in simpler terms without
invoking that assumption. Even when the inherent strength of adhesion
is assumed to be independent of the rate of strain, it is possible
to conceive the dependence of the criterion of failure on the strain
rate, because the stress field which causes the failure depends upon
the viscoelastic deformation of rubber. It is therefore not necessary
to assume that the failure occurs within the rubber or that it is
in any way related to the ultimate properties of rubber.

        There are no such difficulties involved in the author's model[26].
Unlike many previously described models, the qualitative details of

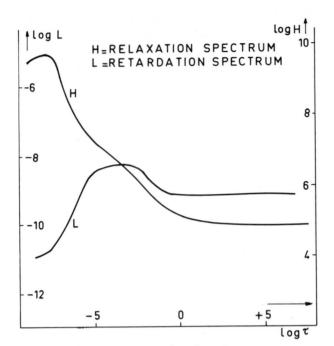

Typical Spectra for Gum Rubber

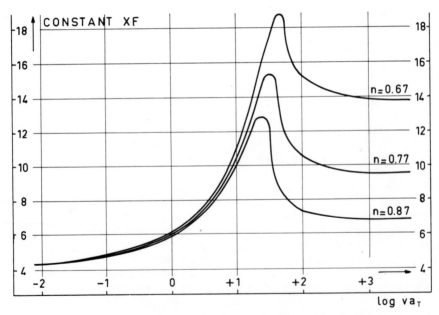

Calculated from Above Viscoelastic Spectra

Fig. 11

the process of failure are compatible with the observed physical processes which are related to these details. The quantification of the problem of failure is however not totally satisfactory and there is still scope for further development on the lines indicated in the section dealing with the fundamental aspects. In that context, it appears that the work of Knauss[29] on viscoelastic fracture seems to offer a promising outlook.

It is at present too much to expect that the absolute magnitude of friction can be predicted to a reasonable degree of accuracy from any available theories. On the other hand, a good measure of the success of a theory may be obtained from the extent to which friction varies with speed. The essential feature of the details of the model may thereby be verified. The reason for this statement is that the absolute magnitude of friction depends strongly upon the absolute values of the less tractable parameters viz. strength of adhesion and real area of contact; the variation of friction with speed is strongly related only to the more tractable viscoelastic properties.

Schallamach[24] has raised two main points of criticism regarding the author's model. The first point is how a bonded asperity can support a force before it has become strained. The point is irrelevant because the constitutive relation (elastic or viscoelastic) defines the relation between force and deformation at any instant irrespective of which of the two quantities are prescribed. The second point relates to the assertion of Schallamach that a bond breakage is determined by a force rather than an energy criterion. The point cannot be resolved as it is philosophic in nature. In fact, our preference is towards a description based upon energy. It should be remembered that the failure of adhesion involves on one hand the mechanics of materials but on the other hand the surface properties. It is obvious that the latter cannot be evaluated easily in terms of forces. Whereas the former property is more conveniently described by forces it can easily be related to the energy. In any case, the choice can be considered to result from convenience of application rather than from a physical principle. Description of failure in terms of stress intensity is convenient to describe a local failure; the stress analysis involved is usually very complex; moreover, it is sensitive to the minute details of boundary conditions.

The other difficulty of obtaining the variation of real area with speed also needs more attention. The kinematic condition of Green[3] has not yet been applied to describe the contact problem of viscoelastic bodies. Such a study is contemplated by the author.

In the forthcoming part, a quantitative theory will be given, which is perforce approximate on account of the reasons mentioned above. It still offers a clear picture of the process; moreover it has been found to be useful in the prediction of frictional properties during the development of rubber compounds for specific purposes.

A QUASI-QUANTITATIVE THEORY

In our previous theoretical work[26], although a phenomenological description is given, molecular parameters have been used in the calculations. It is now proposed to describe an extension of that approach so as to be able to make use of tractable phenomenological quantities. In order to bring out the role of the various concepts discussed so far, the various details of the process are depicted by means of simple models. It is also mentioned how each of these details may be developed further, to the purpose of obtaining a more accurate quantitative theory.

Friction of Asperities

A typical situation of sliding is considered in which an asperity of rubber makes contact with an asperity on the countersurface of a hard solid; a real area of contact is formed. It is convenient to idealize the asperity on the hard solid as a rigid body. According to our interpretation of adhesion, the two asperities adhere over the real contact area. During the subsequent relative motion of the bulk of the bodies, the material near the contact surface is subjected to large shear strains. For the sake of simplicity, it is convenient to assume that the shear strains are concentrated in a distinct layer of material adjacent to the surface. The layer may be called as "top layer".

Schallamach[1], in his experiments on the change of electrical impedance, found that such a top layer may be envisaged. In reality such distinct layer cannot be demarcated. In a detailed analysis based upon the stress field in a semi-infinite body, the author[30] has obtained a truer description. It is,however,too involved to be described here. The process of shear deformation of the top layer, is accompanied by a process of normal deformation of asperities. The relative motion causes the asperities to approach tangentially and to penetrate. This process causes the contact area to increase. In order to describe this process,it is convenient at first to assume that both the asperities are identical, spherically capped. This description is a special case of convex surfaces. The assumption simplifies the study because the area may conveniently be described as a circular area of contact. At this stage,it is necessary to model the approach of the asperities in order to describe the growth of the contact area during the relative motion. The correct condition which describes the approach is the kinetic condition of Green. The asperities come into contact at the sloping sides rather than at the summits. Then they penetrate in such a way that the relative motion of the bodies is parallel to the plane of contact. It is not

difficult to describe the process of penetration which causes the
contact area to grow if there was no adhesion. In the presence of
adhesion, the description of the deformations is very complicated.
An additional difficulty arises on account of the viscoelastic
properties of polymers. The problem is, however, well defined; the
difficulty is of a mathematical rather than a physical nature.

With a view to estimate the growth of contact area during the
process of penetration, it was necessary to make a sweeping assumption
which simplifies the calculations without any serious loss of the
effect. It is assumed that the growth of the area of real contact
may be approximately equated to the growth of real area which may
occur if the asperities approach normally and share a part of the
total normal load as soon as they make contact. The size of the
contact grows in that case, with the time of contact. There is,
however, some loss of effect owing to the assumption. Since the
speed dependence of the tangential approach is not considered, the
expression describing the growth of size does not involve an explicit
term containing speed. The effect as far as the time dependence of
growth is concerned, is described reasonably.

Finally, at some stage the two processes described above must
end because the adhesion has a finite strength. The problem of
describing the condition for the occurence of failure is a formidable
one. It may be recalled from the sections dealing with the fundamental aspects of the theoretical models that an energy approach is
chosen here. The failure is envisaged to take place at once over
the real contact area. It would be more realistic to treat the
failure as local. Such an analysis is not currently available. The
general features however may not be very much different from those
described here.

During the tangential relative motion, it is not unreasonable
to assume that the component of energy arising from the normal
elastic field remains constant. The change in energy arising from
a change of shear stress field occurs when failure takes place. If
a global energy balance, which is similar to that proposed by
Griffith[31] is considered, it would appear that the change in energy
may be equated to the energy required to produce the new surface.
In the case of viscoelastic materials, a part of the work done by
an external force is dissipated within the material itself. The
part that is lost, cannot therefore play any role in the energy
balance. Since the failure is assumed to occur at once over the
whole surface, it is reasonable to equate the component of the energy
stored in the shear stress field within the top layer to the energy
needed to form new surface area which is equal to the real area.
A similar criterion has been proposed by Rabinowicz[32] to describe
the process by which a loose wear particle is formed. An important
feature of the criterion proposed by the present author (cf. reference
26) to describe the failure of adhesion is that the surface energy

term is of major importance in determining the force of friction,
even though its magnitude may be small in comparison to the energy
dissipation within the material. It is of interest to note that a
similar conclusion has been put forward in the theories which deal
with the fracture of viscoelastic materials. While examining the
energy balance for the propogation of a crack in a viscoelastic
material, Williams[33] noted that the term corresponding to the
dissipation of energy disappears from the equation. Rice[34] has
pointed out the important role played by the surface energy term
in determining the fracture strength of inelastic materials, even
when the term is negligible in comparison to the inelastic dissipation.
He argues that the conclusion follows as a consequence of the Griffith
type assumption.

Details of Calculations

The surface of an asperity of rubber is taken to be spherical.
The increase in the area of contact with the time of contact may be
expressed in terms of creep properties, on the assumption that normal
load is borne by the asperity as soon as it makes contact with the
rigid countersurface. It is convenient to relate the area of contact
A_r (t) after a time t, reckoned from the instant of first contact to
the area A_R corresponding to the rubber elastic state. The latter
may be determined from considerations of surface statistics and
Hertz's theory.

$$A_r~(t) = A_R \cdot G_o{}^n \left\{ \frac{1}{G_\infty} + \int_{-\infty}^{+\infty} L~(1 - e^{-\bar{t}/\tau})~d\ln\tau \right\}^n \quad (1)$$

G_o is the modulus in the rubber elastic state.

G_∞ is the glassy modulus, L is the retardation
spectrum with respect to the logarithm of time.

n is an index to specify the elastic relation
describing the contact according to the surface
statistical data. $n = {}^2/_3$ for a Hertzian contact;
n may approach unity depending upon the distribution
of heights of micro-roughness present on the surface.

When the bulk of the elastomer slides at a speed V on a static
countersurface of rigid material, the average shear strain rate $\dot{\varepsilon}$
at which the top layer of thickness h is deformed may be written as

$$\dot{\varepsilon} = \frac{V}{h} \quad (2)$$

The average shear stress σ at time t becomes

$$\sigma(t) = \left\{ G_o\ t + \int_{-\infty}^{+\infty} \tau H\ (1 - e^{-t/\tau})\ d\ln\tau \right\}, \tag{3}$$

where H is the relaxation spectrum with respect to ln. If a mean area A_d of the top layer undergoes shear deformation on account of this shear stress, the tangential force F is given by

$$F\ (t) = \sigma(t) \cdot A_d \tag{4}$$

An expression for the stored energy per unit volume at time t, for a constant rate of strain has been given by Ferry[21].

Making use of that expression and equation (2), the stored energy per unit volume of the top layer adjoining the asperity is given by

$$U(t) = \frac{V t^2\ G_o}{2h^2} + \frac{V^2}{2h^2} \int_{-\infty}^{+\infty} H \tau^2\ (1 - e^{-t/\tau})^2\ d\ln\tau. \tag{5}$$

It has been assumed that the part of the work done by the tangential force is dissipated isothermally. From the previous discussion regarding the criterion of failure, the stored energy is to be equated to the energy required for the formation of new surface equal to the real area.

$$U\ (\bar{t})\ A_d\ h = A_r\ (\bar{t}) \cdot \gamma, \tag{6}$$

where γ is the energy required for a unit area. The tangential force at time t from (2), (3) and (4) is

$$F\ (t) = \frac{A_d}{h}\ V \left\{ G_o\ t + \int_{-\infty}^{+\infty} H\ (\tau)\ (1 - e^{-t/\tau})\ d\ln\tau \right\}. \tag{7}$$

The average tangential force from the instant of first contact to failure is given by

$$F\ (V) = \frac{1}{t} \int_{0}^{t} F\ (\bar{t})\ d\bar{t}. \tag{8}$$

Using equations (1) and (5) in equation (6), $\bar{t}$ is defined for a given V. Then making use of equations (7) and (8), the average tangential force may be found. It is numerically more convenient instead, to prescribe t and obtain V and F. The calculations, although tedious, are otherwise straightforward.

Friction of Surfaces

Once the details of the process are known for an asperity of arbitrary dimensions, it is possible to obtain the friction of surfaces from the surface statistical data. A usual assumption is that the processes taking place at the various asperities do not interact. The statistical model may be made as elaborate as one desires. For certain distributions of heights and peak radii, an attempt has been made to calculate the friction-speed relation; the work is reported elsewhere[35]. There is however, no new physical information contained which may alter the essential nature of the relation. A simpler description of the statistical process would be to imagine a typical asperity of the surface as an average representation. Assuming a narrow band spectrum of asperity dimensions, it is not difficult to visualize the process as a stationary ergodic process. The mean value of the total force generated by the ensemble at any instant, may be equated to the time average of the force. At any given instant, it may be depicted that the various asperities are at different stages of the deformation process.

The expected force of friction may therefore be given by the product of the total number of contacting asperities and the average value of the force given by equation (8).

The Friction-Speed Relation

Since some of the constants involved in the equations are not known, it is not possible to obtain the absolute magnitude of the force of friction. Lumping the unknown constants, it is possible to obtain the nature of the variation of friction with speed if the viscoelastic properties are known. Some typical results are shown. Figure 11 shows the viscoelastic spectra H and L and the friction-speed relation derived by making use of that data. Figure 12 shows the calculated friction-speed curve for a SBR compound, which was obtained from the viscoelastic properties.

Such calculations have been made using the viscoelastic data of various rubber compounds. It appears that the dependence of friction-speed relation upon the viscoelastic properties is clearly exposed. It has also been found[8] that changing the values of γ primarily amounts to scaling the friction-speed relation.

The Magnitude of the Force of Friction

It may be recalled that the magnitude of friction in a given viscoelastic state depends upon certain specific properties of the solids. The roughness of surfaces which are nominally smooth, does

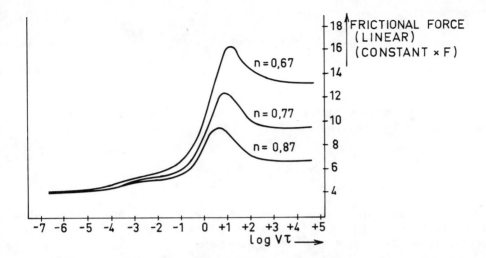

Fig. 12. Friction-Speed Curves Calculated for SBR 913% Styrene)
Spectrum for Storage Modulus

not seem to have a strong influence on the magnitude. It has also
been mentioned that the presently available theories of the physical
nature of contact are quite advanced. It is possible to quantify
the real area by using the statistical theories and the theory of
Hertz.

The physico-chemical nature of the countersurface is the major
factor related to the magnitude. In order to make some progress in
quantifying the relation it is necessary to consider the relation
between the strength of the interface and some tractable parameter.
In the experimental study, it has been shown that the contact angle
is a convenient parameter. The theory of wetting describes the
relation between the contact angle of a drop of liquid on a surface
and the surface energy.

If a drop of liquid of surface energy γ_{lv} rests upon a solid
of surface energy γ_{sv}, a contact angle θ is obtained. From Young's
equation[36],

$$\gamma_{sv} = \gamma_{sl} + \gamma_{lv} \cos\theta ,$$

γ_{sl} is an interfacial energy and the subscripts denote, s-solid,
l-liquid, v-vapour.

Following Good et al.[37] the interfacial energy may be expressed independently.

$$\gamma_{sl} = \gamma_s + \gamma_l - 2 \phi \sqrt{\gamma_s \cdot \gamma_l} \,,$$

ϕ is an interaction parameter which characterizes the surface interactions between s and l.

Fowkes[38] considered the interactions of dispersion type and described the interfacial energy.

$$\gamma_{sl} = \gamma_s + \gamma_l - 2 \sqrt{\gamma_s^d \cdot \gamma_l^d} \,.$$

The superscript d denotes a dispersion type interaction. The equation described by Good may be written in a similar form if the interaction parameter is split up as follows.

$$\phi^2 = \phi_s \cdot \phi_l,$$
$$\gamma_{sl} = \gamma_s + \gamma_l - 2 \{(\phi_s \cdot \gamma_s) \cdot (\phi_l \cdot \gamma_l)\}^{1/2} \,.$$

Combining the equations of Young and Good, using the above relation,

$$\phi_s \cdot \gamma_s = \left[\frac{\gamma_l^2 (1 + \cos\theta)^2}{4 \phi_l \gamma_l} \right] \,, \tag{9}$$
$$= \text{Liquid parameters} \times (1 + \cos\theta)^2$$

In the equation described by Fowkes, the term $\phi_s \cdot \gamma_s$ is replaced by γ_s^d.

It has been pointed out that the strength may be represented by a surface energy. A simple account of the relation between surface energy and the strength which is expressed in terms of an ultimate stress is given by Cottrell[39]. If the surface energy in the friction theory is represented by $\phi_s \cdot \gamma_s$, the contact angles of drops of a given liquid on various solids may be used to describe the strength of the interface.

In the criterion of failure, the stored energy has been equated to the surface energy. The former varies with the square of the shear stress if a given viscoelastic state is considered. Consequently, the force of friction approximately varies with the square-root of the surface energy. From the aforementioned relation between $\phi_s \gamma_s$ and $\cos\theta$, the relation between the forces of friction on various solids and the cosines of the contact angles of a given liquid may be written:

$$F \propto (1 + \cos\theta) \,.$$

This relation has been found in our experiments of Figure 10.

In view of the various assumptions and approximations made, the result may be regarded as tentative. Despite the many weak links in the knowledge of the properties of strength and the equivocality of the wetting theory, the agreement between the theory and experiments is deemed more than merely fortuitous.

A more detailed study is clearly desirable to the purpose of understanding the physical significance of the value $\phi_s \gamma_s$. If it represents the surface energy of the boundary layer, the experimental results obtained so far would suggest that the process of failure occurs within the boundary layer on the countersurface of the hard solid. On the other hand, it would not be easy to explain why the friction of rubber on a rubber countersurface is high. Instead of assuming that the plane of failure lies within the boundary layer, it may be assumed to lie in between the two boundary layers (superfacial failure). These considerations may modify the relation between $Cos\theta$ and friction. Nevertheless, the aforementioned relation seems to be useful in characterizing the potentiality of a surface in respect of the magnitude of the force of friction.

A practical scheme for calculating the frictional properties from the various physical data is presented in Figure 13.

OTHER FACTORS

Of the many factors which have not been considered until now, some work has been done in our laboratory on two important factors. The influence of the presence of a film of liquid (water) is for various reasons regarded as an important factor. The mechanism of boundary lubrication described elsewhere has been more confined to metallic contacts and special lubricants.

Some results have been obtained for sliding of rubber on countersurfaces wetted by water. The sliding speeds are so low that there is little effect, if any, of hydrodynamic lubrication. It is observed that the level of the friction-speed curve on a wet surface is lower than that on a dry surface. The decrease in friction caused by the presence of water is not uniform on the various countersurfaces. In particular,the reduction in friction of rubber on hard polymers was small. On many other solids, there was a significant reduction in friction. A typical result is shown in Figure 14.

The investigation of the effect of an electrical potential on friction has been studied in order that the mechanism may be understood better if this effect is explained. An increase in friction occurs, on application of an electrical potential across the contact plane. Some experiments have been performed using rubber compounds

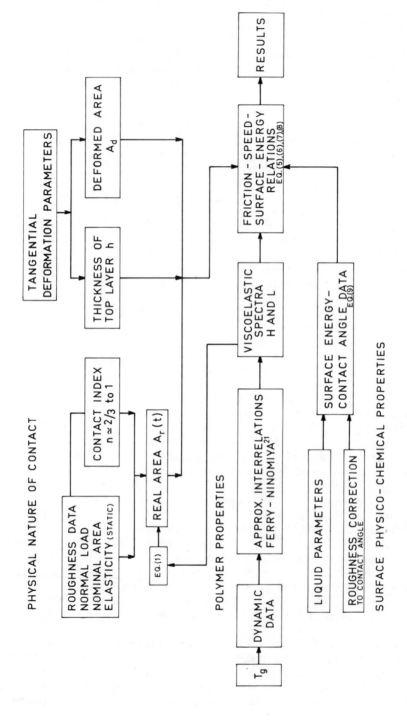

Fig. 13.   Scheme for Estimating Friction Characteristics of Elastomers

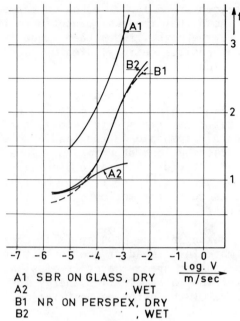

A1 SBR ON GLASS, DRY
A2   , WET
B1 NR ON PERSPEX, DRY
B2   , WET

Fig. 14.   Friction-Speed Curves on Wet and Dry Surfaces (20°C)

which were conductive to different degrees. It appears that the
increase in friction may be explained by taking the forces of
attraction into account. It does not seem that any other process
of fundamental nature is involved. The actual normal force is
determined not only by the mechanical normal load, but also by the
force of attraction. The latter force seems to be exerted in such
a manner that it has a magnitude which is approximately proportional
to the real area of contact. A more detailed description will be
given in a separate paper[40].

DEFORMATION FRICTION

        A substantial portion of the text dealt with the details of the
mechanism of friction which owes its existence to adhesion. In the
process of overcoming the constraint of adhesion, the elastomer
deforms inelastically. The additional resistance corresponding to
the dissipated energy may in effect be described as deformation
friction; it is,nevertheless,dependent upon the strength of adhesion.
For that reason,it is customary in the literature to regard the total
resistance arising from this mechanism as adhesive friction. The
contribution arising from the inelastic deformation is therefore not
separately identified as deformation friction.

        The term "deformation friction" or "hysteresis component" is
usually used to denote a secondary energy dissipation mechanism

which may occur even when there is no adhesion between the surfaces. Tabor[41] et al. have shown that the resistance to rolling may occur by this mechanism. Furthermore, it has been suggested that an appreciable frictional resistance to sliding may occur even when the surfaces are well lubricated.

The basic physical picture may be described as follows. During the motion, various points within a body experience different loading histories. For inelastic materials with memory, the deformation response of the material depends upon the loading history. This leads to asymmetric distributions of normal pressure and deformation even if the initial geometry of the static problem is symmetric with respect to an axis normal to the direction of motion. The asymmetry gives rise to a resistance to the motion. The resistance is known as the deformation friction.

Another way of looking at the problem is to consider that the energy fed into the material, ahead of the contact area is not fully recovered behind the contact area; energy is dissipated on account of the inelasticity of the material. The energy loss may be con-sidered to be the cause of the deformation friction.

It is possible to conceive that the mechanism occurs at two scales of deformation. In the case of concentrated contacts, the mechanism may arise mainly from bulk deformation of bodies. On the other hand, the asperities on extended surfaces may also give rise to such a mechanism during their relative motion. The terms such as ploughing and grooving (non-permanent) of polymers may be thought to describe the latter. The importance of this kind of deformation friction is not yet established rigorously. It is not easy to carry out experiments in which adhesion is totally absent, with a view to obtain the contribution resulting from deformation friction. Neither is any analysis of the problem available currently. The viscoelastic stress analysis involves time dependent, mixed boundary conditions. It should be possible to solve this problem with the help of recent techniques for analysing viscoelastic contact problems. At present, however, neither experimental nor theoretical estimates are available.

Remarkable advance has been made in the solution of problems which deal with the concentrated contact of rolling bodies. A rigorous analysis of the problem of a rigid cylinder which rolls on a viscoelastic half-space has been given by Hunter[42]. He found that the contact area shifts relative to the centre of the cylinder, in the direction of rolling. This gives rise to one asymmetry in the problem. Furthermore, the normal pressure distribution is also asymmetric with respect to the centre of the contact area. Moore[43] arrived independently at a similar conclusion using an approximate, physically motivated reasoning. Hunter's analysis has been limited to a simple viscoelastic response of a standard linear solid. His analysis is unfortunately inadequate to the purpose of accommodating a more complex linear viscoelastic response of real materials. Moreland[44] used another mathematical formulation so as to overcome the difficulties in Hunter's approach. The theory is now capable of handling a finite relaxation spectrum. In a recent paper[45], he has extended the analysis to deal with the rolling contact of two viscoelastic cylinders.

From the explanation concerning the origin of deformation friction, it is not surprising that the frictional resistance is closely related to the viscoelastic properties. It has been established that the coefficient of rolling friction exhibits a maximum at a certain speed of translation. It may also be expected that temperature has a profound effect on the coefficient of rolling friction. The speed-temperature equivalence may again be expected. The various simpler theories of deformation friction have been presented by Moore[43].

Since the physics of the phenomenon is well understood, recent efforts have been directed at obtaining quantitative estimates of rolling friction for specific problems. The problem of a rigid sphere rolling on a viscoelastic body has attracted some attention. The shape of the contact area is circular only at very low and very high speeds. At intermediate speeds it has been found by Halaunbrenner[47] and Birek[48], that the contact area is not circular in shape. Using their experimental results, Moore has described an approximate analysis of this complex problem. The rolling of a rigid cylinder on a thin viscoelastic layer has recently been studied by Alblas and Kuipers[46]. In the aforementioned theories, the problem is considered to be linear. Hunter has pointed out that the pressure distribution obtained from the linear theory assumes that it is normal to the undeformed plane; the tangential force of friction cannot be obtained directly from the corresponding components of traction. Instead, it has to be deduced from the consideration of moment equilibrium. The formalism of the theories is therefore not entirely satisfactory. In this context, it is interesting to note that Batra and Ling[49] have indicated another analytical approach. They base their definition of deformation friction upon the differential work done in moving a pressure profile over an infinitesimal distance. Yandell[50] has described an interesting mathematical model for the viscoelastic bodies. He uses framework of elastic and frictional elements in his analysis. It would be interesting to evaluate the various approaches, to the purpose of consolidating our understanding of deformation friction.

Some efforts have been made to study the deformation friction which occurs during sliding over well lubricated surfaces. Tabor[51] found that the coefficient of rolling and sliding friction were almost equal. A friction coefficient of the order of 0.1 was found to be representative of both the modes of motion at very low speeds. Moore[43] has expressed some reservations as to the order of magnitude

of friction that can be attributed to deformation friction at very
low speeds of sliding. He suggested that the lubricant may not be
very effective at these low speeds; some adhesive mechanism may be
present. On the other hand, in the absence of the deformation
component it is possible to obtain a coefficient of friction of 0.1
in the rolling situation, if pure rolling does not occur. It should
be possible to check if any slip occurs during the rolling experiment.
In view of the experimental difficulties, it would be interesting to
study Hunter's problem for the case of sliding contact. The results
may then be used to compare with his results obtained for the rolling
contact problem.

It is indeed a fortunate situation that the problem of skid
resistance of automobile tyres provides an interesting viewpoint of
looking at the mechanism of deformation friction. In contrast to
many other technologically motivated problems, the aim of a tyre
designer is to find ways and means of increasing the tyre to road
friction on wet roads. It is interesting to know the magnitude of
the deformation friction on well lubricated wet roads; the elastomeric
material to be used for the tread compound may be selected optimally.
It is desirable to obtain an appreciable deformation friction on the
asperity scale in order to increase the wet skid resistance. On
the other hand, the deformation friction which may occur mainly from
inelastic deformation of the bulk material should be minimized so
that a free rolling tyre will have a lower rolling resistance and
a smaller heat build up.

CONCLUSIONS

Major ideas in the field of friction (non-hydrodynamic) of
elastomers have been critically evaluated. While much of the
material presented describes new results, it has been tied up in an
overall review of the field. For the sake of brevity, well-known
topics which have attained definitive forms have been briefly
mentioned in order to prevent loss of continuity. Attention has
been focused on various experimental and theoretical facets which
have remained either unnoticed or less satisfactorily understood.
The following list describes the major contributions of this paper.

1) The basic mechanism of sliding friction has been discussed
   elaborately. The details of the mechanisms of adhesive and
   abrasive friction differ only so far as the final stage of
   failure is concerned. The two mechanisms are not complimentary;
   it is not correct to add the two in order to describe the total
   frictional resistance. The two mechanisms occur on different
   types of surfaces. The topographical feature of surfaces relating
   to the slopes is an essential concept. Nominally smooth surfaces
   yield high friction and low wear whereas sharp textured surfaces
   give a lower level of friction and a high degree of wear.

2) The concept of adhesion as a constraint to relative motion enables to clarify misgivings regarding the adhesive origin of friction. The fact that the normal separation is effected by very small forces whereas the tangential separation requires appreciable forces does not disprove the adhesion hypothesis. A plausible explanation of the anomaly based upon investigations concerning the mechanics of failure has been offered. Two modes of separation have been indicated. The failure may either be superficial or may involve shearing off the boundary layer.

3) The factors which govern the coefficient of friction may be described in terms of a constitutive relation. Various experimental and theoretical aspects have been considered. A good estimate of the static coefficient of friction may be obtained from the value of the kinetic coefficient at a low speed of sliding, at a high temperature. A method of obtaining the friction-speed curve in the region of negative slopes has been presented. The significance of the non-uniform temperature distribution on the sliding friction at higher speeds has been discussed. The factors such as the normal load, nominal area, pressure and surface roughness scale the friction-speed curve by a small amount.

4) The viscoelastic properties of elastomer govern the shape of the friction-speed curve on a countersurface of hard solid. Various details have been described in the text. The influence of the physico-chemical nature of the countersurface in scaling the friction-speed curves may be estimated from the contact angle of a drop of liquid on the surface.

5) Significant theoretical models have been evaluated. It is found that a previous model of the author[26] is essentially consistent with the diverse aspects of the phenomenon of friction. An extended theory which can make use of tractable phenomenological parameters has been presented. A scheme for obtaining an approximate quantitative estimate of friction has been shown. The analysis makes use of relaxation and retardation spectra of elastomers. The estimation of the magnitude of friction has not yet been attempted. However, a major step towards that goal has been shown. In particular the wetting theory has been found to be useful in describing the strength of a countersurface in adhesion.

6) The major ideas pertaining to deformation friction have been described. The deformation friction occurs on two different scales. The significance of these results has been discussed.

ACKNOWLEDGEMENT

The author wishes to express his thanks to the members of the Vehicle Research Laboratory at Delft; Professor ir. H. C. A. van Eldik Thieme for his interest and encouragement and Ir. J. A. Zwaan and Ir. T. Ruyter for their assistance. The cooperation of the Royal Dutch Shell Polymer Laboratories at Delft is also appreciated.

REFERENCES

1. A. Schallamach, Wear, 1, 384 (1958).
2. J. F. Archard, J. Appl. Phys., 32, 1420 (1961).
3. A. P. Green, Proc. Roy. Soc. (London), A 228, 191 (1955).
4. F. P. Bowden and D. Tabor, The Friction and Lubrication of Solids, Clarendon Press Oxford, Vol. I (1950), Vol. II (1964).
5. K. L. Johnson, K. Kendall and A. D. Roberts, Proc. Roy. Soc. (London) A 324, 301 (1971).
6. A. R. Savkoor, Techn. Hogeschool Delft, Lab. voor Voertuigtechniek, Report P 159 (1972), (1973).
7. J. S. Courtney-Pratt and E. Eisner, Proc. Roy. Soc. (London), A 238, 529 (1957).
8. A. R. Savkoor, Tech. Hogeschool Delft, Lab. voor Voertuigtechniek, Report P 093 (1968).
9. K. A. Grosch, Proc. Roy. Soc. (London), A 274, 21 (1963).
10. K. C. Ludema and D. Tabor, Wear, 9, 329 (1966).
11. G. M. Bartenev and A. I. Elkin, Wear, 8, 79 (1965).
12. V. V. Lavrentev, Plaste u. Kautschuk 6, 282 (1962).
13. J. A. Zwaan, Tech. Hogeschool Delft, Lab. voor Voertuigtechniek, Report P 113 (1969).
14. E. Rabinowicz, Proc. Phys. Soc., Vol. LXXI, 668 (1958).
15. A. I. Elkin, Proc. Fifth Intern. Congress on Rheology, Kyoto, 1968, Univ. of Tokyo Press (1970).
16. H. Blok, S.A.E. Journal, 2, 54 (1940).
17. T. Ruyter, Tech. Hogeschool, Delft, Lab. voor Voertuigtechniek, Report 710 (1973).
18. J. B. Hunt, I. Torbe and G. C. Spencer, Wear, 8, 455 (1965).
19. H. Rieger Diss. T.H. München (1968).
20. J. A. Greenwood and J. B. P. Williamson, Proc. Roy. Soc. (London) A 295, 300 (1966).
21. J. D. Ferry, Viscoelastic Properties of Polymers, Wiley (1970).
22. R. P. Steijn, Metals Engng. Quart. 9, May (1967).
23. M. H. Walters, M.O.T. Conference Proc. Waltham Abbey, Paper 3, 30 (1969).
24. A. Schallamach, Rubber Chem. Techn. 41, 209 (1968).
25. A. Schallamach, Wear, 6, 375 (1963).
26. A. R. Savkoor, Wear, 8, 222 (1965).
27. H. W. Kummer, Engng. Res. Bull. B-94, The Penn. State Univ., 149, July (1966).
28. H. Rieger, Kaut. Gummi Kunstst. 20, 293 (1967).

29. W. G. Knauss, Intern. J. Fracture Mech. 1, 2, (1970).
30. A. R. Savkoor, Tech. Hogeschool Delft, Lab. voor Voertuigtechniek Report P098 (1968).
31. A. A. Griffith, Proc. 1st Intern. Congress of Appl. Mech. Delft, 55 (1924).
32. E. Rabinowicz, J. Appl. Phys. 32, 1440 (1960).
33. M. L. Williams, Proc. First Intern. Conf. Fracture Mechanics, Sendai Japan, Vol. 1, Ed. T. Yokobori, T. Kawasaki, and J. L. Swedlow (1966).
34. J. R. Rice, same conference as above.
35. A. R. Savkoor, Tech. Hogeschool Delft, Lab. voor Voertuigtechniek, Report P 111 (1969).
36. W. A. Zisman, Ind. Engng. Chem. 55, 18 (1963).
37. R. J. Good and L. A. Girifalco, J. Phys. Chem. 61, 904 (1957).
38. F. M. Fowkes, Contact Angle, Wettability and Adhesion, Am. Chem. Soc. Appl. Div. (1962).
39. A. H. Cottrell, The Mechanical Properties of Matter, Wiley (1964).
40. A. R. Savkoor and T. Ruyter, in Advances in Polymer Friction and Wear, Ed. L. H. Lee, Plenum Press, New York (1974).
41. D. Tabor, J. Greenwood and H. Minstrall, Proc. Roy. Soc. (London) A 259, 480 (1961).
42. S. C. Hunter, Trans. A.S.M.E. 12, 611 (1961).
43. D. F. Moore, Friction and Lubrication of Elastomers, Pergamon Press (1970).
44. L. W. Moreland, J. Appl. Mech., 28, 611, (1961).
45. L. W. Moreland, Quart. J. Mech. and Appl. Math. XX Pt. 1, 74 (1967).
46. J. B. Alblas and M. Kuipers, Intern. J. Engng. Sci., 8, 363 (1970).
47. J. B. Halaunbrenner, Wear, 8, 30 (1965).
48. B. Birek, Rheologica Acta 5, 3, 235 (1966).
49. S. K. Batra and F. F. Ling, A.S.L.E. Trans 10, 294 (1967).
50. W. O. Yandell, Australian Road Research 1, 3, (1967).
51. D. Tabor, Engineering 186, 838 (1958).

Notes

* (from Page 89) Limited surface reproducibility makes it difficult to detect the small vertical shift required by WLF equation.

* (from Page 100) This theory ignores the influence of the sliding speed upon the real area of contact that ought to follow from the viscoelastic behavior.

DISCUSSION OF PAPER BY A. R. SAVKOOR

J. J. Bikerman (Case Western Reserve University): The hypothesis
advanced by the speaker, namely that the adherence between two solids
can be overcome by zero force when the separation takes place in the
direction normal to the interface, whereas separation in the tangential direction requires a strong force, has been tentatively advanced
by William Hardy over 50 years ago. It is impossible to accept this
idea. The difference between the ease of raising the slider and the
difficulty of dragging it along is common to all solids; thus, an
extreme anisotropy of molecular forces would become a common properties of all substances. There is no theoretical suggestion or
experimental evidence for such an anisotropy. The large experimental
material on the breaking stresses of adhesive joints proves its
absence. Perhaps, the difference between adhesion and friction
phenomena may be made clearer by the two following sketches. Let,
say, a pin go round and round over a plane, and let plot the force
F needed to maintain this motion (at a constant normal load N) as a
function of the path length L. At low speeds v and low values of N,
usually a straight horizontal line will be obtained,

Fig. 1. If v, or N, or both, are large, the dotted
line of Fig. 1 may result. Its curvature may be due
simply to gradual rise in temperature, and an identical
curve may be obtainable by slow heating of the whole
system. Thus, it is easy to admit that both these
behaviors (F constant, and F gradually varying) have
identical mechanisms.

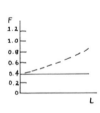

Fig. 2 is quite different from Fig. 1. The initial
force F is about 2500 times as great as the initial
F in Fig. 1. Instead of remaining constant or mildly
changing during the travel, F suddenly (when the L
is still less than 1 micron) drops to a value similar
to those seen in Fig. 1 and becomes independent of L.
What happened? The pin was glued to the plane before
the experiment started.

E. Southern (Malaysian Rubber Producer's Research Association):
The use of a damper to suppress stick-slip, although convenient,
would seem to be open to the objection that it will produce an
average value of the periodic force observed in an undamped system.
Our observations reported in our paper on the friction of rubber on
ice indicate that the perk value of the frictional force is more
appropriate. Without repeating our reasons for believing this, there
is the experimental observation that increasing the transducers
strong stiffness increases the minimum value of the periodic force
without affecting the maximum. It would affect, therefore, that an
infinitely stiff mechanical system would produce a non-periodic force
equal to the maximum value obtained in a stick-slip measurement.

A. V. Savkoor: When a periodic oscillation occurs such as the stick-slip oscillation at speeds just exceeding that at which the friction-speed curve shows a peak, the maximum value of the measured force corresponds approximately to the peak value. The value of the frictional force, which is of interest is obviously lower than that value since the friction decreases as speed increases in that range of speeds. It is, therefore, incorrect to assume that the maximum value is equal to the value of the actual force of friction. From the above arguments it is also evident why you find that the maximum value is not appreciably altered when you increase the stiffness. At higher sliding speeds, the maximum value corresponds to a frictional force which occurs at speed lower than the speed of peak friction. The speed depends upon the system.

It is necessary to note that the sliding speeds are varying over a wide range of speeds; when the inertia forces are not known it is not possible to obtain the value of the force of friction at the instant when the sliding speed equals the imposed speed. The use of a damper does not necessarily give an average value of the force; the oscillation is generally not harmonic. Furthermore the sliding speed equals the imposed speed when stick-slip is eliminated so that you can measure the force directly. There are no complex instrumentation and data processing involved. The additional theoretical advantage of our method is that any possible complications arising from terms involving higher derivatives of displacements are out of question.

The Interactions of n-C$_5$F$_{12}$ (a Model Compound to Partially Represent P.T.F.E.) on a Clean Iron Film

M. O. W. Richardson and M. W. Pascoe

Loughborough University, Loughborough, Leics., England

Brunel University, Uxbridge, Middx, England

As an initial exploratory experiment dodecafluoro n-pentane was chosen as a model compound for P.T.F.E. to simulate part of the perfluorinated molecular chain and allowed to adsorb at 0.13 µPa pressure and ambient temperature on a freshly prepared iron film produced by an evaporation process in situ from an iron wire wrapped around a tungsten filament. The events inside the vacuum chamber were followed with an MS10 mass spectrometer and (where appropriate) a Pirani gauge. These experiments suggest that the 'clean iron' acts as a catalyst for the breakdown of the perfluorinated alkane producing a variety of transiently reactive fragments, such as CF$_2$, that in turn may react with the borosilicate (Pyrex) glass walls of the reaction chamber. Generally speaking, perfluorinated compounds of the type used are exceptionally stable and do not break down except under extreme conditions (such as reaction with chlorine at 1073k) and so any apparent degradation products at such a low temperature are particularly interesting, especially since no mechanical shearing was involved.

INTRODUCTION

Freshly exposed metal surfaces can be continually generated by such processes as wear and metal fatigue. It is, therefore, reasonable to suppose that during such a process the interface sites involved can be potentially very reactive for at least a few nano-seconds. In fact it has been reported by Morecroft[1,2,3] and others that, under conditions of ultra-high vacuum, oxide free iron will promote the breakdown of long chain alkanes and aliphatic acids into smaller molecules and more recently a comprehensive survey of this type of reaction has appeared[4] in tabular review form. It was decided, therefore, to establish whether a material such as P.T.F.E. (used in bearing manufacture) might reasonably be expected to degrade under such conditions of exposure to a catalytic surface of 'clean iron'. Such an investigation becomes particularly important when considering the use of materials in the context of a spacecraft environment where there is not sufficient oxygen in the atmosphere to reoxidize the clean iron surface generated by a wear process and thus effectively reduce its catalytic effect.

As an initial exploratory series of experiments dodecafluoro-n-pentane ($n-C_5F_{12}$) was chosen as a model compound to simulate part of the P.T.F.E. chain and allowed to come in contact at 0.13 µPa pressure and ambient temperature with a freshly prepared iron film on borosilicate glass (Pyrex) produced by an evaporation process in situ from an iron wire wrapped round a tungsten filament. The events inside the reaction chamber were followed with an M.S.10 mass spectrometer and a modified Pirani gauge.

EXPERIMENTAL

Reaction Vessel

This consisted of a $5 \times 10^{-4} m^3$ Pyrex glass flask inside which were positioned a pair of stainless steel electrodes culminating in a 280 µm tungsten wire filament wrapped with fine iron wire. Wire thicknesses were critically adhered to in order to promote iron vaporization without alloying with the tungsten. Connected to the glass chamber was a modified Pirani gauge and a cold trap for removing condensible vapours when necessary. Sample vapour ($n-C_5F_{12}$) introduction to the reaction chamber was restricted initially by a glass break-seal and later in time, after this seal had been broken, controlled by a high vacuum leak valve. The ultra-high vacuum conditions required were accomplished using a getter-ion pump and the analyses of the gases within the main reaction chamber were carried out with an A.E.I. MS.10 mass spectrometer modified to cover the mass range m/e 12-200.

Materials

The dodecafluoro-n-pentane ($n-C_5F_{12}$) sample was obtained from Fluoro-Chem. Ltd. A comprehensive analysis of the material was made using an A.E.I. M.S.9 high resolution mass spectrometer and a gas/liquid chromatograph. This confirmed a 95% pure sample with some impurities due to undecafluoropiperidine and C_5F_{12} isomers which were difficult to remove due to their boiling points in some cases being within 0.7 deg. C of the sample. Other potentially purer samples of longer or shorter perfluorinated alkanes were either not available or possessed unsuitable vapour pressures for this particular experiment.

RESULTS AND DISCUSSION

Essentially the present study involved making (under initially U.H.V. conditions) mass spectrometer analyses of the $n-C_5F_{12}$ sample vapour either in the presence of a 'clean' iron film or in the absence of a 'clean' iron film. For simplicity and clarity only the major carbon/fluorine relative abundance peaks have been presented and these have been drawn up in histogram form for easy comparison in Fig. 1.

Firstly, diagram A shows the cracking pattern quoted in the literature[5,6,7] for a pure $n-C_5F_{12}$ sample. Diagrams B and C are control analyses of the $n-C_5F_{12}$ sample used. B stems from an A.E.I. M.S.9 high resolution mass spectrometer external to the reaction chamber and C from an analysis made on an A.E.I. M.S.10 mass spectrometer connected directly into the glass reaction chamber without any 'clean' iron present. Thus, it can be seen that the cracking pattern of the sample used was fairly typical of the fragmentation of an $n-C_5F_{12}$ molecule under these analysis conditions.

Secondly, diagrams D and E show the cracking patterns obtained when the same $n-C_5F_{12}$ vapour was allowed to adsorb onto the surface of a 'clean' iron film. D was obtained after 20' adsorption and E 60' later after the gases in the reaction chamber had been deliberately condensed and then revaporized in the glass vapour trap.

Thus when a 'clean' iron mirror was present in the apparatus and $n-C_5F_{12}$ was introduced on to it the cracking pattern saliently changed from C to D. Presumably this must have meant that one was now observing the cracking pattern of $n-C_5F_{12}$ along with at least one major new type of gas molecule produced by interaction with the clean iron. Clearly one of three processes could be occurring to produce this effect. Since the $C_2F_5^+$ fragment has increased the most relative to CF_3^+ then the new gas might be assumed to have its base peak at m/e 119. But, checking the literature tables for

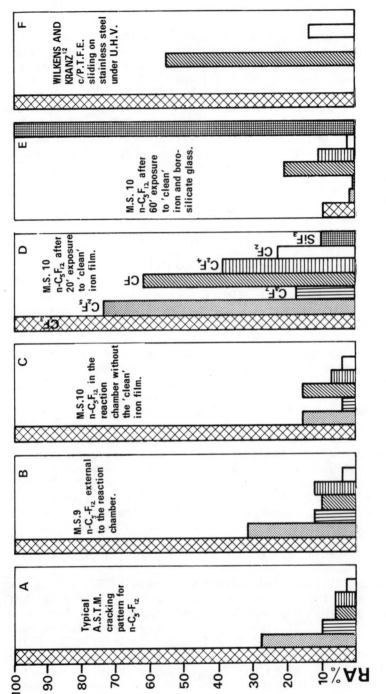

Fig. 1.   Comparative Mass Spectrometer Analysis

molecules producing base peaks at m/e 119 there are no cracking patterns that fit closely enough to confirm the hypothesis. Alternatively it may be argued that as $C_2F_4^+$ has increased the most relative to its own original concentration then the new molecule might have its base peak at m/e 100. In this case octafluorocyclo-butane (C_4F_8) is a possibility but unfortunately even allowing for a mixture with nC_5F_{12} the average cracking pattern does not success-fully account for the high proportion of $C_2F_5^+$ and $C_3F_7^+$ in the analysis. The most likely explanation of the D cracking pattern, however, is that by some process a cracking pattern very similar to nC_5F_{12} has been produced but with the overall proportion of CF_3 being reduced by interactions in the spectrometer head. The coreactant with the CF_3^+ ion in this case could very likely be CF_2 that is known to be stable for at least a few seconds. Thus the following reaction sequence is possible following the principles cited by Bond for formic acid degradation[8].

$$nC_5F_{12} + 1^* \longrightarrow C_4F_{10} + \quad :CF_2 \quad + \quad 1^*$$

$$CF_2 + CF_3^+ \longrightarrow C_2F_5^+$$

N.B. 1^* is the notation for a catalyst ('clean' iron) reaction site. Obviously other combinations of CF_2 with fragmentation ions are possible but since CF_3^+ is the most abundant species obtained from the cracking of $n-C_5F_{12}$ in the spectrometer head then the generation of $C_2F_5^+$ ions seems to be, on balance, more probable.

The cracking pattern for C_4F_{10} when considered in conjunction with nC_5F_{12} and the CF_3^+ removal process compares reasonably well in terms of the order of the most abundant fragments. The fit is certainly not a perfect one but in the light of fragmentation patterns quoted in the literature for perfluorinated alkanes and alkanes C_1-C_5 and higher this explanation seems the only one at all feasible.

The third major type of cracking pattern to be interpreted is that shown by E. Once again a complete change has taken place, presumably as a result of the condensation and revaporization of the clean iron reaction products in the vapour trap. Between m/e 85-87 a new parent peak has appeared that can be resolved by reference to associated peaks. This is necessary because of the limited resolution of the M.S.10 instrument when dealing with the substantial peaks in the range m/e 85-87 which are close together and obviously swamped by one very large peak in the middle. This is probably due to the base peak associated with a SiF_4 cracking pattern and would give rise to a parent peak at m/e 85 and subsidiary peaks at m/e 87, 28, 33, 47 and 19. All of these occur in experiment E but not all in experiment C. Thus it would be more reasonable to assume that reaction products from the clean iron/$n-C_5F_{12}$ are attack-ing the borosilicate glass than for a very large increase in the

unstable C_4F_2 molecule (F-C≡C-C≡C-F) found at m/e 86. It is
conceivable that the reactive fragment interacting with the glass
is a diradical difluoromethylene (F-Ċ-F) or difluorocarbene (:CF_2)
that can appear more easily due to the lower activation energy
encountered in removing itself from the end of a perfluorinated
alkane[9]. Using the concept of thermal bond dissociation energies,
once a carbon/carbon bond has been broken [for uncatalyzed nC_5F_{12}
requiring at least (87 k cals/mole) 365 k J mol^{-1} (9)] then it can
be shown that the subsequent energy for removing one CF_2 unit can
be as low as (44 k cals/mole) 185 k J mol^{-1} (9)

$$\begin{array}{ccccc} 86 & 86 & 85 & 78 & 44 \end{array}$$
$$\text{e.g.} \quad R - CF_2 - CF_2 - CF_2 - CF_2 - CF_2.$$

Thus assuming that the clean iron is functioning as a catalyst
to thermal degradation and assuming equal lowering of activation
energy all along the nC_5F_{12} molecule then elision of a CF_2 fragment
is highly likely and indeed has been postulated for the thermal
breakdown of P.T.F.E. and other perfluorinated molecules[9,10,11].
A chain reaction of "unzipping" process, however, is likely to be
hindered here because unlike P.T.F.E. when the chain length is small
the bond energies increase quite dramatically.

$$\text{e.g.} \quad CF_3 \xrightarrow{\quad 122 \quad} F \gg CF_3 \xrightarrow{\quad 94 \quad} CF_3 \rangle CF_3 \xrightarrow{\quad 91 \quad} CF_2CF_3$$

$$\rangle CF_3CF_2 \xrightarrow{\quad 88 \quad} CF_2CF_3 \rangle CF_3CF_2 \xrightarrow{\quad 87 \quad} CF_2CF_2CF_3$$

This tentatively supports the earlier hypothesis which proposed (in
the absence of strong evidence to the contrary) the formation of
C_4F_{10}. After elision of one CF_2 unit from the nC_5F_{12} molecule
recombination of the remaining fragments could occur leaving behind
the shortened perfluorinated alkane.

The reaction mechanism could follow one of two routes according
to the nature of the transition state formed in conjunction with the
clean iron surface (See Fig. 2).

This reaction scheme includes a metastable transition of
$C_2F_4 \xrightarrow{+} CF_2^+ + CF_2$ and is feasible in the light of the results
obtained in analysis D and providing one postulates that the high
level of $C_2F_4^+$ ions observed in that experiment is partially due to
the recombination of CF_2 radicals.

Whilst acknowledging that one can never compare directly the
results and conclusions of different experiments performed under
different conditions nevertheless when there are at least a few
common factors present then the exercise can be a useful and
stimulating pointer in planning future investigations. This is

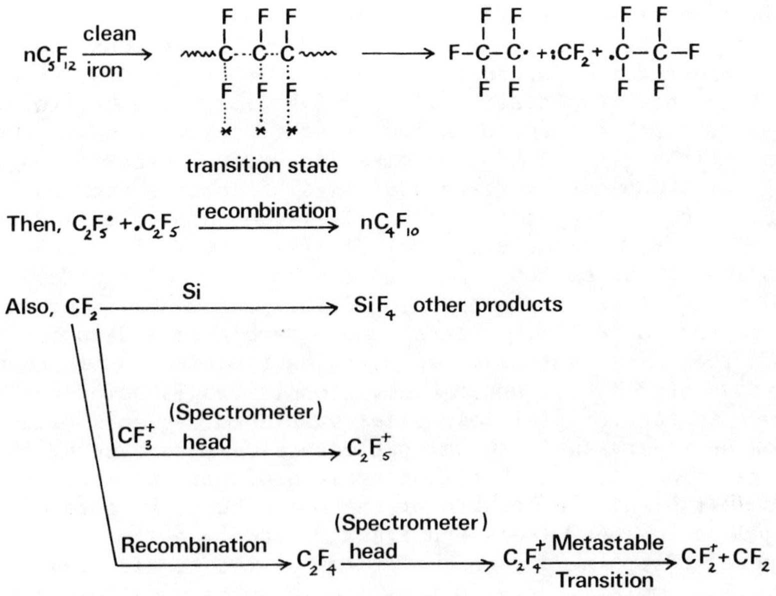

Fig. 2. A hypothetical reaction path for the catalytically
         induced breakdown of $n-C_5F_{12}$ on a 'clean' iron film
         (Route 1)

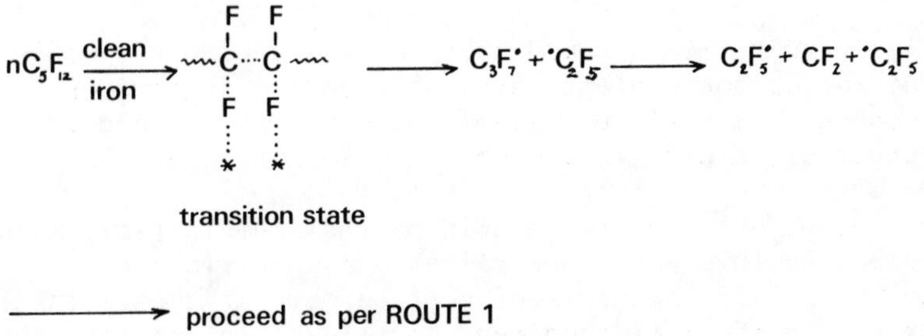

proceed as per ROUTE 1

Fig. 3. Route 2.

especially relevant when attempting to assess how far one can
extrapolate the results of experiments using model compounds to
predict the likely behaviour of the materials they are intended
(albeit imperfectly) to represent. For instance it is interesting
to note that in the excellent work carried out by Wilkens and Kranz[12]
they have reported similar relative concentrations of CF_3, CF and
CF_2 wear fragments from carbon filled P.T.F.E. sliding on stainless
steel in U.H.V. as were found in the present work with nC_5F_{12} on
clean iron. This can be seen by comparing D and F cracking patterns
in Fig. 1. In D however certain additional fragments such as C_2F_5,
C_3F_7, C_2F_4, etc. also appeared in fairly large quantities and so
comparisons are very imperfect. Nevertheless the similar relative
proportions of these common fragments can hardly be fortuitous and
are open to one of two explanations. Either the mode of mechanochemical rupture of P.T.F.E. bears some resemblance and common
factors to that of a thermally degrading perfluorinated alkane or
the breakdown of P.T.F.E. under U.H.V. conditions is being catalyzed
by the worn surface of stainless steel unable to replenish its
natural oxide layers that are being progressively removed by the
tribological process. The first of these explanations would seem
to be the most plausible because of the conspicuous absence of C_2F_5,
etc. fragments in the Wilkens and Kranz[12] experiments. Thus it
can be argued that, in very general terms at least, any weak links
in the P.T.F.E. molecule are just as likely to be open to rupture
by energy from mechanical agencies as thermal. The alternative
possibility of the stainless steel acting as a catalyst under these
U.H.V. conditions can only be ascertained by further experimentation
that would involve repeating both experiments in the same vacuum
chamber. One of the most important considerations would be that
the same mass spectrometer be used in such a fashion that the flight
paths of carbon/fluorine fragments en route to the cracking head
were both the same. This would reduce the risk of any integrating
or disintegrating short life time fragments producing different
cracking patterns merely because of the different flight times
involved in reaching the ionization chamber.

Similarly adjustments would have to be made for the differences
in chromium and carbon content, etc. between the surface of the
'clean' iron and the stainless steel assuming that it had been
polished under ultra high vacuum conditions.

Thus the authors are of the opinion that the initial exploratory
work described in this paper underlines the significance and
importance of defining the role of surface catalytic effects (if any)
when interpreting chemical phenomena at sliding interfaces especially
when the experiments are performed under U.H.V. conditions. Obviously
to establish the existence or exact nature of any individual polymer/
iron interactions (where the metal has been stripped of its oxide)
would require extensive further work with the actual systems involved
and it is hoped that this paper will stimulate research in this area.

CONCLUSIONS

1. Oxide free 'clean' iron appears to act as catalyst for the breakdown of nC_5F_{12} at room temperature and under U.H.V. conditions.

2. A hypothetical reaction path is proposed involving the elision of a CF_2 fragment after the nC_5F_{12} molecule has passed through a transition state on the clean iron surface.

3. One of the proposed reaction products, $:CF_2$, is highly reactive and may be capable of removing Si from the Borosilicate (Pyrex) glass walls of the reaction chamber.

ACKNOWLEDGEMENTS

The authors would like to thank Mr. D. W. Morecroft, Dr. P. T. Davies, and Dr. H. Naylor of Shell Research Ltd. for their advice and kindness in making available the reaction vessel and associated equipment; Prof. W. A. Holmes-Walker (Brunel University) in whose department most of the experiments were carried out; Prof. G. C. Bond (Brunel University) for helpful discussions and the Science Research Council for providing the finance.

REFERENCES

1. D. W. Morecroft, Wear, 18, 333-339 (1971).
2. R. W. Roberts, Trans, Fara. Soc., 58, 1159 (1962).
3. P. A. Redhead, Can. J. Phys., 42, 886 (1964).
4. J. R. Anderson (Editor), Chemisorption and Reactions on Metallic Films, Vols. 1 and 2, Academic Press (1971).
5. A. Cornu and R. Massot, Compilation of Mass Spectrometer Data, Heyden (1966).
6. Eight Peak Index of Mass Spectra, M.S. Data Centre, A.W.R.E., Aldermaston, Vol. 2, Table 3 (1970).
7. A.S.T.M. Index of Mass Spectral Data, STP 356 (1969).
8. G. C. Bond, Catalysis by Metals, Academic Press (1962).
9. L. A. Errede, J. Org. Chem., 27, 3425-3430 (1962).
10. R. K. Steunenberg and G. H. Cady, J. Am. Chem. Soc., 74, 4165 (1952).
11. G. H. Cady, Proc. Chem. Soc., April, 133 (1960).
12. W. Wilkens and O. Kranz, Wear, 15, 215-227 (1970).

DISCUSSION OF PAPER BY M. O. W. RICHARDSON

D. T. Clark (Durham University, U. K.): In terms of the energetics
of the situation, I find your postulate of fragmentation of C_5F_{12}
into $2C_2F_5$ and CF_2 without specific involvement of the surface
rather strange. Would not an insertian reaction preceeding fragmentation be more reasonable. The relatively unreactive nature of
difluorocarbene would also tend to instigate against your proposal
for formation of SiF_4.

M. O. W. Richardson: Firstly, if you refer to Fig. 2 on p. 236
and to the third to last paragraph on p. 232 of our preprint you
will see that we do indeed postulate the involvement of the clean
iron surface in assisting the fragmentation process. Extending
general principles of metallic catalysis outlined by Prof. G. C. Bond
(Ref. 8), we feel that the perfluorinated alkane goes through a
transition state on this surface as a precursor to C - C bond
rupture stimulated by the metal acting as an energy sink.

Secondly, regarding the comments on the CF_2 fragment there are in
fact a number of literature references to it being stable for only
a few seconds. Thus I would have thought that our proposal of it
reacting with the glassware was not unreasonable in the light of
the mass spectrometry evidence we obtained. Furthermore, Dr. Lansland
of the Naval Research Station tells me that he has obtained evidence
of SiF_4 formation when P.T.F.E. is degraded in the presence of
borosilicate glass. Since P.T.F.E. is said to thermally degrade by
the successive elision of these same CF_2 fragments then presumably
here is another example of CF_2/Si interaction to back up our own
tentative proposal. The exact mechanism of this type of reaction,
however, we are unable to comment on without carrying out more
experiments.

H. Gisser (Frankford Arsenal): Did you observe any "fragments" having
more than five carbon atoms? Would you expect any?

M. O. W. Richardson: Answering the last part first, yes in theoretical terms at least, the formation of perfluorinated alkanes with
chain lengths longer than the original C_5 are possible by recombination of only the larger free radicals. For example taking our
'Route 2' example in Fig. 2, two $C_3F_7\cdot$ radicals could produce C_6F_{14}
as a side reaction. In practice, however, examination of cracking
patterns from the literature for perfluorinated alkanes longer than
C_5 and comparing them with those we actually obtained did not reveal
any evidence for this.

Fracture Mechanics Applied to Rubber Abrasion

D. H. Champ, E. Southern and A. G. Thomas

Malaysian Rubber Producers' Research Association

Welwyn Garden City, England

A novel theory has been developed which directly relates the abrasion of rubber by a knife edge to its crack-growth characteristics. The approach is based on fracture mechanics which has been successfully used to interpret other failure processes in rubbers. The theory treats the removal of the rubber when a steady state has been reached in the development of the abrasion pattern in the rubber surface. It is suggested that crack growth occurs into the rubber from stress concentrations in the abrasion pattern. An important point is that once a steady state is established the detailed way in which the particles of rubber are finally detached is unimportant. The crack growth behaviour of the rubber is measured independently using established techniques.

The experimental apparatus is described and results have been obtained at various loads and sliding speeds. Good agreement with the theory is found for most rubbers but departures are noted for natural rubber which crystallizes on straining. The reasons for these departures are discussed.

133

INTRODUCTION

Although a considerable amount of work has been done on the mechanism of rubber abrasion[1-8], it is still not clear what the fundamental process is. It has been proposed that in some cases the rubber is removed by a tensile rupture process, and in other cases by a 'fatigue' mechanism. However, no quantitative theory has been developed to relate the rate of abrasion to independently measured fundamental strength properties. In this paper we propose a theory which relates the abrasion produced in a particularly simple abrasion process, the passing of a blade over the rubber surface, to the crack growth behaviour of the material. The approach used is that which has been referred to as 'fracture mechanics'. It has been successfully applied to the tear[9,10], crack growth[11,12] and fatigue[13] behaviour of rubbers.

During abrasion of rubber, in many cases the surface develops a pattern of ridges perpendicular to the direction of abrasion. This phenomenon, which has been described and discussed by Schallamach[14], clearly plays a very important part in the abrasion process. The detailed way it is developed, and the precise way it is deformed during abrasion is in general complicated. It is these complications that have perhaps been the main difficulty in producing a precise theory.

Schallamach[15] studied the damage produced when a needle is scratched across a flat surface of rubber, and this work was inform-ative as to the nature of the deformations produced by a rubbing asperity. However, virtually no rubber is removed in this operation so that it is not possible to examine the practically significant quantity, the amount of rubber removed by abrasion. We therefore considered what may be thought of as a one dimensional analogue of this case, namely the abrasion produced by a line contact, physically a blade, moving perpendicular to its length across the rubber. This method removes a measurable amount of rubber, and produces an abrasion pattern essentially similar to that developed on other abrasives.

BLADE ABRASION MACHINE

A schematic diagram of the machine is shown in Figure 1. The abrading razor blade bears on the rubber sample which is in the form of a wheel 6.25 cm diameter and 1.2 cm wide. The blade is clamped at one end of a freely pivoted arm which has a simple dashpot damper at the other end. The whole arm is mounted in a spring cantilever the movement of which, and hence the applied force, is detected by a transducer. The geometry of the machine is such that this force is the frictional force developed, and the normal reaction on the blade is equal to the applied vertical load. The abrasion loss is obtained from the weight change of the rubber sample. Abraded rubber

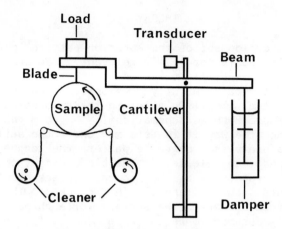

Fig. 1. Schematic diagram of blade abrasion machine

clings to the surface of the sample and this debris is easily removed
using transparent adhesive tape which is driven independently of
the rubber wheel. Subsidiary experiments have shown that the use
of the tape does not itself cause any significant loss of rubber
from normal unfilled rubbers. If very weak rubbers are used, however,
there is a danger that the tape will pull off significant amounts.
Care has been taken to ensure that vibration from the motors used
to drive the sample and cleaner do not induce vibrations in the arm
and blade assembly. The damper reduces the vibrations which arise
from the passage of the blade over the sample. These vibrations
can be quite large in an undamped system if the abrasion pattern is
coarse. All the measurements have been carried out with a stiff
single edge razor blade which is clamped over most of its major
surface so that it is not significantly deflected by the forces
involved.

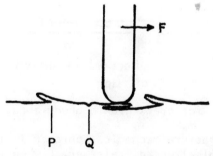

Fig. 2. Schematic diagram of abrasion pattern and its deformation by blade. Force F applied to blade. Points such as P
and Q are where crack growth is assumed to occur. P is the
re-entrant angle in the undeformed abrasion pattern, and Q
represents a similar element deformed by the blades.

THEORY

The actual development of the abrasion pattern is difficult
to follow, and indeed is not really relevant. Measurements on all
abrasion machines are usually taken only after a steady state has
been reached and the pattern fully developed. Under these circumstances a substantial simplificiation in the theoretical approach
is possible. Figure 2 shows schematically the form the abrasion
pattern takes, and the way it is deformed under the action of the
blade. As the blade passes over the pattern the tongue of rubber
is pulled back and then released as the blade passes on. The stress
produced in this process is assumed to cause crack growth in the
re-entrant corners such as P and Q. This crack growth will have a
component Δx perpendicular to the surface for each pass of the
blade. If, as is assumed to be the case, the surface has reached
a steady state with the abrasion pattern maintaining a constant
overall appearance, the surface of the rubber must be lowered on
average by Δx at each pass. Thus the volume of rubber removed per
unit area of surface is Δx. An important point is that this conclusion is not dependent on the detailed mechanism by which the
rubber is finally removed, but it is merely assumed that a steady
state is reached and that a crack growth mechanism is operative.

In order to apply the necessary fracture mechanics analysis
the deformation of the abrasion pattern is somewhat idealized as
shown in Fig. 3. The pattern is taken to be uniform across the
surface of the sample, which observation indicates is a fair
approximation, and the frictional force F to be wholly sustained by
the tongues of the pattern. For the present, we will also assume
that the crack growth is perpendicular to the surface. This
assumption is not strictly necessary, and the refinements will be
dealt with in a later paper. They do not however affect the general
conclusions.

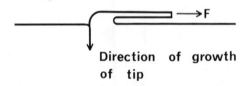

**Direction of growth
of tip**

Fig. 3. Model for crack growth under abrading force F.

To apply the fracture mechanics approach it is necessary to
calculate the mechanical energy released dU when the crack grows
an incremental amount dc. This energy release rate, which has been
termed the tearing energy, and is usually denoted by T in the rubber
work[9], is defined by

$$T = \frac{1}{h} \frac{dU}{dc} \qquad (1)$$

where h is the length over which F is applied (in this case the width of the specimen). It has been found from extensive tear and crack growth studies that the T value determines the rate of crack propagation independently of the overall shape of the specimen and the detailed way the forces are applied[9,10]. Thus if the tearing energy versus crack growth relation is known for a particular material, the behaviour in any type of specimen can be predicted provided the T value developed in that specimen can be calculated from the measurable applied forces.

Considering Fig. 3, it can be seen that if the crack tip advances into the rubber a distance dc, the force F will move by a similar distance relative to the rubber, provided the extension of the tongue is not large.

The situation is in fact very similar to that existing for a tear test (the 'simple extension' or 'trouser test' piece) described by Rivlin and Thomas[9]. For the conditions stated above it is easy to show, following a similar analysis, that the tearing energy is given by

$$T = \frac{F}{h} \qquad (2)$$

The crack growth behaviour of a number of rubbers under repeated stressing has been examined using the tearing energy approach, and the results expressed as the crack growth per cycle r as a function of the maximum T value attained during each cycle[11,12]. For many materials over a fair range of T values (approximately $10^5 < T < 10^7$ dynes/cm) it is found that this function is of the general form

$$r = BT^{\alpha} \qquad (3)$$

The exponent α varies from about 2 for natural rubber to 4 or more for non-crystallizing unfilled rubbers such as SBR[16]. With the above model of the abrasion process the rate of abrasion A, in cm^3 per revolution of the wheel, will therefore be given by

$$A = rhs \qquad (4)$$

where s is the circumference of the abrasion wheel and h its width. Equations (2), (3) and (4) thus determine the abrasion loss in terms of the frictional force F on the blade and the independently measured crack growth behaviour, exemplified by equation (3).

RESULTS AND DISCUSSION

It is noteworthy that the sharpness of the abrading blade does not appear in the theory. Clearly, it must be sufficiently sharp so that only one element of the abrasion pattern is deformed at any one time, but this should not be a severe limitation. This conclusion is broadly confirmed by experiment. The difference in abrasion between a new razor blade and one used for several thousand revolutions is negligible. Cutting thus plays little part in this abrasion measurement. Also, the normal load itself does not appear in the theory, being relevant only insofar as it controls the frictional force F. The rate of revolution used was equivalent to a linear abrading velocity of 3.41 cm/sec. This velocity is not critical and changing it by a factor of 2 produces an inappreciable effect.

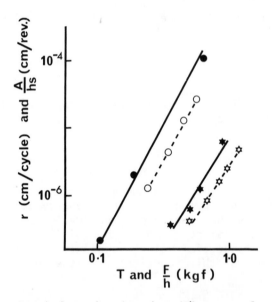

Fig. 4. Depth loss in abrasion A/hs as a function of F/h and crack growth rate[12] r as a function of T
isomerized natural rubber;   ○ abrasion   ● crack growth
butadiene-acrylonitrile;   ✦ abrasion   ★ crack growth

According to the theory given above, plots of abrasion results, (A/hs vs. F/h) and crack growth results (r vs. T) from ref. 12 should superimpose. Figure 4 shows such results for two of the four non-crystallizing materials examined, gum (unfilled) rubbers of isomerized natural rubber[17] (43% cis and 57% trans polyisoprene) and butadiene acrylonitrile. It can be seen that complete superposition is not found. For all four rubbers the abrasion results lie to the right of the crack growth curves, but the slopes are similar and the materials are rated in the same order. Agreement can be improved

by multiplying all the abrasion results by a constant factor of 0.63
in T as shown in Figures 5a and 5b. The agreement between the
abrasion and crack growth results is now quite good, probably within
the accuracy of the measurements and reproducibility of the materials.

The substantially linear curves are consistent with the approximate relation (3). Because of this, it is not possible with the
present results to distinguish between a shift factor in T and A. The
necessary factor is not large considering the simplifications in the
theory, and indeed can be explained in terms of the crack growth not
being accurately perpendicular to the surface. (This will be considered in detail in a later paper). The general agreement is
sufficiently good to suggest that the mechanism of abrasion in this
case has been correctly identified.

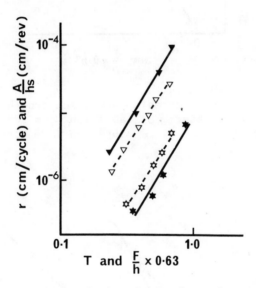

Fig. 5a. Comparison of abrasion and crack growth data.
Butadiene acrylonitrile;   *   abrasion   ★   crack growth
Cis-polybutadiene;   ∇   abrasion   ▼   crack growth

All of the above rubbers are non-crystallizing. Similar
measurements taken on a normal natural rubber compound, which is
strain crystallizing, give the results shown in Fig. 6. Clearly
there is a substantial difference in behaviour. The crack growth
resistance is much greater than for the isomerized NR but the abrasion
losses are surprisingly little different. This strongly suggests
that under the present abrasion conditions crystallization is either
suppressed or is ineffective in strengthening the material. A
possible reason for this is the relatively rapid deformation of the
rubber during abrasion, perhaps giving insufficient time for

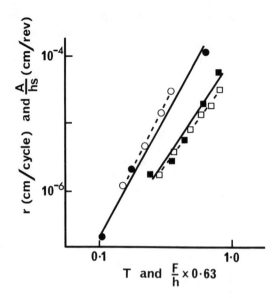

Fig. 5b. Comparison of abrasion and crack growth data.
Isomerized natural rubber;   o   abrasion   ● crack growth
Butadiene styrene;   □   abrasion   ■ crack growth

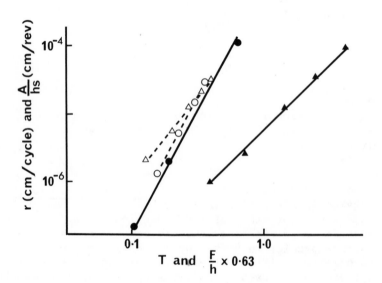

Fig. 6. Comparison of abrasion and crack growth data.
Isomerized natural rubber;   o   abrasion   ● crack growth
Natural rubber;   △   abrasion   ▲ crack growth

crystallization to occur, although crack growth and abrasion measurements at widely varying rates have so far given little support to this. Alternatively, the substantial compression undergone by an element of rubber as the blade approaches it, which is not reproduced in the crack growth measurements used here, may impede crystallization in the subsequent rapid extension.

CONCLUSIONS

The theory put forward, which relates the abrasion by a blade to the crack growth behaviour of the rubber is in good accord with results on non-crystallizing gum rubbers, suggesting that the mechanism proposed is essentially correct. Strain crystallizing natural rubber abrades more than would be expected on the basis of its crack growth behaviour, in fact very similarly to non-crystallizing isomerized NR, indicating that crystallization is inhibited or ineffective in the present abrasion process.

ACKNOWLEDGEMENT

This work forms part of the Research Programme of the Malaysian Rubber Producers' Research Association.

REFERENCES

1.   A. Schallamach, Wear, 1, 384, (1958).
2.   K. A. Grosch and A. Schallamach, Rubber Chem. and Technol., 39, 287 (1966).
3.   I. V. Kragelskii, Soviet Rubber Technol., 18, 22 (1959).
4.   I. V. Kragelskii and E. F. Nepomnyashchii, Wear, 8, 303, (1965).
5.   S. B. Ratner, Dokl. Akad. Nauk USSR, 150, 848 (1963).
6.   G. I. Brodskii, N. L. Sakhnovskii, M. M. Reznikovskii and V. F. Evstratov, Soviet Rubber Technol., 19, 22 (1960).
7.   A. Schallamach, Proc. NRPRA Jubilee Conf. Cambridge, L. Mullins, ed., Maclaren, London (1965).
8.   A. Schallamach, J. Appl. Polymer Sci., 12, 281 (1968).
9.   R. S. Rivlin and A. G. Thomas, J. Polymer Sci., 10, 291 (1953).
10. A. G. Thomas, J. Appl. Poly. Sci., 3, 168 (1960).
11. A. G. Thomas, J. Polymer Sci., 31, 467 (1958).
12. G. J. Lake and P. B. Lindley, Rubb. J., 146, (10) 24 and (11) 30 (1964).
13. A. N. Gent, P. B. Lindley and A. G. Thomas, J. Appl. Poly Sci., 8, 455 (1964).
14. A. Schallamach, Trans. Inst. Rubber Ind., 28, 256 (1952).
15. A. Schallamach, J. Poly. Sci., 9, 385 (1952).
16. P. B. Lindley and A. G. Thomas, Proc. 4th Rubb. Technol. Conf. London, 428 (1962).

17. J. I. Cunneen and G. M. C. Higgins, Chemistry and Physics of
     Rubber-Like Substances, L. Bateman, ed., Chapter 2, Maclaren
     (1963).

DISCUSSION OF PAPER BY E. SOUTHERN

J. R. Beatty (Goodrich Rubber Co.): We have done a comprehensive study of the effect of temperature on crack growth of rubber.

E. Southern: At low Temperatures?

J. R. Beatty: At high temperatures. Our review on the fatigue of rubber was published in Rubber Chemistry and Technology (J. R. Beatty and A. E. Juve, <u>37</u> (5) 1341 (1964).

S. Bahadur (Iowa State University): Did you estimate the error involved that results due to ignoring the stretching of the highly elastic thin rubber filament that gets partially detached from the bulk material due to abrasion?

How can you consider the process as steady state in which the rubber filaments are stretched and broken repetitively giving rise to humps in the layer of material deposited due to abrasion?

E. Southern: A more vigorous approach to the problem may be made to include the effect of stretching the rubber torques following the method of Rivlin and Thomas (Reference 9) for the "trouser" testpiece. Unfortunately we do not know by how much the torques of rubber are extended but the inclusion of a term to take account of this effect would make agreement between theory and experiments worse rather than better. The approximation of neglecting the extension of the rubber is normal practice in calculating the tearing energy.

The steady state to which we refer is the steady rate of loss of rubber from the whole surface of the rubber as assessed by weight loss measurements. Furthermore, the appearance of the rubber surface is not changing with time (i.e. the abrasion pattern is fully developed) although, of course, individual elements are not in a steady state.

D. Tabor (Cambridge University): Could the author first tell us how he thinks the tongues of rubber are first formed? Secondly, is it possible that the stretching of the tongues is one of the reasons for the factor 0.63? Thirdly, has the author considered the effect of temperature? This might give some insight into the role of recrystallization of the rubber in comparing abrasion with tear-strength.

E. Southern: The formation of the tongues is difficult to understand, but if our proposed mechanism is correct, it suggests that the first stage is the formation of cracks perpendicular to the rubber surface. These cracks are likely to be initiated by natural flaws in the rubber. Once the cracks start to grow, the deformation of the crack lips by the blade will cause the direction of crack growth to become more nearly parallel to the rubber surface. Further growth of the crack will produce the observed tongues of rubber. Although this plausible explanation arises from our proposed mechanism of abrasion, perhaps I should re-emphasize that our theory is concerned with the steady state after the abrasion pattern has been formed.

We have carried out the rigorous analysis which includes a term taking into account the extension of the tongues. The problem is that we do not know what the extension of the tongues is in practice, and we have,therefore,made the same assumption that is usually made in fatigue and tearing experiments that the extrusion is negligible. Nevertheless the form of the accurate equation is such that if the effect of tongue extension is included there would be greater disagreement between theory and experiment. Preliminary measurements suggest that the factor 0.63 will largely, if not completely, be accounted for when we have determined the angle at which cracks grow relative to the rubber surface.

Unfortunately our equipment does not readily lend itself to temperatures variation. I fully agree that measurements of this sort would be informative and we hope to carry them out in the future.

PART TWO

Polymer Properties and Friction

Introductory Remarks

Thomas L. Thourson

Xerox - Xeroradiography

125 N. Vinedo Street, Pasadena, California   91107

I'm always amazed to run into lively discussion on technological topics that I once thought put to bed. In our earliest introduction to the study of the physical science in high school we addressed the problem of friction - measuring the force required to move a slider over a flat surface. We were taught that the ratio of the force required to keep the slider in uniform motion to the force holding the slider to the surface was a constant called the "coefficient of friction". What our teachers often neglected to tell us was that the coefficient of friction wasn't always a constant and that there was a lot of physics buried in that constant which was obscured from our view. And the problem is especially complicated when the materials of concern are as complicated as polymers.

My own field of expertise is not friction nor polymers but xerography. Nonetheless, in our efforts to improve the reliability and performance of xerographic devices we keep encountering friction problems. Xerographic devices must run automatically and reliably and must do many things that are not related to making and developing a latent image. For example, the device must handle paper, piece by piece, over a range of environmental conditions and the friction between paper and paper handling system is important. The visible image is made up of powder toner which is transferred to a sheet of paper. To repeat the process requires that the residual toner be removed from the photoreceptor surface. There are several ways to do this including wiping with brushes or with a polymeric blade. In either case we run into friction problems. In this session Dr. Ganesh Harpavat will describe a portion of his work which is an outgrowth of this concern. Specifically, he will describe the heating that occurs when a polymeric blade wipes over a photoreceptor to get rid of the residual toner. It is important to know the amount of heat generated because local heating of the photoreceptor can result in a change of state which would render it unusable for xerographic purposes.

Friction problems are complicated because they involve so many factors and may have many different outcomes. This is what forces those of us whose interests lie in other technologies to seek the advice and counsel of those who have studied the friction problems in detail. In addition to the generation of heat, friction problems may involve a lubricant as well as a slider and surface; that is, we may be confronted with a three body problem. Friction may be accompanied by the deformation of the slider and/or the plane. And, referring again to the xerographic example, friction may be affected by the electrostatic state of the components.

In this session we will hear some of the complexities of the problem of friction and, in particular, those related to friction between polymers and other materials. J. J. Bikerman will discuss the role of adhesion (or the lack of the role of adhesion) in friction. Dr. Shyam Bahadur and Mr. M. A. Saleem will discuss the effect of the deformation of a polymer on friction or perhaps we should say the effect of friction on deformation since he will speak of this in the context of the forming of polymers by extrusion. Dr. L. C. Towle will tell us of his studies in solid film lubricants and how we might apply his findings to more effective desigh of bearings. Professor Tabor, Dr. B. J. Briscoe and Miss Christine M. Pooley will discuss studies of relating polymeric molecular structure to friction. Drs. E. Southern and R. W. Walker will describe studies on the effect of velocity and temperature on the coefficient of friction between rubber and ice and relate it to existing theories of the behavior of rubber sliding on other smooth surfaces.

These talks amply illustrate the wide variety of effects that occur in friction and the wide variety of technologies that require attention to frictional effects.

The Nature of Polymer Friction

J. J. Bikerman

Consultant

15810 Van Aken Blvd., Cleveland, Ohio 44120

The sliding friction between a solid polymer and another solid (polymer, metal, glass) usually is determined by the work of deformation of the polymer(s). The ratio F/N of frictional force to normal load commonly decreases for polymers when N increases because F is proportional to N^{m+n} and $m+n < 1$; m and n_m are defined by the expressions $w = k_1 N^m$ and $d = k_2 N^n$; w and d are the width and the depth of the disturbed material; k_1 and k_2 are proportionality constants. Friction in reproducible sliding is never caused by adhesion. The absence of adhesion is proved by the ease of normal separation of slider and support; and the cause of this absence is the presence of weak boundary layers (air, moisture, etc.) between the two. When these layers are eliminated, adherence of slider to support occurs and no usual sliding is possible. Attempts to correlate F with the true area S of contact are misleading because work must be performed to achieve this S. The common determinations of S are unsatisfactory. Theoretical and experimental determinations of the work done during frictional deformation of viscoelastic polymers is needed to account for the observed friction.

If one solid body (slider) is dragged over the surface of another (support), resistance has to be overcome and work has to

be spent. The nature of these resistance and work depends on the
nature and shape of the two solids, the normal load N, and other
circumstances; see, for instance, [1]. As long as N is small and
the solids are rigid, the main process consists in lifting the
slider over the surface hills of the support. In this region,
frictional force

$$F = \mu N; \tag{1}$$

μ is the coefficient of friction, approximately equal to the tangent
of slope of a few steepest hills on the two surfaces.

Equation (1) usually is not valid for polymers, no doubt because
organic polymeric solids are far from being rigid. Most investigators
[2-4] believe that adhesion between slider and support is the main
cause of polymer friction. If the attraction between the superficial
atoms of the two solids is A per unit area, and if S is the area
of atomic contact, then the total attractive force between slider
and support is AS, and it has to be overcome to start or continue
the relative tengential motion of the two solids. Hence,

$$F = AS. \tag{2}$$

The adhesion theory of friction is incorrect [5]. In the first
place, attraction A is not achieved in usual friction experiments.
If slider (A) is placed on support (B) in air, a film of adsorbed
air (plus moisture and numerous impurities) remains between the two
solids and no true contact between atoms of A and atoms of B occurs
anywhere along the surface of apparent contact. In the modern theory
of adhesive joints [6], this film is an example of weak boundary
layers [7] which so often lower the final strength of adhesive
assemblies. Only because a solid cannot displace air (or another
gas) from the surface of another solid, it is necessary to apply
adhesives, that is materials which are liquid in the moment of
application, are (therefore) able to eliminate the gaseous weak
boundary layers, and which later solidify thus giving rise to an
unbroken 3-layer solid consisting of first adherend - adhesive -
second adherend.

Then, the absence of attraction in common friction is easily
proved by lifting the slider, that is, moving it in the direction
normal to the geometrical area of contact; the force needed for
this motion is equal to the weight of the slider, and no force
component attributable to adhesion can be detected. In a few special
experiments the adhesion component was measured, but then ordinary
friction was absent. Ordinary adhesion and ordinary friction are
mutually exclusive. A typical test is described here to render the
meaning of this rule clear.

Two strips of poly(methyl methacrylate) were pressed [8] to
a metal at about 130°C for 20 min., and the system was cooled to 95°.
The force needed (after cooling) to displace the strips tangentially
was about 3×10^7 dynes, independently of the normal pressure
(between 80 and 900 kilodynes/sq. cm) during the application of this
force. In other words, μ in equation (1) appeared to be inversely
proportional to N. When the normal pressure was made negative, i.e.
a vertical pull upward was applied to the slider, the force required
for tangential motion still was positive, so that "the coefficient
of friction" was negative. Evidently, this behavior is very different
from that observed in true friction tests. On the other hand, it
is exactly as expected from a poor adhesive joint. When the polymer
was pressed to a metal at a relatively high temperature, a part of
the polymer was liquid enough to displace air from the support so
that a (very incomplete) adhesive joint formed; and the above
tangential force was the shear strength of the joint; as usual, it
was independent of the pressure applied to the joint during the
shear. After the rupture of the joint, the force needed for further
motion of the slider over the support was of the order of 10^5 dynes,
i.e. only a small fraction of the breaking stress. This force
corresponded to real friction.

Adherence can be achieved also at a constant temperature. If
the two solids are degassed before, and kept in a good vacuum during
the contact, the protective gas film between the two cannot form,
and the solids seize. Work is required to separate them either
normally or tangentially, and the breaking stress is independent of
N in a wide interval of loads. Consequently the system can simulate
large or small, positive, negative, or infinite "coefficients of
friction"; an infinite coefficient is calculated when N = 0, i.e.
$\mu = F/N = \infty$. In these instances again the strength of a joint
rather than true friction is determined.

As long as the slider can be moved along the support in a
reasonably steady manner, no adhesion can be noticed, that is, only
the weight of the slider must be overcome to lift it from the support.
It was repeatedly attempted to explain this absence of adhesion by
invoking elastic stresses. When two solids are pressed together,
elastic stresses are created in them. When the pressure is removed,
elastic bodies tend to regain their initial shapes; in some instances
this means also their initial positions. However, this tendency is
suppressed by adhesive forces: whenever adhesion was achieved, it
can be measured also after the removal of the initial load. This
was seen in the experiments described above [8], in the vacuum tests
of the preceding paragraph, and in the innumerable adhesive joints
prepared daily on our planet. If no "normal" adhesion can be
measured, it does not exist or is too weak to account for the resistance to tangential motion. Moreover, the above elastic stresses
would lower also this resistance.

Even if the adhesion A existed, force F of equation (2) would
not be proportional to area S of contact. This is so because the
stress pattern in an adhesive joint is never uniform and because
brittle fracture starts at a point where local stress first exceeds
local strength. For instance, in single lap joints, Fig. 1, which
are geometrically similar to systems used in sliding friction tests,
the force needed for rupture, see arrows, is not proportional to the
area of the adhesive - adherend contact; it is possible to remove
the central half of the adhesive layer, as in Fig. 2, almost without
altering this force [6].

Since equation (2), or one of its close relatives, was considered
paramount for polymer friction, experimental determination of S was
repeatedly attempted. These attempts were unsuccessful because the
notion of contact area was not clarified. In order to achieve atomic
contact (at which molecular attractions are strong), no other molecule
should be present between the atoms of A and atoms of B. The area
of this atomic contact may be - and usually is - very much smaller
than other (estimated or measured) contact areas.

The area most commonly determined is that of "frustrated optical
reflexion". Total reflexion occurs (in a range of incidence angles)
when light reaches a body of a low refractive index from a body of
a high index, for instance, when a light beam passes a polymer and
is refracted

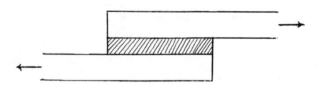

Figure 1. A simple lap joint. The adherends are white,
           the adhesive layer is shaded.

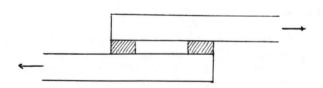

Figure 2. A simple lap joint in which the central half of
           the adhesive layer (shaded) is absent.

at the boundary of the latter with air. In other words, when air is
present between two polymeric solids, it may, at a suitable angle of

incidence, "intercept" the beam. When no air is present, the effect
is impossible. Thus it is feasible to determine the area over which
the air interlayer is absent. Unfortunately, this absence is very
relative. The sensitivity of this optical method is so small, that
a gas layer whose thickness is equal to the wave length of the light
employed would not be registered [9]. A gas film of this thickness
would easily reduce molecular attraction to practically nothing.

All these remarks well support the above statement that friction
is not caused by adhesion. In all probability, the work of deforming
polymer slider and support during their tangential motion is the
main component of frictional work in polymer systems. The existence
of this deformation is clearly shown, e.g., by experimental values
of the optical contact area. For instance, when the base of a thin
cylinder of a modified polycarbonate was pressed against optically
flat glass, this area increased by a factor of 100 when pressure
was raised from a low value to, say, 20 megadynes per sq. cm.
Obviously the cylinder had to be deformed to adapt itself to the
profile of the glass.

It is remarkable that this deformation is generally disregarded
in the theory popular at present. When A is pressed to B, the
change in their shapes results in area S. When now tangential motion
is enforced, the adhesion along S is said to be overcome and frictional force is supposed to be needed for this process. But what about
the force required for the above change in shape? If a polymer
filament is extended, it will finally break. The work performed
during this extension is essentially work of deformation because
the work of tearing a few chains over the rupture area is negligibly
small. It would be absurd to claim that the whole work of extending
the filament (perhaps by 300%) should be counted as the work of
fracture. The neglect of the work spent on forming S is equally
unconvincing.

The deformation theory of polymer friction has been advocated
before [10,11] but without much success. Here some additional
arguments supporting it are presented.

                              NORMAL LOAD

As a rule, the ratio F/N in polymers decreases when normal
load N or the average normal pressure N/S_o increases. S_o is the
geometrical area of contact; for instance, when a cube of 1 cm edge
slides over a plane, $S_o = 1$ cm^2. Examples of this rule can be seen
in [12,13] and other publications.

A simple explanation of it seems possible. The volume significantly deformed by a moving slider may be expressed as a product
wdl; w is the width of the disturbed material, d is its depth, and

l is the length of the trajectory. Consequently the work of deformation may be approximated as the product kwdl, k being a constant
$(g/cm.sec^2)$ which depends on the mechanical properties of the two
solids. Frictional force F is the ratio of frictional work and the
path length. Hence,

$$F = kwd \qquad (3)$$

Both w and d are functions of N (or N/S_o) and in many systems may
be represented as $w = k_1 N^m$ and $d = k_2 N^n$; k_1 k_2, m, and n being
constants. Writing K for the product kk_1k_2, equation

$$F = KN^{m+n} \qquad (4)$$

is obtained. Thus, ratio F/N is proportional to N^{m+n-1}. It
decreases on an increase in N whenever

$$1 > m + n \qquad (5)$$

The classic theory (Hertz 1881) gives the external shape of
the two bodies pressed to each other or the depth d_1 and width w_1
of the depression produced in an elastic plate by an elastic sphere.
If the dependence of the above w and d on N is similar to that of
w_1 and d_1, some information on the exponents m and n is obtained.
According to Hertz, m = 1/3 and n = 2/3 so that m + n = 1. However,
the theory disregards surface roughness. Because of roughness, the
area (the "bearing area" of rugosity measurements) supporting the
slider increases more rapidly with load, see Fig. 3, than for ideally
smooth solids. Consequently the increase of d with N is smaller than
would be for an ideal system, and n for rough surfaces must be smaller
than 2/3, so that m + n must be less than 1. Unfortunately, the
irregular nature of surface roughness renders a more precise statement
impossible at present. It is also regrettable that the research on
the contact of two rough surfaces [3] dealt with the area of contact
rather than with d and w.

A better idea of the effective values of w and d would be
obtained if the stress distribution in the bodies in relative
tangential motion were better known. Distribution of stresses in
the static case has been calculated [14,15] but is insufficient for
finding the strain energy of the system in sliding. Moreover, the
classic theory neglects the friction between slider and support
during the indentation and supposes that the diameter of the contact
circle is very small compared with the diameter of the spheric
indenter; this latter assumption is invalid for rough surfaces in
which each hill acts as a plunger.

In several systems, e.g., [16], the ratio F/N was constant in
a wide range of pressures. The belief in the adhesion theory of
friction prevented the authors from looking for, and finding the

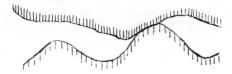 

Figure 3.   Apparent contact between two rough surfaces at a
            low (left) and a high pressure.

true cause of this insensitivity. A review of recent experimental
results concerning the dependence of F/N on N for polymers will
soon be available [17]. Experimental determination of the volume
deformed and of the stresses and strains in this volume would be
highly welcome. Instructive observations of this kind have been
published [18,19] but more quantitative results, accounted for by
the theories of elasticity and viscoelasticity, are still not
available.

SLIDER's SHAPE AND SIZE

     Unfortunately, the effect of the shape of the slider on the
ratio F/N is almost unknown. However, stress patterns produced by
a wedge dragged over a plate of transparent rubber was not significantly different from that observed when the slider was a hemisphere;
also, when the wedge was asymmetric, it did not matter whether the
more or the less inclined face of it was advancing [18]. Evidently,
these observations are in accord with the deformation theory of
polymer friction: as long as w and d remain constant, it matters
little what the actual profile of the indenter is.

     A few results could be found in the literature for the relation
between F/N and the radius of a hemispherical or spherical slider.
When this radius R was [20] 0.285, 1.33, and 10.0 cm, respectively,
F/N was 0.4, 0.46, and 0.57. The slider was of glass, the polymer
was Nylon-66, and N was constant at 10^6 dynes. From Hertz's theory
for ideally elastic materials, the product wd is independent of R
(w is proportional to the cubic root of R and d is inversely proportional to this root). It is seen that F/N also was little affected
by the radius: when R rose in the ratio of 35 to 1, F/N increased
only in the ratio 1.4 to 1. Presumably it was not quite constant
because Nylon-66 was not a Hookean solid. The change of the track
width with R also seems to confirm the poor applicability of Hertz's
equation to nylon: this width increased less steeply than did the
cubic root of R.

     The F/N for <u>starting</u> friction of a hard hemisphere on a plastic
plate was 2-5 times as great for a large slider (R = 4 cm) as for

a small slider (R = 0.24 cm) [10]. It is not known why in this instance the dependence of F/N on R was more pronounced than in the above experiments.

A metal sphere was rotated between 2 polymer plates pressed to the sphere by a constant normal load. When the sphere radius was altered from 0.3 to 0.63 cm, the torque required to maintain rotation was proportional to r^3, r being the radius of the indentation left in the plate by the rotating sphere [21]. Thus the frictional torque was proportional to the volume visibly deformed; this seems to be another confirmation of the deformation theory of friction. Perhaps the strain pattern in the plates will once be determined and the strain energy calculated from it; then a better test of the theory will be feasible.

RATE OF MOTION AND TEMPERATURE

Perhaps the clearest indication of the deformation nature of the usual polymer friction is afforded by the fact that the time - temperature transformation valid for various deformations of polymeric solids is valid also for the results of many friction experiments [22]. If the ratio F/N is measured at a constant temperature T at different speeds u cm/sec, a curve (often with a maximum) is obtained. If the measurements are repeated at several other temperatures, a family of curves results. It has been repeatedly found [22-24 etc.] that all curves of F/N versus log u could be combined to a single curve by shifting them along the axis of log u. In other words, if F/N is plotted against log a_T + log u instead of log u, the experimental data obtained at different temperatures almost coincide.

The value of log a_T is not purely empirical. For ordinary deformations it was established long ago that log a_T could be expressed as

$$\log a_T = \frac{-8.86(T - T_s)}{101.5 + T - T_s},\qquad (6)$$

T_s being a temperature characteristic for each polymer. It is this value of log a which permits construction of a "master curve" for F/N against the rate of sliding. Another transformation [24] uses the equation

$$\log \frac{u_1}{u_o} = \frac{Q}{R}\left(\frac{1}{T_o} - \frac{1}{T_1}\right);\qquad (7)$$

u_o is the velocity at the "standard" temperature T_o. For experiments

performed at another temperature (T_1) the corresponding velocity
is u_1; Q is the "activation energy" of the deformation and R is the
gas constant. The value of Q calculated from friction experiments
was almost identical with that obtained from undoubtedly mechanical
deformations for a given polymer; thus also in this procedure no
effect of any adhesion was noticed.

Nevertheless the authors [24] managed to consider their results
a confirmation of the adhesion theory of friction. The reasoning
was essentially as follows. When "adhesive junctions" between slider
and support are broken, the two bodies are deformed in the process,
and this deformation proceeds according to the rules of the time -
temperature superposition. This hypothesis implies that the work
of rupturing adhesive bonds is negligible in comparison with the
work of deformation which precedes the rupture. Thus it agrees
with the present author's view [25] and the statements on the sixth
page of this paper; but it renders the idea of adhesion even more
elusive. Of the four processes considered in the adhesion theory
of friction, namely (1) deformation leading to the final contact
area, (2) formation of an adhesive joint over this area, (3) rupture
of this joint, and (4) deformation associated with rupture, processes
1 to 3 are implicitly assumed not to affect the measured ratio F/N.
This is a strange assumption. In reality, processes 2 to 4 do not
occur in ordinary friction, and process 1 dominates the experimental
values of the frictional work of polymers.

The importance of deformation for friction is confirmed also
by the observation that the coefficient F/N suffers marked changes
in the narrow temperature interval in which the mechanical properties
of the polymer exhibit rapid changes; this occurred [26], for instance,
in a linear polyethylene between 115° and 120°.

In conclusion, two essential statements are repeated:

(1) Contact of a solid polymer with another solid in a gas does
    not give rise to adhesion, as is proved by direct tests.
(2) The work of sliding friction between a solid polymer and
    another solid depends on the rate of sliding, temperature,
    etc. in the same manner as does the work of mechanical
    deformation of slider and support.

Consequently, the adhesion theory of friction is incorrect and
the deformation theory of polymer friction is the most probable at
present.

REFERENCES

1. J. J. Bikerman, Physical Surfaces. Academic Press, New York, 1970, p. 444.
2. G. M. Bartenev and V. V. Lavrent'ev, Friction and Wear in Polymers [Russian], Khimiya, Leningrad 1972.
3. D. F. Moore and W. Geyer, Wear 22, 113 (1972).
4. V. A. Belyi et al., Friction of Polymers [Russian], Nauka, Moscow 1972.
5. J. J. Bikerman, Rev. Mod. Phys. 16, 53 (1944).
6. J. J. Bikerman, The Science of Adhesive Joints. Academic Press, New York, 1968.
7. J. J. Bikerman, Ind. & Eng. Chem., Sept. 1967, p. 40.
8. A. I. El'kin and V. K. Mikhailov, Mekh. Polim. 1970, 688.
9. R. S. Longhurst, Geometrical and Physical Optics, p. 475, Wiley, New York, 1967.
10. K. Tanaka, J. Phys. Soc. Japan 16, 2003 (1961).
11. W. O. Yandell, Wear 17, 229 (1971); 21, 313 (1972).
12. Sh. M. Bilik et al., Mekh. Polim. 1969, 850.
13. G. V. Vinogradov et al., Wear 23, 33 (1973).
14. J. L. Lubkin in Handbook of Engineering Mechanics (W. Flugge, ed.) p. 42-1. McGraw-Hill, New York, 1962.
15. S. P. Timoshenko and J. N. Goodier, Theory of Elasticity, p. 413. McGraw-Hill, New York 1969.
16. K. V. Shooter and D. Tabor, Proc. Phys. Soc. B 65, 661 (1952).
17. J. J. Bikerman, J. Macromol. Sci.- Rev. Macromol. Chem., in press.
18. A. Schallamach, Wear 13, 13 (1969).
19. A. Schallamach, Wear 17, 301 (1971).
20. S. C. Cohen and D. Tabor, Proc. Roy. Soc. (London) A 291, 186 (1966).
21. N. M. Mikhin and K. S. Lyapin, Mekh. Polim. 1970, 854.
22. K. A. Grosch, Proc. Roy. Soc. (London) A 274, 21 (1963).
23. K. C. Ludema and D. Tabor, Rubber Chem. Technol. 41, 462 (1968).
24. S. Bahadur and K. C. Ludema, Wear 18, 109 (1971).
25. J. J. Bikerman, SPE Trans. 4, 290 (1964); J. Paint Technol. 43, No. 9, 98 (1971).
26. G. V. Vinogradov, Yu. G. Yanovskii, and E. I. Frenkin, Brit. J. Appl. Phys. 18, 1141 (1967).

DISCUSSION OF PAPER BY J. J. BIKERMAN

L. C. Towle (Naval Research Laboratory): As I understand your
comments today and your preprint you state that the friction force
observed with polymers is not given by the expression $F = \mu N$. My
understanding of the definition of the coefficient of friction is
that it has an operational definition, i.e. to measure the friction
between solids A and B one forces A to slide against B under measured
load N and simultaneously measures the tangential force F. The
friction coefficient is then defined to be the ratio $\mu \equiv F/N$. If
this is the case, then it would seem that one <u>always</u> has $F = \mu N$
being valid. If this is not the case, then I wonder how does one
define the friction coefficient?

J. J. Bikerman: If the ratio of every tangential force F to every
normal force N is called "coefficient of friction", then the term
<u>friction</u> would cover such a multitude of phenomena that no reasonable
discussion of the mechanism of friction would be possible. For
instance, shearing an adhesive butt joint would also be accepted
as an experiment on friction, and slicing a bread loaf would be
another. Over thirty years ago I suggested the term <u>resistance to</u>
<u>sliding</u> as a general notion; in true friction the ratio F/N
is independent of N or is a steady function of N; also, the value
of F/N is reproducible in a given sample. Only when the meaning
of <u>friction</u> is restricted in this, or a similar manner, is it possible
to discuss causes of friction. The situation is similar to that in
which medicine was 200 years ago. As long as people referred to
every epidemics as "plague" and to many illnesses as "fever", as
long as different diseases have not been mentally separated from
each other, almost no progress was possible. Only when it was
realized that each malfunction requires its own remedy, that pills
against scurvy will not combat cholera, real advance in treatment
became feasible. Different types of resistance to sliding must be
treated in different manners.

K. C. Ludema (University of Michigan): You criticize the adhesion
theory of friction on the ground that a thin film of fluid (gases,
contaminants, etc.) exists on all surfaces and this thin film
prevents adhesion. You then shift your argument to say that friction
is due to interaction of asperities and in polymers this means a
higher hysteresis loss than in the case of metals. Would your
argument not be strengthened if you would do an analysis and experiment in which you are open to the possibility that viscous drag in
the fluid film is one cause of friction resistance?

J. J. Bikerman: The adsorbed air films are so thin and so mobile
that their deformation requires very little work compared with that
spent on the deformation of the polymer. When the polymer is not
extremely rigid and when the normal load is not too small, frictional
work of polymers is needed for this deformation rather than for the

lifting of asperities over other asperities; this lifting is important
only as long as the classic law of friction is valid, that is, the
coefficient of friction is independent of load.

G. Steinberg (Memorex Corp.): On the subject of temperature-time
superposition, one must define the system (as in all cases). When
adhesion can occur, e.g. in <u>thermoplastic</u> systems, the frictional
force will increase wide sliding and load. More and more adhesion
(fusion) will take place as more and more "new" (nascent)surface
is exposed, and "friction" will increase. This results in stick-slip.
Such systems can also "weld" when in contact after sliding stops:
adhesion. Rupture then occurs on restarting sliding. Please
comment.

J. J. Bikerman: As long as the temperature of the slider-support
interface is moderate, there is no difference between thermosetting
and thermoplastic polymers. No fusion will be expected as long as
this temperature is below the softening range of the polymer material.
Usually, the "nascent" surface becomes covered with adsorbed gases
before it comes in contact with another solid: when a nail is
driven into a polymer plate, a layer of air is still present between
the two. The "stick-slip" kind of motion depends above all on the
inertia of the measuring instrument.

D. Dowson (University of Leeds): I would be grateful if Dr. Bikerman
could elaborate on his view that a major source of frictional resistance arises from the work done in drawing asperities over each other,
since this classical seventeenth and eighteenth century view is
frequently refuted on the grounds that this is a non-dissipative
process.

J. J. Bikerman: When asperities of solid A are lifted over those
of solid B and then fall into the valleys of the latter, the work
performed during the first half of the process is transformed into
heat during the second half because surface roughness is random.
For instance if two asperities climb a hill in the direction from
South to North, one of them may fall (along the North slope of the
hill) toward the North-West, whereas the other may descent into a
North-East valley. These and analogous irregularities of the path
are the reason why the momentum acquired during ascent is not
preserved during the following descent.

D. Dowson (University of Leeds): Dr. Bikerman restricted his remarks
to friction, but I would welcome his/views on the mechanism of
'adhesive wear' and the evidence of cold-welding of asperities
followed by shearing and the production of wear particles.

J. J. Bikerman: In true friction, no adhesion occurs. Hence, there
is no "adhesive wear". Wear takes place wherever an outside force
is strong enough to break or tear off a piece of the slider or the

support. This force may act through a layer of a gas or a liquid.
A strong wind removes sand grains from a sand pile, and water digs
canyons out; thus no solid-to-solid adhesion is needed for wear
processes. Wear particles of solid A may remain attached to solid B
also when some air is present between the two solids; asperities on
the surface of B act as hooks.

E. Sacher (I.B.M. Corp.): Can one have sliding friction on polymers
without wear? If not, is this why the frictional coefficient may
decrease with increasing normal load?

J. J. Bikerman: As long as no excessive speeds or loads are used,
sliding friction of polymers without wear readily occurs for long
periods of time.

G. M. Robinson (3M Company): I find it hard to believe your theory
of asperities riding over one another as being the cause of friction.
Your figure 3 in the preprint shows only two or three asperities on
both the slides and fixed surface in contact with one another, but
in the real world, thousands of asperities on both surfaces are in
contact. If both surfaces are reasonably smooth then they will
have similar distributions of peak heights. Then both surfaces
should have enough high asperities always in contact with its contacting surface that no up and down motion will occur. This seems
to me to leave a frictionless situation according to your theory
which is contrary to facts.

How does your theory explain friction between two smooth surfaces?
How does it explain the function of a boundary lubricant? How does
it explain stick-slip?

As a final comment, I think you are confusing the differences between
a continuous and discontinuous adhesion joint. If the joint is
discontinuous as the adhesion theory of friction states, a slightly
off-normal force removing the slides would break the asperity bonds
individually rather than collectively. This situation would be very
difficult to measure.

J. J. Bikerman: Every solid surface is covered with hills of various
heights and various slopes. When the slider is placed on the support,
the average distance between the two surfaces is determined by the
height of relatively few hills; in sliding along a tilted plate,
sometimes the "contact" occurs in one "point" only. If H_i means
the distance of a contact spot from the main plane of the slider,
and h_i is the corresponding magnitude for the support, the above
clearance (average!) between the two surfaces will be determined by
the lengths $H_1 + h_1$, $H_2 + h_2$, $H_3 + h_3$, and a few others. The
probability that, after a small tangential shift, the new values of
$(H_1 + h_1)$, $(H_2 + h_2)$, $(H_3 + h_3)$, etc., all will be equal to the
earlier set of values is extremely small. Your objection reminds

me of the objections raised against the probability theory of
Brownian movement; I hope, you will find time to get acquainted
with this item of history of science. Several good books on surface
roughness also are available. By the way, the theory you flatteringly
attribute to me is over 200 years old.

The only smooth surfaces we know are those of liquids at rest. No
sliding friction is observed on them. Boundary lubricants fill the
depressions on the rough surfaces and thus lower the effective angle
of slope; usually, they have also other functions which are of no
interest to the problem at hand. Frictional vibrations (or "stick-
slip" friction) has been explained long ago by the difference between
static and dynamic friction. Perhaps you will be interested to
consult my paper in Rev. Modern Physics (1944).

The strength of incomplete adhesive joints has been discussed many
times before. My book "The Science of Adhesive Joints" (1968) may
be a suitable source of information.

S. Bahadur (Iowa State University): Do you believe in the removal
of adsorbed layers of foreign material from the interface during
the sliding process?

J. J. Bikerman: Adsorbed layers are not removed during sliding.
Usual sliding is a very slow process compared to evaporation,
condensation, and tangential motion of adsorbed molecules; thus,
these three processes are not markedly influenced by friction.

S. Bahadur: What other than adhesion is responsible for the
resistance to sliding in the case of very smooth polymeric surfaces?

J. J. Bikerman: No polymer surface is smooth on atomic or molecular
scale. However, the friction of polymers (at not too small pressures)
usually is determined not by surface roughness but by deformation
of the bodies in tangential motion; if this deformation is extensive,
surface roughness ceases to be significant.

D. Tabor (Cambridge University): I am so charmed by Dr. Bikerman's
personality as well as by his sophistry that I do not know how to
treat the criticisms he has raised of the adhesion theory of friction.
His main point is that the friction of polymers is due to deformation.
I completely agree that a large part of friction arises from this
source. Indeed, we can simulate the deformation process in the
following way. Instead of sliding a sphere, we can roll it over the
polymer and in this way we are able to determine the deformation
losses without any appreciable contribution from adhesion. The
observations agree well with those calculated from the deformation
losses of the polymer so that the point raised by the author this
morning is in complete agreement with what I said yesterday. However,

if we now slide the ball over the polymer, the friction is perhaps ten times higher than the rolling friction. We would say that this is because adhesion is involved at the interface and that we have to expend work in overcoming this adhesion.

It is at this point that Dr. Bikerman introduces his major disagreement. He says that as soon as there is adhesion there is no true friction. Clearly if you define sliding as a condition in which there can be no interatomic interaction between the surfaces of the type we call adhesion, obviously the force of friction as so defined does not arise from adhesion. If Dr. Bikerman lays down his ground rules in these terms, we must all be forced to agree with him, but I cannot say that this is friction in the way that engineers and ordinary people understand it.

Finally, may I add the following? During the last few years, we have been studying the contact between molecularly smooth surfaces using mica. We have been able to obtain contact between them and from our refined optical measurements have been able to show the presence of adsorbed monolayers of water between the surfaces. We are also able to measure both friction and adhesion between such surfaces and these are not zero. Undoubtedly, the water monolayer reduces the adhesion force but it is completely erroneous to say that adhesion does not occur unless such films are removed. On this specific point, some of Dr. Bikerman's comments have been outdated by more recent experiments.

Polymer Friction under Conditions of Deformation Processing

M. A. Saleem and S. Bahadur

Department of Mechanical Engineering and Engineering
Research Institute

Iowa State University, Ames, Iowa 50010 (USA)

The sliding friction behavior of low
density and high density polyethylenes with
normal pressure and velocity is investigated
for unlubricated conditions using the
conventional friction measuring arrangement.
Apart from its dependence on velocity, the
coefficient of friction is found to vary
significantly with normal pressure. The
investigation of friction at very high
contact pressures, as encountered in
deformation processing, has been carried
out for the case of direct extrusion using
an instrumented extrusion die with movable
head. The variation of extrusion and die-
head force with ram travel is recorded for
particular ram speeds. The extrusion data
has been analyzed assuming von Mises' yield
criterion and spherical velocity fields
and using the adhesion and constant shear
theories of friction. The frictional shear
stress and coefficient of friction are
calculated for the state of stress involved
in the extrusion process. The frictional
shear stress of low density polyethylene
is found to be of the order of its bulk
shear strength which is in agreement with
the adhesion theory of friction.

INTRODUCTION

In deformation processes the polymeric material is visco-elastically deformed where deformation in the early stages is mostly elastic and in the latter stages is mostly viscous. Here the polymeric material slides in contact with a metallic tool/die surface and the stresses at the interface vary over a wide range. The rate of deformation processing also affects the conditions at the interface. This investigation is therefore directed towards the study of sliding friction under dry sliding conditions with contact pressure and velocity for the conventional case as well as for a deformation process which happens to be extrusion in this work.

The variation of friction with normal load has been studied[1-4] but no definite trend has been reported. According to Steijn[1] and Fusaro[2], the coefficient of friction decreases with increasing load while the investigations of Barlow[3] show otherwise. The theoretical analysis of Gupta and Cook[4] shows that for elastic contact the coefficient of friction should decrease while for plastic contact it should increase with increasing load. The anomalous trend described above is probably because the normal load is not an independent parameter. The latter is instead the contact pressure which is a function of the normal load and the area of real contact. An investigation of the variation of the sliding friction of bulk polymers with contact pressure has recently been carried out[5]. The friction of grossly deformed thin polymer films against hard surfaces has been studied by Towle[6] and Bowers et al[7]. The friction in metal working processes has been studied[8-12], but no such direct study has been reported for polymers. Shaw et al.[13] have shown that for metals the results of conventional studies on sliding friction do not seem to apply to deformation processing situations.

EXPERIMENTAL

The variation of sliding friction with normal pressure and velocity was measured using a variable speed lathe and a strain gage tool dynamometer. The instrumentation used and the experimental procedure are described in detail elsewhere[5]. Here a ground and polished disc of AISI 4340 steel, which had earlier been austenitized, water quenched, and stress relieved, was mounted in the lathe chuck. The polymeric indentor with a flat, polished tip was secured to the dynamometer which was capable of measuring the normal force and the frictional force simultaneously. The dynamometer output from strain gages was fed into power amplifiers, strain gage balancing units and then into a 2-channel recorder.

For the measurement of sliding friction during the course of deformation of the polymeric billet in the extrusion process, a direct extrusion set-up (Fig. 1) with a movable die-head was designed

and instrumented. The container and the die-head were both made of
AISI 4340 steel, austenitized, water quenched, and stress relieved,
and the surfaces encountering sliding contact with the polymeric
material were ground and lapped to provide approximately the same
degree of surface finish as in the case of metallic disc used for
friction measurement above. The movable die-head which slides
freely in the container is supported on a tube carrying strain gages.
The latter provide a measure of the force F_d exerted on the die-head
during extrusion. The ram compressing the billet material in the
container is also mounted with strain gages which provide a measure
of the force F_t needed to extrude the billet. These two measurements
along with the other data lead to an estimate of friction under the
extrusion conditions. The temperature rise during extrusion at the
sliding interface as well as within the billet material was monitored
with the help of thermocouples installed in the set-up.

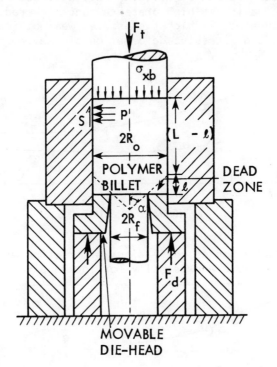

Fig. 1.   Schematic arrangement of the instrumented extrusion die.

The investigation was carried out on commercial low density and
high density polyethylenes. These materials were bought from
Cadillac Plastic and Chemical Co. in the form of 1 in. diameter melt-
extruded rods. For friction studies, the cold extrusion under dry
conditions of low density polyethylene was performed at room

temperature and that of high density polyethylene at 110°F. For the
comparison of surface finish of extrudates, both the materials were
extruded under dry conditions at elevated temperatures and under
lubricated conditions at room temperature, but no friction data was
taken for these conditions.

RESULTS

Using the flat-tip polymer indentor and metal disc arrangement,
the variation of coefficient of friction at room temperature with
normal pressure and sliding velocity was obtained. The data is
plotted in Figs. 2 and 3 for low density and high density polyethylenes, respectively. With increasing normal pressure and decreasing
velocity, the coefficient of friction initially decreases and the
curves flatten out in the high pressure range covered. The measurement of friction at still higher pressures was not feasible because
of the buckling and excessive deformation of the indentor.

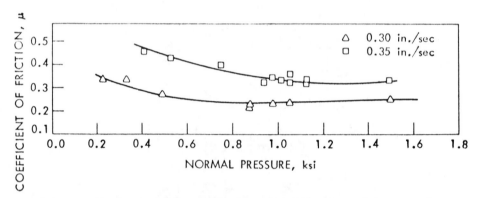

Fig. 2.   Variation of coefficient of friction with normal
          pressure and velocity for low density polyethylene.

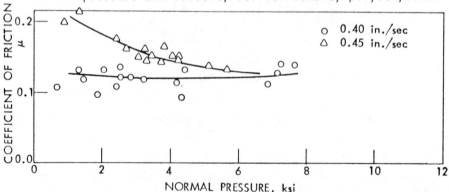

Fig. 3.   Variation of coefficient of friction with normal
          pressure and velocity for high density polyethylene.

The data obtained from the extrusion of 1 in. diameter and
1-3/4 in. long billets into 1/2 in. diameter rods of low density
polyethylene at room temperature and of high density polyethylene
at 110°F is shown in Figs. 4 and 5 respectively. The friction force
F_f is plotted as the difference of F_t and F_d. The dead zone semicone
angle by repeated measurements was found to be 45 ± 3°.

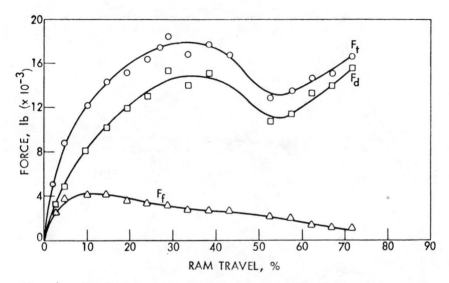

Fig. 4.   Variation of extrusion force F_t, die-head force F_d,
          and friction force F_f with ram travel for low density
          polyethylene; ram travel 0.2 in./min; extrusion at
          room temperature.

From the adhesion theory of friction, F = AS, where F is the
friction force, A the real area of contact and S the shear stress
of the interface. The coefficient of friction μ may thus be
expressed as

$$\mu = \frac{F}{W} = \frac{A}{W} \cdot S = \frac{S}{P} \tag{1}$$

where W is the normal load and p the average contact pressure at
the polymer-metal interface. It assumes that due to the flow of
material under extrusion condition the areas of real and apparent
contact will be approximately the same. Using von Mises' yield
criterion and assuming spherical velocity fields, the free body
equilibrium approach leads to the following equation[14] for extrusion

$$\frac{\sigma_{xb}}{\sigma_o} = -1 + \left\{ \exp < \frac{2\mu}{R_o} (L - \ell) > \right\} \left\{ 1 - 2 \left[f(\alpha) \cdot \ln\left(\frac{R_o}{R_f}\right) \right. \right.$$

$$\left. \left. + \frac{1}{\sqrt{3}}\left(\frac{\alpha}{\sin^2\alpha} - \cot\alpha\right) + \frac{1}{\sqrt{3}} \cot\alpha \cdot \ln\left(\frac{R_o}{R_f}\right) \right] \right\} \qquad (2)$$

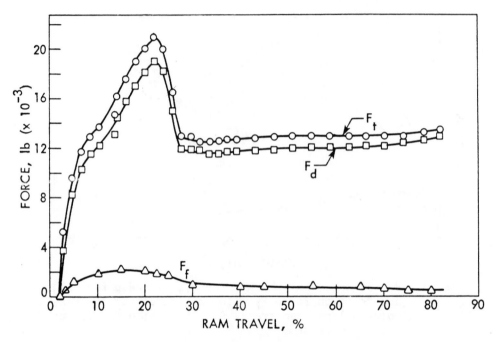

Fig. 5.   Variation of extrusion force F_t, die-head force F_d, and friction force F_f with ram travel for high density polyethylene; ram travel 0.4 in./min; extrusion at 110°F.

where

$$f(\alpha) = \frac{1}{\sin^2\alpha} \left[1 - (\cos\alpha) \sqrt{1 - \frac{11}{12}\sin^2\alpha} \right.$$

$$\left. + \frac{1}{\sqrt{132}} \ln\left(\frac{1 + \sqrt{11/12}}{\sqrt{\frac{11}{12}}\cos\alpha + \sqrt{1 - \frac{11}{12}\sin^2\alpha}}\right) \right],$$

α is the dead zone semicone angle, σ_{xb} the external pressure on the billet, σ_o the material flow stress, L the axial length of the billet, and ℓ the length of the dead zone in the axial direction.

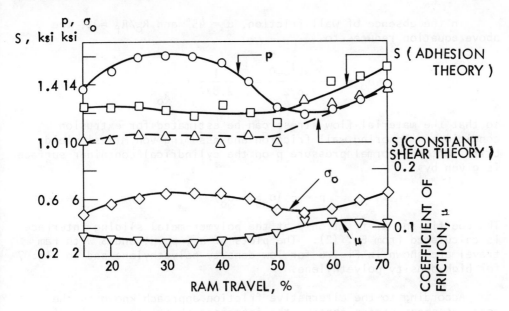

Fig. 6. Variation of flow stress σ_o, normal pressure p,
coefficient of friction μ, and interface shear stress
S with ram travel for low density polyethylene.

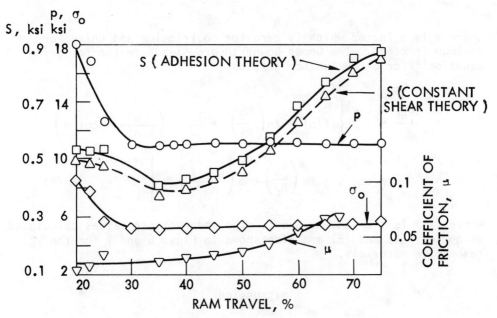

Fig. 7. Variation of flow stress σ_o, normal pressure p,
coefficient of friction μ, and interface shear stress
S with ram travel for high density polyethylene.

In the absence of wall friction, $\alpha = 45°$ and $R_o/R_f = 2$, the above equation reduces to

$$\sigma_o = - \frac{(\sigma_{xb})_{\mu=0}}{2.872} = - \frac{1}{2.872} \cdot \frac{F_d}{\pi R_o^2} \tag{3}$$

so that the material flow stress can be estimated for extrusion conditions. Ignoring wall friction and assuming von Mises' yield criterion, the normal pressure p on the cylindrical container surface is given by

$$p = - (\sigma_{xb} + \sigma_o) \tag{4}$$

The coefficient of friction μ at the polymer-metal sliding interface is calculated from Eq. (2). The plots of σ_o, p, μ, and S with ram travel are shown in Fig. 6 for low density polyethylene and in Fig. 7 for high density polyethylene.

According to the alternative friction approach known as the constant shear stress theory, the interface shear stress S is given by

$$S = \frac{m\sigma_o}{\sqrt{3}} \tag{5}$$

where m is a factor which is zero for no friction and unity for maximum friction. The upper bound theorem leads to the following equation[14] for extrusion for this case

$$\frac{\sigma_{xb}}{\sigma_o} = - 2 \left\{ f(\alpha) \cdot \ln\left(\frac{R_o}{R_f}\right) + \frac{1}{\sqrt{3}} < \left(\frac{\alpha}{\sin^2\alpha} - \cot \alpha\right) \right.$$

$$\left. + \cot \alpha \cdot \ln\left(\frac{R_o}{R_f}\right) + m\left(\frac{L - \ell}{R_o}\right) > \right\} \tag{6}$$

where $f(\alpha)$ is the same as above. The shear stress values calculated using Eqs. (5) and (6) are also shown in Figs. 6 and 7 for the respective materials.

DISCUSSION

From the variation of friction measured using the indentor and the disc arrangement (shown in Figs. 2 and 3), it is found that the coefficient of friction decreases initially with normal pressure and then remains fairly unchanged. There is some scatter observed

in the data. This is attributed to the heating of the indentor tip,
irregularities on the surface and some unnoticeable buckling of the
tip when subjected to compressive load.

Figures 4 and 5 show that the friction force increases rapidly
with ram travel in the early stage of deformation of the extrusion
billet. This deformation of the billet is mostly of elastic nature.
At about 10 to 15 percent of the ram travel which corresponds to the
yield strain of materials, a drop in the friction force is observed.
Thereafter the friction force decreases almost linearly with ram
travel because of the continuously decreasing area of sliding contact.
The variation of the material flow stress and normal pressure with
ram travel, shown in Figs. 6 and 7, is in general similar to that of
the extrusion pressure. The interface shear stress increases rapidly
when the dead zone is approached. It also results in the increased
value of the coefficient of friction. The actual increase in both
the cases would be smaller than that obtained by calculation, because
some sliding and cross flow in the dead zone is likely to take place
in the latter part of the extrusion process.

The variation of the interface shear stress and coefficient of
friction with normal pressure is shown in Fig. 8 which was plotted
from Figs. 6 and 7. It is noted that the interface shear stress
increases with normal pressure. It is similar to the variation of
tensile mechanical properties with hydrostatic pressure, as observed
by a number of workers[15-17]. Analogous to the case of conventional
friction measurement, the coefficient of friction decreases with
normal pressure even under actual extrusion conditions. The decrease
is in agreement with the adhesion theory of friction. The interface
shear stress seems to be about equal to the bulk shear strength of
the material, because the shear strength of low density polyethylene
at room temperature and at a strain rate of 30 min^{-1} was measured as
1200 psi. The temperature rise in the billet of this material during
extrusion was measured as about 20°F. Thus the shear strength of the
bulk material under extrusion conditions would have been lower and is
definitely in the range of the interface shear stress calculated.

It is seen from Figs. 2 and 3 that the coefficient of friction
increases with velocity. This is because, with increased velocity,
the strain rate associated with the rupture of adhesive bonds at the
interface increases. This causes an increase in the interface shear
stress without any significant effect on the area of real contact.
The variation of friction with normal pressure determined from both
the conventional test and the extrusion test tallies, but the
magnitudes cannot be compared due to the large difference in velocity
and normal pressure range. It is also noted from Figs. 2 and 3 that
a 12-17% change in velocity produces a significant change in friction
at low pressures, but the difference in friction decreases at higher
pressures. In spite of the fact that the velocity in the die is very
different from that in the pin-on-disc arrangement, there is not a

great deal of change in friction values obtained from the two types
of tests. This is perhaps because of the vastly different contact
pressures prevailing in the two tests.

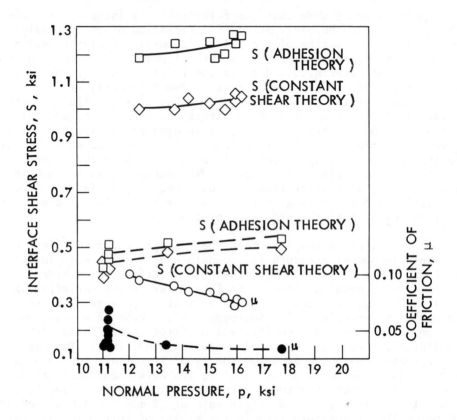

Fig. 8. Variation of coefficient of friction μ and interface
         shear stress S with normal pressure p; continuous line
         for low density polyethylene, dotted line for high
         density polyethylene.

     In the case of the dry extrusion of low density polyethylene
at room temperature, the surface of the extrudate was very rough.
There was also an indication of severe adhesive fracture occurring
during the extrusion process. When the extrusion was performed
under lubricated condition or under dry conditions at high temperature, the surface was smooth. High density polyethylene could not
be extruded satisfactorily under dry conditions at room temperature,
but its extrusion at elevated temperature resulted in a smooth surface
extrudate. Thus, from the point of view of surface finish, the
extrusion of polymeric materials is recommended only under lubricated
conditions or under dry conditions at elevated temperatures.

ACKNOWLEDGMENTS

The authors gratefully acknowledge the generous support of this work by a grant from the National Science Foundation. This work was also supported by the Engineering Research Institute at Iowa State University, Ames, Iowa.

NOMENCLATURE

A	real area of contact
F	friction force
F_d	force on die-head in extrusion
F_f	friction force in extrusion
F_t	total extrusion force
ℓ	length of dead zone parallel to the axis of billet
L	length of cylindrical billet
p	average contact pressure at interface
R_f	radius of extrudate
R_o	radius of billet
S	shear stress of interface
W	normal load
α	dead zone semicone angle
σ_o	flow stress of material
σ_{xb}	external pressure on billet
μ	coefficient of friction

REFERENCES

1. R. P. Steijn, Metals Eng. Quarterly, ASM, 9 (May 1967).
2. R. L. Fusaro, "Friction and Wear Life Properties of Polimide Thin Films", NASA TN D-6914 (August 1972).
3. D. A. Barlow, Wear, 20, 151 (1972).
4. P. K. Gupta and N. H. Cook, Wear, 20, 73 (1972).
5. S. Bahadur, "Mechanism of Dry Friction in Deformation Processing of Polymers", ERI Project 896-S, Final Report, NSF Grant GK-27845 (May 1973).
6. L. C. Towle, ASME Paper No. 70-WA/PT-1.
7. R. C. Bowers, W. C. Clinton, and W. A. Zisman, Lubrication Eng., 9, 204 (1953).
8. F. F. Ling, R. L. Whitley, P. M. Ku, M. B. Peterson (Ed.), Friction and Lubrication in Metal Processing, ASME (1966).
9. M. B. Peterson and F. F. Ling, J. Lubr. Tech., ASME Trans., 92, 4, 535 (October 1970).
10. J. A. Newnham and J. A. Schey, "Investigation of Interface Friction between Tool and Workpiece Materials under Conditions of Plastic Deformation", Contract No. NOw-66-0503-d, Final Report (August 31, 1967).

11. J. A. Newnham and J. A. Schey, J. Lubr. Tech., ASME Trans.,
     91, 2, 351 (April 1969).
12. V. Depierre, J. Lubr. Tech., ASME Trans., 92, 3, 398 (July 1970).
13. M. C. Shaw, A. Ber, and P. A. Mamin, J. Basic Eng., ASME Trans.,
     D, 342 (1960).
14. B. Avitzur, Metal Forming: Processes and Analysis, McGraw-Hill,
     New York (1968).
15. L. Holliday, J. Mann, G. Pogany, H. Ll. D. Pugh, and D. A. Green,
     Nature, 202, 381 (1964).
16. D. Sardar, S. V. Radcliffe, and E. Baer, Polym. Eng. & Sci.,
     8, 290 (1968).
17. K. D. Pae and D. R. Mears, J. Polym. Sci., Polymer Letters, B,
     6, No. 4, 269 (1968).

DISCUSSION OF PAPER BY S. BAHADUR

J. J. Bikerman (Case Western Reserve University): Is the contact pressure you mention the macroscopic pressure?

S. Bahadur: Yes. In case of such high pressures as used in this work, the difference between the real and apparent areas of contact will be negligibly small.

L. H. Lee (Xerox Corporation): This paper is very interesting especially to those who are involved in polymer processing. To me, this is the first time that I see the application of a friction theory to polymer extrusion. For example, the need of a lubricant during extrusion is clearly explained by the adhesion theory of friction. Though the adhesion theory was challenged by J. J. Bikerman, it is further verified by Dr. Bahadur through shear strengths determined with the constant shear theory. For low density polyethylene, the frictional shear stress was found to be in the order of its bulk shear strength.

This work also verifies the relationship between pressure and the coefficient of friction as discussed by Dr. Towle. Dr. Bahadur and his co-author should be congratulated for this fine paper which helps, directly or indirectly, clear up many confusions about the theories of friction.

Discussion of Paper B- STEGMAIER

Shear Strength and Polymer Friction

Laird C. Towle

Naval Research Laboratory

Washington, D. C. 20375

The friction coefficient of thin
polymer sheets under high loads can be
accounted for by a simple extension of
the adhesion theory of friction in which
the pressure dependence of the shear
strength of the polymer is taken into
consideration. Recent measurements on
polypropylene films, as well as earlier
data on several other polymers, are used
to illustrate these ideas.

INTRODUCTION

Solid film lubrication is becoming an important aspect of modern
technology. The application of such lubricants in the form of bonded
films, as suspensions in conventional oils and greases, or impregnated
in porous substrates is becoming widespread. The better known solid
lubricants, such as graphite and MoS_2, have been extensively investigated; here we are concerned with polymer films which are now finding
increasing use.

The frictional properties of bulk polymers were studied in an
important series of paper by Tabor, Shooter, and coworkers[1-4] in the
1950's. Their fundamental studies showed that these properties of
bulk polymers can be explained in terms of the adhesion theory of
friction which was developed for metals some years earlier, primarily
by Bowden, Tabor, and coworkers[5-9]. According to this theory, the
frictional resistance to sliding arises from the shearing of cold-
welded junctions which form spontaneously at the microscopic areas

179

of contact when two solids are brought together. The cross-sectional
area of the junctions is thought to increase or decrease in proportion
to the load, thereby producing a frictional force proportional to the
load as required by Amontons' classical laws of friction.

There is scattered evidence in the literature which shows that
Amontons' classical laws frequently do not apply to solid film
lubricants in load ranges of practical importance. This subject has
been reviewed elsewhere[10]. The failure of these classical laws
has recently been demonstrated for several polymer films by Towle[10-14]
and Bowers[15]. It has been found that for both organic and inorganic
solid film lubricants the friction coefficient decreases markedly
with increasing load approaching a small asymptotic value at very
high loads. The decrease observed is sometimes as much as an order
of magnitude.

In order to account for the very small and variable friction
coefficient observed with solid film lubricants, Bowers and Zisman[16]
suggested that is was necessary to take into account the pressure
dependence of the shear strength of the solid lubricant, and proposed
the empirical equation

$$\mu = \frac{S(p)}{p} \tag{1}$$

where μ is the coefficient of friction, $S(p)$ is the pressure
dependent shear strength of the solid lubricant, and p is the normal
pressure acting on the microscopic areas of contact where the shearing occurs. One can readily show that this equation follows in a
very simple way from basic definitions, and that it reduces to
Amontons' classical laws at low loads where there is only partial
contact between the solid lubricant film and the substrate or
superstrate[10,11,17].

It is known from the work of Bridgman[18-21] and others[22-24]
that the strength of solids generally increases smoothly and slowly
with pressure. Hence one can usually write

$$S(p) = S_o + \alpha p + \beta p^2 + \gamma p^3 + \cdot \cdot \cdot \cdot \tag{2}$$

where the coefficients S_o, α, β, γ, etc., may be temperature and
strain-rate dependent quantities. Equations (1) and (2) combine
to give the following general expression for the pressure dependence
of the friction coefficient

$$\mu = S_o/p + \alpha + \beta p + \gamma p^2 \cdot \cdot \cdot \cdot \cdot \cdot \tag{3}$$

Since a solid has a finite shear strength at p = 0, by definition,
the first term in Eqs. (2) and (3) cannot be ignored as was done in
Ref. 16; however, it is usually sufficient to retain only the first

two terms in the expansion[10,11]. This has been found to be the case
in recent measurements on the crystalline polymers polyethylene (PE),
polytetrafluoroethylene (PTFE), polyvinylidenefluoride (PVF), poly
(chromium phosphinate) (PCP), and the copolymers ethylene-tetrafluoroethylene (ETFE) and hexafluoroethylene tetrafluoroethylene (FEP)[10-15].
Those results are summarized here and new data are presented on the
crystalline polymer polypropylene (PP).

EXPERIMENTAL PROCEDURE

The measurements were made with a rotating anvil shear press.
In this type of device a thin sheet of sample material is compressed
between the flat, circular faces of a pair of tungsten carbide anvils.
One of the anvils is then rotated relative to the other about their
common axis of symmetry. The measured axial load, N, and torque, M,
are converted to friction coefficient and shear data using the
equations

$$\mu = (3/2a)M/N = (3/2\pi a^3)M/P \qquad (4)$$

$$f = (3/2\pi a^3)M \qquad (5)$$

In these equations μ is the coefficient of friction, a is the measured
radius of the anvil faces, P is the nominal applied pressure, i.e.
the axial load divided by the macroscopic area of the anvil face,
and f is the friction force per unit area. At high loads where there
is complete contact between the sample and the anvil faces P = p, and
f can usually be equated with the shear strength, S, of the sample
material; at low loads P < p, and f < S. Details regarding the design
and operation of our rotating anvil shear press have been given
earlier[10,11].

The PP samples used in these tests were from two sources. The
bulk of the data were taken on sheet stock prepared by Bowers[15]
from commercial powder (Hercules Powder Co.). A few measurements
were also taken on a commercial film stock (Diamond Shamrock Chemical
Co., No. 8602L). The former samples were initially about 10 mils
thick, but compressed and extruded laterally down to a thickness of
~1 mil under loading and shearing. The latter samples were only
about 1 mil thick initially and underwent only a little such extrusion.
From the observed macroscopic deformation of the samples it was
evident that the samples underwent bulk shearing rather than the
surficial shearing above pressures of about 500 bars[10-12]. The
samples were circular discs punched from the sheet or film stock,
and were degreased by washing in ethyl alcohol. They were stored in
a vacuum desiccator prior to testing.

All measurements reported here were made at room temperature
using anvils with ~0.25 in. diameter faces. The anvil motion was

oscillatory through an angle of 90° at a rate of about 30°/sec.

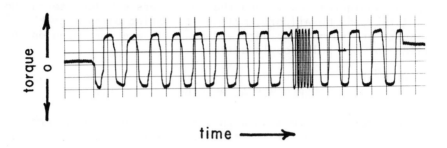

Figure 1. Typical torque versus time curve.

The torque-time curve was initially somewhat irregular due to the
lateral extrusion of some of the sample material during the first
few cycles of anvil oscillation. Data was taken after the torque-
time curve had stabilized and approximated a square wave. A typical
tracing obtained on PP is shown in Fig. 1. Note that during the
experiment the rotation speed was increased about 5-fold with no
apparent residual effect on the sample. There was only a small
increase in torque associated with the higher strain-rate or sliding
speed. This is typical of what has been observed with other polymers
studied[10].

RESULTS AND DISCUSSION

The observed shear strength of PP is shown in Fig. 2 as a
function of pressure. The circles show results obtained on the
samples prepared from Hercules powder stock (type 1). Each point
is an average of several measurements. Standard deviations calculated from the range of the data are shown by flags. The crosses
represent results of a few individual measurements made on the
Diamond Shamrock sheet stock (type 2). There is good agreement
between the two types of samples.

The data in the range from 1 to 4 kbars shows a linear dependence
on pressure. A least squares fit of the type 1 data in that pressure
range to the straight line in Fig. 2 yielded the parameters S_0 = 33
bars and α = 0.110. Note that there is an apparent break in the
data near 1 kbar which suggests some anomaly in the behavior of the
material. Because the PP samples consisted of a mixture of an
amorphous component and a crystalline component, there are two
obvious possible sources of the anomaly; (1) a glass transition in
the amorphous component, or (2) a polymorphic transition in the
crystalline component.

Passaglia and Martin[25] measured the pressure dependence of the glass transition temperature in PP at pressures up to about 700 bars. They found that it increased with pressure from about -28°C at zero pressure at a rate of about 2°C/kbar. By a rather lengthy extrapolation one can estimate from their data that the glass transition temperature of PP is increased to room temperature by a pressure of about 2.6 kbars. The transition to the glassy state should be accompanied by an increase in strength. Thus, the break in the data near 1 kbar may be due to the glass transition, although it appears to occur at a lower pressure than anticipated. It is also possible that the break is due to a polymorphic transition; PP is known to exist in at least three different crystalline forms[26]. In particular the triclinic γ phase crystallizes from the melt under pressures of a few kilobars. This suggests that the break in our data may be due to a transformation from the normal α phase to the γ phase, but confirmation of this possibility is lacking.

The mechanical properties of PP have been studied under pressures up to about 7 kbars by Mears et al.[27] They observed a linear increase in peak yield stress with pressure in a series of tensile tests. Their data were taken at large pressure increments which precluded the detection of small effects such as the break shown in Fig. 2.

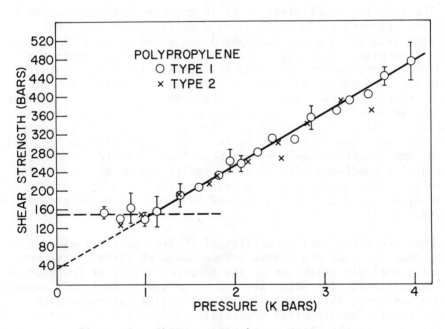

Figure 2. Shear strength versus pressure

From their data one can estimate that the atmospheric pressure shear
strength of PP is $S_o \sim 150$ bars in reasonable agreement with Fig. 2.
In addition they determined the parameter α to be 0.114 in good
agreement with the present result. The observed agreement is no
doubt partly fortuitous, because their tests involved strains that
were two or three orders of magnitude smaller than ours, and there
were probably morphological and structural differences in the sample
materials as well.

The friction coefficient of PP as deduced from our measurements is shown in Fig. 3. Note the sharp drop in the pressure range
up to about 1 kbar and the near plateau in the higher pressure region.
This general shape is typical of what we have presviously observed
with several other polymers[10-14]. The sharp drop results from the
S_o/p term in Eq. (3), while the plateau at high pressures results
from the fact that all the higher order terms in that equation are
negligible except the constant α. In the present case the steepness
of the drop is accentuated because of the break in Fig. 2, and the
near horizontal character of the curve at high pressures results
from the fact that the linear portion in Fig. 2 extrapolates back
nearly through the origin, i.e, $S_o \sim 0$. The solid curve in Fig. 3
was calculated from Eq. (3) using the parameters S_o and α determined
from the high pressure linear region in Fig. 2 with the higher order
terms being neglected. The dashed and dotted curves in Fig. 3 denote
the break shown in Fig. 2 by similarly dashed or dotted segments.

The friction coefficient of PP in the pressure range up to 8
kbars was measured by Bowers[15] using a conventional friction
apparatus in which a steel hemisphere was drawn along a PP film on
a soda lime glass substrate. The low pressure region where the
curvature occurs was not examined in detail. In addition, the
sample pressure was calculated using the Hertz equation for the
area of elastic contact between a sphere and a plane. In calculating
the contact area the presence of the polymer film was neglected.
Because the slider must indent the film to some extent, the calculated
areas may have been significantly in error, especially at low loads,
and the contact pressures may have been over-estimated. Hence,
meaningful comparisons with Bowers' results cannot be made at low
pressures. In the high pressure region where these limitations do
not apply he found $\mu \sim 0.125$ in good agreement with the present
results.

Thus the frictional properties of PP are readily explained by
taking cognizance of the pressure dependence of its shear strength
and using a simple extension of the adhesion theory of friction as
represented in Eq. (3) with only the first two terms in the expansion
being necessary. This result is in qualitative agreement with
earlier results on six other crystalline polymers[10-15]. In the
present case the situation is slightly complicated by the existence
of a break in the shear strength versus pressure curve which is

indicative of some type of transition in the mechanical properties of the material. Similar breaks have previously been observed in PTFE[10] and ETFE[14].

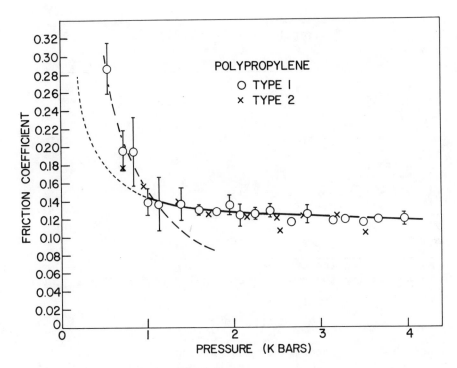

Figure 3. Friction coefficient versus pressure.

The values of S_0 and α applicable to the various polymers studied to date are summarized in Table I. The shear strength data on PP, PVF, FEP, and PCP were well represented by linear fits with the parameters shown in the Table[10-13]. The pressure dependence of the shear strength of PE ($\alpha = 0.020$) is very low, being comparable to the effect seen in metals[24], while the other polymers show a much larger effect. Consequently the friction coefficient of PE at high loads approaches a very small value. It should be noted, however, that significant variations are observed in different PE samples[12]. In ETFE a small discontinuity in shear strength was observed, near 2.8 kbars, but disregarding that, it behaved very much like the other polymers[14]. The PTFE and PP films exhibited distinctly anomalous behavior at low loads, but in the higher pressure range also showed a linear strength relationship. Disregarding the perturbations caused by phase transformations or other extraneous effects, all the polymer films examined have shown a linear increase in shear strength with pressure, and a friction coefficient which drops sharply with increasing load approaching a small asymptotic value at high contact pressures.

TABLE I.   Shear Strength and Friction Parameters
of Several Crystalline Polymers

Material	S_0	α
PE	27	0.020
PVF	65	0.113
PTFE[a]		
P < 1.8 kbar	~ 23	~ 0.010
P > 1.8 kbar	~ 30	0.051
ETFE[b]	76	0.084
PP		
P < 1.0 kbar	~ 150	~ 0.000
P > 1.0 kbar	33	0.110
FEP	41	0.049
PCP	99	0.070

a) Neglecting the curvature of S vs P at low pressure.
b) Neglecting the small discontinuity in S vs P near 2.8 kbars.

CONCLUSIONS

Since the shear strength of most polymers is significantly increased by pressure, this property becomes important in analyzing the behavior of polymer films used as solid film lubricants where high contact pressures are commonly encountered. Indeed one finds that it may be advantageous at times to design bearings with high unit loading in order to make use of the lower friction coefficients which may prevail under those circumstances. In any event the friction of polymer films under high contact pressures is readily accounted for by the simple extension of the adhesion theory of friction discussed here.

ACKNOWLEDGEMENTS

Mr. E. R. Carpenter, Jr. provided technical assistance and Mr. R. C. Bowers and J. R. Degenfelder provided sample materials.

REFERENCES

1. K. V. Shooter and D. Tabor, Proc. Phys. Soc. (London) B65, 661 (1952).
2. E. Rabinowicz and K. V. Shooter, Proc. Phys. Soc. (London) B65, 671 (1952).
3. M. W. Pascoe and D. Tabor, Proc. Roy. Soc. A235, 219 (1956).
4. S. C. Cohen and D. Tabor, Proc. Roy. Soc. A291, 186 (1966).
5. F. P. Bowden and D. Tabor, The Friction and Lubrication of Solids, Oxford Univ. Press, London, Part I, 1950, Part II, 1964.
6. F. P. Bowden and L. Luben, Proc. Roy. Soc. A169, 371 (1939).
7. F. P. Bowden and D. Tabor, Proc. Roy. Soc. A169, 391 (1939).
8. F. P. Bowden, A. J. W. Moore and D. Tabor, J. Appl. Phys. 14, 80 (1943).
9. F. P. Bowden and D. Tabor, J. Appl. Phys. 14, 141 (1943).
10. L. C. Towle, ASLE Proceedings - International Conference on Solid Lubrication 1971, Amer. Soc. Lubr. Eng., Park Ridge, Ill., 1971.
11. L. C. Towle, J. Appl. Phys. 42, 2368 (1971).
12. L. C. Towle, J. Appl. Phys. 44, 1611 (1973).
13. L. C. Towle and P. Nannelli, Asle Journal. (In Press).
14. L. C. Towle, J. Appl. Phys. (In Press).
15. R. C. Bowers, J. Appl. Phys. 42, 4961 (1971).
16. R. C. Bowers and W. A. Zisman, J. Appl. Phys. 39, 5385 (1968).
17. L. C. Towle, J. Appl. Phys. 43, 4807 (1972).
18. P. W. Bridgman, Phys. Rev. 48, 825 (1935).
19. P. W. Bridgman, J. Geol. 44, 653 (1936).
20. P. W. Bridgman, J. Appl. Phys. 8, 328 (1937).
21. P. W. Bridgman, Am. Acad. Arts Sci. 71, 387 (1937).
22. D. T. Griggs, F. J. Turner, and H. C. Heard, Geol. Soc. Am. Memoir 79, 39 (1960).
23. L. F. Vereschagin and V. A. Shapochkin, Fiz. metal. metalloved. 9, No. 2 258 (1960).
24. L. C. Towle and R. E. Riecker, Science 163, 41 (1969).
25. E. Passaglia and G. M. Martin, J. Research Natl. Bur. Standards 68A, 273 (1964).
26. K. D. Pae, J. Polymer Sci.: Part A-2 6, 657 (1968).
27. D. R. Mears, K. D. Pae, and J. A. Sauer, J. Appl. Phys. 40, 4229 (1969).

DISCUSSION OF PAPER BY L. C. TOWLE

S. Bahadur (Iowa State University): It is well known that the mobility of polymer molecules is greatly increased in the glass transition region. How would you then explain in the light of the above the departure in the shear strength of material from the expected shear strength vs. pressure variation towards the low pressure range in terms of the glass transition temperature?

L. C. Towle: The answer to this question was given in conjunction with an earlier question. Namely, the more rapid increase of shear strength with pressure in the range above 1 kbar is probably due to the sample being forced into the glassy state by the high pressure.

J. J. Bikerman (Case Western Reserve University): In order to separate the interfacial effect from the bulk effect, we should prepare samples of different thickness and different heights so we can extrapolate to zero thickness. The zero thickness can hopefully give us interfacial data, otherwise your experiments, however interesting, have nothing to do with friction.

L. C. Towle: Since we have fundamental disagreement on when one can apply a definition of friction coefficient, there is no hope to resolve other differences. In regard to the thickness dependence, it is not possible to do this kind of experiment because once you load the sample and extrude down, the thickness is determined by mechanical properties of the polymer.

D. H. Buckley (NASA - Lewis Research Center): (This comment was made at the end of the presentation of his own paper.) In regard to the discussion on the extrapolation to zero thickness, I would like to say "Don't do it." As you can see, the properties at the interface (or at the zero thickness) are generally different from those in the bulk.

E. Sacher (I.B.M.): Since the frictional coefficient changes with wear, did you find such wear during your experiments, especially at the higher pressures attained?

L. C. Towle: Wear was not specifically studied in our experiments which were all short time tests. In this geometry where the sample material is largely confined, wear does not occur in the usual sense of the word. However, there does tend to be some slow extrusion of the sample material from between the anvils such that the sample becomes thinner as the experiment continues. The sample also becomes thinner with increasing load. Either of these effects can lead to direct anvil-to-anvil contact which can invalidate the measurements. We monitored the electrical resistance between the anvils, which were electrically isolated from the frame of the apparatus, in order to detect such metal-metal contacts. This was rarely a problem in these short term tests. We found that an electrical short through the sample can develop with no noticeable increase in shearing torque so that this is a very sensitive and satisfactory method for detecting the onset of break-down of polymer films.

L. P. Fox (RCA Labs): At higher pressures does the temperature exceed the glass transition point? If so, aren't you really measuring viscosity characteristics of the polymer and interpreting this as an increase in shear strength and a decrease in μ?

L. C. Towle: The experiments were run at room temperature and we do not believe any significant heating of the sample material occurred because: (1) the sliding speeds were below those where frictional heating is generally observed; (2) the samples were very thin and in intimate contact with good thermal conductors; (3) the shearing torque did not decrease with increasing sliding speed (see Figure 1). At atmospheric pressure and room temperature, polypropylene is between its melting point and its glass transition temperature in a state with some viscous properties, but which is nevertheless normally described in terms of conventional mechanical properties (see References 26 and 27). Increased pressure is qualitatively similar to a reduction in temperature and moves the material toward the glass transition. In the present case it seems probable that the transition seen near 1 kbar is the glass transition in fair agreement with an extrapolation of the measurements of Passaglia and Martin as discussed in the paper.

Friction and Transfer of Some Polymers in Unlubricated Sliding

B. J. Briscoe, (Mrs.) C. M. Pooley, and D. Tabor

Physics and Chemistry of Solids

Cavendish Laboratory, University of Cambridge, Cambridge,

England

The paper deals mainly with the friction
of PTFE and high density polyethylene sliding
over hard clean flat surfaces. The behavior
of these polymers is somewhat anomalous and
is contrasted with the behavior of "normal"
polymers; that is virtually all other polymers.
The friction and transfer behavior falls into
two groups. (i) At low speeds the friction
is low ($\mu \simeq 0.05$) a thin film of polymer
is drawn out of the slider and adheres to
the counter surface. Its thickness is of
order 50-200Å and the molecular chains
are highly oriented in the sliding direction.
This behavior does not appear to depend on
the crystallinity or band structure of P.T.F.E.
or on the spherulite size of P.E. If, however,
bulky or straggly side-groups are incorporated
into the polymer backbone the friction and
transfer increase and resemble those of
"normal" polymers. The behavior appears
to be connected with a smooth molecular
profile. (ii) At high speeds or low
temperatures these "low-friction" materials
give a high friction ($\mu \simeq 0.3$) and the
transfer is much heavier consisting of
lumps or smeared films of polymer several
1000 Å thick. The behavior in this regime
may be determined by crystallite size, the

band structure of the polymer or the
molecular weight. Experiments on High
Density P.E. containing a mixture of CuO
and Pb_3O_4 show that the filler has no
influence on the friction or wear behavior
at low sliding speeds. But at high
sliding speeds the wear against a smooth
steel counterface is markedly reduced.
The filler apparently functions by
producing a strongly adhering polymer
film on the steel surface.

INTRODUCTION

There are two main processes involved in the friction between
solid bodies: the ploughing of asperities of the harder solid
through the surface of the softer and the shearing of adhesive bonds
formed at the regions of real contact. In the sliding of a thermoplastic over a hard smooth surface the ploughing term is negligible.
The major part must arise from adhesion at the interface and it is
this factor which is considered in this paper. If the adhesive
bonds are weaker than the polymer the shearing will occur truly at
the interface and there will be virtually no transfer of polymer on
to the other surface. This can occur even if the coefficient of
friction is large. For example it occurs over a limited range of
loads and speeds when P.M.M.A. slides over glass although the
coefficient of friction may be as large as $\mu = 0.4$. In general,
however, weak interfacial adhesion is generally associated with
low friction ($\mu < 0.1$). By contrast strong interfacial adhesion
is generally associated with high friction ($\mu > 0.3$) and sliding
occurs at a plane within the polymer.

The friction of P.T.F.E. has long been considered anomalous.
At low sliding speeds it shows a relatively low friction ($\mu \sim 0.06$
at 20°C) and this has often been attributed to poor interfacial
adhesion. However at high speeds the friction of P.T.F.E. against
a clean surface is high ($\mu \simeq 0.3$ at 20°C) and there is appreciable
transfer of polymer. Clearly adhesion can be strong. As we shall
see, relatively strong adhesion may also occur in the low-friction
regime.

The work described in this paper deals briefly with the friction
of P.T.F.E., polythene and related polymers. It includes a detailed
study by optical microscopy, electron microscopy and electron
diffraction of any polymer transfer that is produced on the counter
surface. Mention is also made as to how the adhesion of the film
to the substrate may be improved by using certain inorganic oxide
fillers.

THE FRICTION OF P.T.F.E. SLIDING AGAINST A HARD SURFACE.

Low Speeds

General Behavior. If a hemisphere of P.T.F.E. is slid over a clean smooth glass surface at a speed of order 1 mm s^{-1} under a load of about 1 Kg (though the load is not critical) the following features are observed[1,2,3].

(a) An initial "stick" occurs, and the coefficient of friction at the stick, μ_s, is approximately 0.3. This is accompanied by the formation of a relatively large lump of polymer adhering to the glass surface and smeared material covering one diameter of the contact area. The thickness of the transfer is of order 1 to 20 μm.

(b) Immediately after the "stick" the friction falls to a low value and the coefficient of friction during sliding, μ_k, is about 0.06. This is accompanied by the formation of a thin transferred film of polymer which adheres to the glass. It can be removed easily by wetting the surface with water. The detached film is found to be only 30 to 100 Å thick; it is tenuous with many discontinuities but electron diffraction shows that the molecular chains are highly oriented in the direction of sliding.

(c) If the slider is moved to a fresh position on the glass surface and sliding commenced in the same direction, there is no initial high static friction. The friction and transfer are similar to those in (b).

(d) If the slider is slid at right angles to the original direction (e.g. by rotating the slider about its vertical axis through 90°) it reverts to condition (a), that is an initial high static friction and lumpy transfer, followed by low friction and thin film transfer (Figure 1).

(e) Repeated sliding over a thin transferred film always gives a low friction ($\mu_s = \mu_k \simeq 0.06$) and there is little increase in the thickness of the film, i.e. sliding occurs essentially at the interface between the slider and the transferred film.

We deduce from these experiments that once the first high-friction stick has occurred the P.T.F.E. in the contact region is pulled into a direction which favors easy drawing of the polymer chains: These adhere to the surface as sliding occurs. The results of (e) suggest that the force to draw out the polymer chains from within the bulk of the specimen is essentially the same as the force required to slide the polymer, now itself oriented, over an oriented film. Thus additional drawing of fibre, i.e.

thickening of the transferred film may not occur: on the other
hand the friction will be the same as when drawing <u>does</u> occur.

(f) Similar results are observed in the sliding of P.T.F.E. on
clean smooth surfaces of steel or aluminum.

(g) If the counter surface is roughened then, for a surface roughness
greater than about 0.1 μm, high friction is observed and torn-out
fragments are formed. For this degree of roughness the easy
drawing process is inhibited and the low-friction thin-film-
transfer regime is no longer observed.

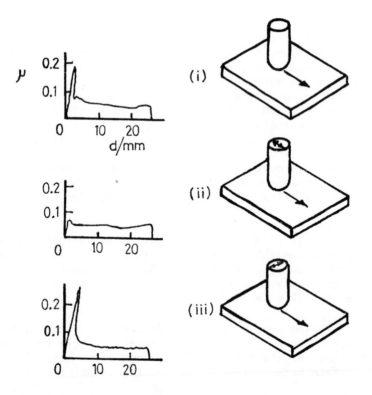

Figure 1.   High and low friction in P.T.F.E.. Coefficient of
friction plotted against distance d travelled along
glass (i) first traversal; high static friction
followed by low kinetic value. (ii) Second traversal
parallel to first; static value of friction is about
the same as the kinetic value. (iii) Another traversal
after slider has been turned through 90° about its axis.
High static friction again.

(h) If the hard smooth surface is contaminated the transfer may
almost completely disappear but the friction does not fall below
about $\mu \simeq 0.05$. Thus in the absence of adhesion the coefficient
of friction is only a little less than that observed when reasonably strong adhesion occurs. Clearly in the low friction regime
on clean smooth surfaces, the low friction is not due to poor
adhesion: it is due to the formation of a conveniently oriented
layer on the polymer and/or on the counter-surface. However as
we shall note later, the transferred film is not firmly attached
to the substrate and may be removed by the slider under more
extreme experimental conditions.

The Effect of Speed and Temperature in the Low-Friction Regime.
For speeds less than about 2 mm s^{-1}, a 50-fold increase in sliding
speed produces approximately a 2-fold increase in friction and no
visible change in transfer. Evidently in the low friction regime
involving the sliding of chains length-wise over one another there
is a small shear-rate dependent factor. If, however, the speed is
increased to 100 mm s^{-1} the friction rises to a high value $\mu \simeq 0.3$
and the transfer changes to the heavy type observed in the "stick"
region at lower speeds.

The effect of temperature is more marked. If the temperature
is raised from 20° to 150°C the friction falls from 0.06 to 0.024.
If an estimate is made of the area of contact A from the width d of
the friction track ($A \simeq \pi d^2/4$) an estimate may be made of the shear
strength s of the interface ($s = F/A$). It is found that, over this
temperature range s falls by a factor of 7 implying an "effective"
activation energy of about 16 Kcal/mol. This is of the same order
as that observed in the viscous flow of long chain organic liquids
and in the shear of thin organic films[4]. Although increasing the
temperature causes a reduction in contact pressure the effect of
the pressure itself on s is not large[4].

The Effect of Crystal Texture. Three samples of P.T.F.E. were
examined. One was sintered and slowly cooled (band width ~3000Å).
The second sintered, then quenched (band width ~500Å). The third
was preformed under pressure but not sintered: it had no band
structure but the original polymerization grains were discernible
(about 2000Å diameter). Although the friction values varied a
little due to difference in hardness of the samples and hence contact
pressure all three showed the same behavior: a high static friction
with heavy transfer followed by a low kinetic friction and light
transfer. Three samples of dispersion grade P.T.F.E. were also
studied and again showed similar behavior.

The Effect of Side-Groups. The effect of tetrafluoroethylene
(TFE) and hexafluoropropylene (HFP) copolymers was studied. In these
materials a bulky perfluoromethyl (CF_3) group replaces a fluorine
atom at intervals along the carbon chain but apart from this the

general structure, including the helical chain, remains substantially unchanged. The results are summarized in Figure 2. For a H.F.P. content of less than about 2%, i.e. about one (or less) CF_3 group per 30 carbons the behavior resembles P.T.F.E. With the larger concentrations of CF_3 groups the behavior resembles that of a "normal" polymer. The friction is high ($\mu>0.2$) there is lumpy transfer and there is no low-friction low-transfer regime. Similarly polychlorotrifluoroethylene (P.C.T.F.E.) giving the repeat unit $-CFCl - CF_2 -$ is a "normal" high friction polymer.

All these results support the rather simple view that the main factor responsible for the low friction of P.T.F.E. is the presence of a "smooth" molecular profile.

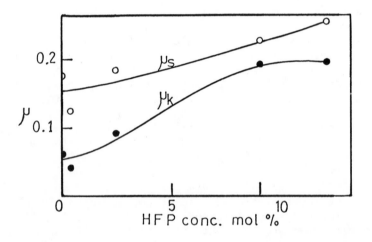

Figure 2.   Coefficients of static and kinetic friction, μ_s, and μ_k respectively plotted against concentration of HEP (mol percent) in a range of TFE - HFP copolymers. Load 1kgf; sliding speed 1mm s^{-1}; temperature 20°C.

THE FRICTION OF POLYTHENES AT SLOW SPEEDS

The behavior of high density polythene is similar to that of P.T.F.E. while low density polythene behaves as a "normal" polymer. Low density polythene has a high coefficient of static friction

($\mu\approx0.3$) with lumpy transfer and there is no low-friction low-transfer regime, i.e. $\mu_k\approx\mu_s\approx0.3$. This type of polythene has many straggly side groups. By contrast high density polythene which has few, or no, side groups has a static friction of order 0.15 which is associated with lumpy transfer and a low kinetic friction of about 0.08. The transfer is even less discernible than is the case with P.T.F.E. but again contains molecular chains well oriented in the sliding direction. Spherulite size, over the range 20 to 200µm, produces a small change in the frictional values but not in the frictional behavior. Similarly extended chain (high density) polythene which has a band structure similar to that of P.T.F.E. shows a high static and a low kinetic friction. The behavior closely resembles that of the spherulitic polymer. By contrast polypropylene behaves like a "normal" polymer. There is no low-friction low-transfer regime and $\mu_s\approx\mu_k\approx0.3$. These results show that molecular profile is more important than crystallinity or spherulite size.

The Effect of Speed and Temperature

Here again in the low friction regime the effect of speed on friction is small. The effect of temperature is more pronounced, a temperature rise from 20 to 100°C causing a reduction of the interfacial shear-strength by a factor of about 2.2.

THE FRICTIONAL BEHAVIOR OF P.T.F.E. AND HIGH DENSITY POLYTHENE IN THE HIGH FRICTION REGIME

We have already seen that in the low friction regime the force to slide chains length-wise over one another increases if the sliding speed is increased or the temperature decreased. At sufficiently high speeds or low temperatures this may reach a stage where the force required to produce such sliding is equal to or greater than the force required to tear out lumps of polymer. In such a case the friction is high ($\mu>0.3$) and the transfer lumpy. Further the shear strength at the interface is very close to the bulk shear strength of the polymer. The behavior resembles the initial "stick" observed in the low speed friction of these materials.

The fact that bulk properties are involved in the high friction regime is shown for P.T.F.E. by the following simple example. If the variation of interfacial shear strength s with temperature T is plotted as log s against 1/T a straight line is obtained, the slope of which corresponds to an "effective" activation energy of 10 KJ mol^{-1}. This is the same as the activation energy involved in the bulk shear strength of the polymer. By contrast a plot of log s against 1/T for the low friction regime gives a very different activation energy, or an order of 16 KJ mol^{-1} (Figure 3).

The factors which determine the amount of transfer in the high friction regime have not yet been established. With P.T.F.E., Tanaka[5] suggests that it is the band structure that is important: Jost[6] suggests that it is the crystallite size. Other workers have noted the importance of molecular weight[7,8]. The subject merits a detailed study.

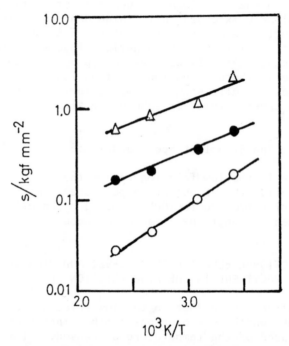

Figure 3.   Arrhenius plots s= s exp (Q/RT) for the variation
             of shear strengths with temperature. ○,s calculated
             from kinetic friction, Q=15.5kJ/mol; ● ,s calculated
             from static friction, Q=10kJ/mol; △, bulk shear
             strength, Q=10kJ/mol.

THE EFFECT OF LEAD ADDITIVES ON THE HIGH-SPEED FRICTION
AND WEAR OF HIGH DENSITY POLYTHENE

Earlier work has shown that though P.T.F.E. is often a convenient low friction material for use in porous metal bearings its wear rate is often excessive both in presence and absence of lubricants. A good deal of effort has been expended in recent years to find additives or fillers which reduce the dry wear to more acceptable values. A wide variety of materials have been used including glass fibres[6,7], lead[9,10], lead oxide-bronze[7,9,10,11], and cadmium oxides[9,10]. Recently we have studied an analogous situation, the use of additives to reduce the wear of high density polythene[11].

The polymers investigated were both low and high density polythene (molecular weight approximately 30,000). Specimens of these
polymers were prepared containing 5% by weight of CuO and 30% by
weight of lead oxide (Pb_3O_4). The moulded polymer was used as a
flat specimen and pressed against a flat smooth surface of glass,
or mild-steel.

The results show the following clear-cut features:

(i)   With low density polythene at both low and high speeds
        of sliding the friction is high and the wear (in the
        form of transferred lumps) heavy. No difference is
        observed between the filled and unfilled polymer samples.
        For this reason all the subsequent results refer to high
        density polythene specimens.

(ii)   At low sliding speeds the transfer is very small. The
        transfer and frictional behavior for the filled and
        unfilled polymers are indistinguishable.

(iii)   At high sliding speeds, where appreciable frictional
        heating occurs, the friction and wear are high on glass
        for both filled and unfilled specimens.

(iv)   At high speeds there is a marked difference in behavior
        on a mild-steel surface. If the surface is rough (C.L.A.
        greater than 2.0µm) the filled and unfilled specimens
        behave similarly giving high friction and wear. There
        is a marked change however if the surface is smooth
        (C.L.A. less than 0.5µm). The unfilled polymer continues
        to give high friction ($\mu \approx 0.5$) and heavy wear but the
        filled specimen gives a lower friction and the wear rate
        of the polymer is reduced by a factor of 20 to 100 depend-
        ing upon operating conditions.

        Figure 4 shows the total wear of two specimens, unfilled
        and filled high density polythene, as a function of
        running time for two initial counterface temperatures, T_o
        The sliding velocity was 2.1 ms^{-1} and the normal load 5.0
        Kg for 5 mm diameter pins. The substrate observed tem-
        perature during the experiment is indicated at each data
        point (rubbing on mild steel (C.L.A.$\approx$ 0.5 µm)). Four
        important differences between the samples are seen.
        First the wear of the unfilled sample is greater. Second-
        ly this sample continues to wear throughout the test
        whereas the filled material attains a zero wear rate
        after about 15 min. Thirdly for the unfilled sample the

wear increases with increasing temperature while the
converse is true for the filled sample. Finally the
substrate temperature increases much more for the unfilled
sample during the experiment indicating that the
frictional work is greater for this material.

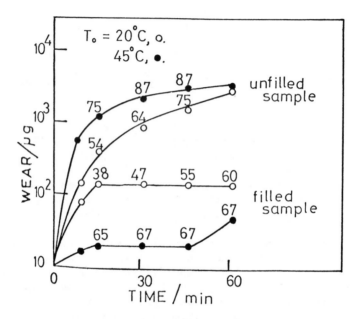

Figure 4.   Total wear of two HDPE specimens, filled and unfilled,
             as a function of sliding time on a steel counterface
             for two initial temperatures. Also shown next to the
             data points are the substrate temperatures during the
             experiments. Load 5.0kg, sliding speed 2.1 m s^{-1}.

(v)   Examination of the steel surface by optical and electron
         microscopy shows that the low-friction low-wear-rate
         regime with the filled polymer is accompanied by the
         formation of a film of polymer strongly adhering to the
         steel surface. By contrast relatively little of such

a film is formed if the unfilled polymer is used.

(vi)   A strongly adhering polymer film may be formed with the
        filled polymer. If unfilled polymer is now slid over
        this surface, the friction and wear remain low for very
        protracted periods and the material behaves like filled
        polymer. Apparently the oxide additives do not appreciably
        affect the intrinsic frictional properties of the polymer.
        Their main function appears to be to produce a film of
        polymer strongly attached to the counter surface. The
        attachment is specific and it occurs on steel but not
        on glass. Further CuO alone and Pb_3O_4 alone do not have
        any effect on the frictional behavior. The precise
        mechanism of polymer adhesion in the presence of the
        oxide mixture is not established and needs further
        investigation.

SUMMARY

These results show that at low sliding speeds and moderate
temperatures P.T.F.E. and high density P.E. give a low coefficient
of friction (of order $\mu \approx 0.06$) and produce a thin adherent film of
polymer on the counter surface. Its thickness is of order 30-100Å
and the molecular chains are highly oriented in the direction of
sliding. The behavior does not appear to depend on the crystallinity
or band structure of the polymer or, in the case of P.E., on spherulite size. On the other hand if bulky side groups are incorporated
into the molecular backbone the friction and transfer increase and
resemble those of "normal polymers". These results suggest that the
low-friction low-transfer behavior is connected with a smooth
molecular profile.

At high speeds (or low temperatures) these "low friction"
polymers give a high friction and the transfer is much heavier,
consisting of lumps or smeared films of polymer several 1000Å thick.
The behavior in this regime may be determined by crystallite size,
band structure or molecular weight or, in the case of polythene by
spherulite size.

The incorporation of a mixture of copper and lead oxides into
high-density polythene produces no change in the frictional behavior
at slow sliding speeds. At high sliding speeds, however, where there
is appreciable frictional heating, the behavior against a smooth mild-
steel counterface is markedly changed. Both the wear and friction
are markedly reduced. It would seem that with the unfilled specimens
of polythene there is transfer of polymer to the counterface but it
is relatively weakly held: the film is readily detached and replaced
so that there is a continuous transfer and removal of polymer from
the counterface. This accounts for the relatively high wear. With

filled polymer, however, the adhesion is stronger and once a continuous film of polymer is formed further transfer is small and the wear is light. The strong adhesion is specific since it does not occur on glass surfaces. Nor is it found if copper oxide or lead oxide are used alone. Further the effect is confined to P.T.F.E. and high density P.E.; no significant improvements in wear resistance are noted with "normal" polymers. The mechanism of adhesion is not understood. Nor is it clear why sliding of polymer against the transferred film gives weak adhesion and correspondingly low wear.

ACKNOWLEDGEMENTS

The authors are grateful to several members of their laboratory, both past and present, who have participated in various aspects of the work presented in the paper.

REFERENCES

1. C. M. Pooley and D. Tabor, Nature, Physical Science, 237, 88 (1972).
2. C. M. Pooley and D. Tabor, Proc. Roy. Soc., Lond., A 329, 251 (1972).
3. K. R. Makinson and D. Tabor, Proc. Roy. Soc., Lond., A 281, 49 (1964).
4. B. J. Briscoe, B. Scruton and F. R. Willis, Proc. Roy. Soc., Lond., A 333, 99 (1973).
5. K. Tanaka, Y. Uchiyama and S. Toyooka, Wear, 23, 153 (1973).
6. H. Jost, S. Mothes, M. Raab, J. Richter-Mendau and C. Staedler, Schmierungstechnik, 4, 69, 109 (1973).
7. G. C. Pratt, "Plastic-Based Materials" in Lubrication and Lubricants, edited by E. R. Braithwaite, p. 403, Elsevier (1967).
8. R. T. Steinbuch, Wear, 5, 458 (1962).
9. G. C. Pratt, Trans. J. Plastics Institute (London), 32, 255-260, (1964).
10. G. C. Pratt, Recent Developments in P.T.F.E. Based Dry Bearing Materials and Treatment, Proc. Conf. Lubric. and Wear, 1967, paper 16, 132-143, (Inst. Mechn. Engineers, London).
11. B. J. Briscoe, A. K. Pogosian and D. Tabor, Wear, 27, 34 (1974).

DISCUSSION OF PAPER BY D. TABOR

G. M. Robinson (3M Company): Would you please discuss friction and wear of a composite containing a metallic oxide and a binder such as an urethane or vinyl.

D. Tabor: We have no experience with these types of materials and we are reluctant to comment on this important area of composite technology.

R. J. Nash (Xerox Corporation): As you have demonstrated, the low friction found for PTFE and polyethylene is associated with the removal of molecules from the polymer sample, and transfer of these molecules to the substrate. I have two questions concerning this removal/transfer process:

1. Have you examined the chain length/s of the extracted moelcules? If not, which of the following hypotheses do you feel is most plausible:

   a) the extracted molecules are identical to those in the polymer sample.
   b) the extracted molecules are the shortest, and hence least entangled, of the polymer sample.
   c) the extracted molecules are chain fragments generated by a mechanochemical process.

2. Have you studied the effect of polymer molecular weight and distribution on the friction process?

D. Tabor: We have not attempted to analyse the chain length distribution of the "extracted" polymer molecules. Such measurements would be very difficult as the amount of material available for analysis is quite small. In our experiments, we estimate that no more than 20 μg of PTFE or high density polythene could be extracted in a simple experiment. It may be possible to carry out a molecular weight analysis for examples using gel permeation chromatography. The results of such an experiment would be very interesting.

We feel that speculating on the change in molecular weight distribution (if any) during the transfer process is a dangerous exercise at the present time since the nature of the transfer process is not understood. Finally, we consider from energy considerations (as well as Buckley and Pepper's work[1]) that extensive chain scission does not occur for PTFE and high density polythene. The main dissipative process involves the sliding of chains over one another.

Concerning the second question, the work of Mrs. Pooley showed that for unfilled specimens of PTFE and PE the friction and transfer were hardly influenced by molecular weight, crystallinity or spherulite size. The essential factor was the "smoothness" of the molecular profile. We have no data for filler materials.

1. See, for example, S. V. Pepper, NASA TN D-7533, (1974).

J. J. Bikerman (Case Western Reserve University): Dr. Tabor still believes that climbing of the slider over the hills of the support cannot explain the dissipation of energy during sliding, that is the transformation of frictional work into frictional heat. But, when the slider is first lifted and then permitted to fall into the depressions between the hills, the work performed by the experimenter in lifting is completely dissipated as heat after the fall.

D. Tabor: The real point of issue in the roughness model of friction is the method of dissipation of energy. If Dr. Bikerman believes that when the slider falls into the depressions between the holes, work is lost by deformation, he is describing a very valid mechanism. However, in a multi-asperity model the rate of fall of each individual asperity has to be extremely rapid if it is to dissipate any apprec-iable amount of energy by this method. A little thought will show that unless the rate of return is extremely rapid the asperities on one surface will simply follow the contours on the other as best they can. Only if all the asperities are shaped like saw teeth with a sharp drop in one direction can one envisage the loss of a large amount of energy by this mechanism.

H. Gisser (Frankford Arsenal): I would like to make a comment on the matter of side chains. In the case of teflon and hexafluoro-propylene, there is no question about the effect of trifluoromethyl group which increases the friction coefficient. When one studies polyalkyl methacrylate, one finds that the effect of side chain is different. Of course, the simplest one already contains a side chain, but when the length of side chain increases, for example, polyethyl methacrylate, polypropyl methacrylate, the friction coefficients decrease. It turns out that we are dealing with different materials. Polymethyl methacrylate is an amorphous (random) material, but the higher methacrylates begin to show crystallinity.

D. Tabor: The results you quote are very interesting. My only comment is that the work I have just described was limited to PTFE and PE and related polymers. It was not meant to provide a general explanation for the friction of all polymers and the influence of side chains.

Frictional Heating of a Uniform Finite Thickness Material Rubbing Against an Elastomer

G. Harpavat

Xerox Research Laboratories

Xerox Square, Rochester, N.Y. 14644

In this paper a general mathematical analysis is presented to study the rise in
temperature due to frictional heating of
two surfaces rubbing against one another.
The problem has been formulated in terms
of a plane of finite thickness rubbing
against rectangular and pyramidal shaped
elastomeric sliders. The plane has been
treated as a rigid solid, but the analysis
is also expected to hold for an elastomeric
plane. Formulas for the average and maximum temperatures in the contact area and
the distribution of the temperature in the
bodies of the slider and the plane are
presented. The parameters investigated
include the coefficient of friction, velocity of slide, normal load, thickness
of the plane, geometric dimensions of the
slider, and the elastic and thermal constants of the plane and the slider. The
results are presented in closed form
analytical expressions which can be easily
evaluated on an electronic computer. The
analytical predictions are correlated with
measurements on a lightly loaded pyramidal
shaped rubber slider moving over a smooth
plane.

I.   INTRODUCTION

When two surfaces slide against each other the energy dissipated in friction appears as heat at the interface. This causes
the temperature of the sliding surfaces to rise. The rise in
temperature of the sliding surface due to dissipated energy can be
quite substantial. For example, experiments[1] have shown that even
at moderate velocities of sliding the heat of friction can easily
cause a thermal softening or local melting of the metals at points
of contact. In thermal non-conductors these local high temperatures
occur even more rapidly.

Many investigators have attempted to predict the temperature
rise in the sliding contact, including Blok,[2] Jaeger,[3] Holm,[4]
Ling & Pu[5] and Kounas, et al.[6] In order that the analytical work
be realistic, appropriate solutions are needed for each geometric
configuration and the sliding conditions. Furthermore, an analytical determination of flash temperature requires a knowledge of
the partition of the heat of friction between the bodies in sliding
contact. Jaeger[3] has presented a detailed mathematical analysis of
a square slider sliding over a semi-infinite medium. In many
practical applications the sliders are rectangular or pyramidal in
shape and the surfaces on which they slide are finite in thickness.
In this paper a detailed mathematical analysis is presented to
study the frictional heating in problems involving the above configurations. The heat loss from the exposed surfaces is considered.
Expressions for the partition of the heat of friction between the
slider and the plane, the average and maximum steady state temperatures in the contact area and the distribution of temperatures in
the bodies of the slider and the plane are derived. These expressions are particularly suited for soft elastomers sliding over a
tough plane.

II.   NOMENCLATURE

A = Real area of contact (cm^2)
W = Total normal load in the contact area (g)
P_m = Mean yield pressure of the irregularities on the con-
tacting surfaces (g/cm^2)
A_∞ = Ultimate contact area (cm)
E = Young's modulus of the elastomer (g/cm^2)
β = Non-dimensional parameter characterizing the geometries
in contact
$2b$ = Width of the real area of contact (cm)
L = Length of the area of contact (cm)
μ = Dynamic coefficient of friction at the sliding surfaces
g = Acceleration due to gravity ($cm/sec.^2$)
v = Velocity of slide (cm/sec.)
N = Normal load per unit length of the contact area (g/cm)

Q_R = Heat of friction per unit contact area per unit time (cal/cm^2-sec.)

J = Mechanical equivalent of heat (4.18x10^7 ergs/cal)

α = Fraction of total heat of friction which flows into the slider

∇^2 = Laplace Operator = $\dfrac{\partial^2}{\partial x^2} + \dfrac{\partial^2}{\partial y^2} + \dfrac{\partial^2}{\partial z^2}$

t = Time coordinate

k_1 = Thermal diffusivity of the plane = $\dfrac{K_1}{\rho_1 C_1}$ (cm^2/sec)

K_1, ρ_1, C_1 = Thermal conductivity, mass density and specific heat of the plane (cal/sec cm°c), (g/cm^3), (cal/g°c)

ℓ = Thickness of the plane (cm)

H_1 = Coefficient of surface heat transfer for the plane (cal/cm^2-°c-sec)

T_a^P = Temperature rise in the contact area (-b<x<b) as shown in Fig. 2 (°c)

T_b^P = Temperature rise in the region behind the contact area (-∞<x<-b) (°c)

T^S = Temperature in the slider (°c)

H_2 = Surface heat transfer coefficient for the blade cal/cm^2-°c-sec)

K_2 = Thermal conductivity of the slider material (cal/sec-cm°c)

T_{ave}^P = Average temperature of the plane in the contact area (°c)

T_c^S = Temperature of the slider in the contact area (°c)

III.   FORMULATION OF MODEL

The two configurations considered in this paper are shown in Fig. 1 (a) and (b). The area of contact between the slider and the plane has dimensions 2b and L in X and Y directions, respectively. We shall assume that the slider is stationary and the plane moves in X direction with a velocity V. The problem of the slider moving and the plane being stationary is equivalent to that of the slider being stationary and the plane moving. Initially, the plane is at rest and the temperature in the contact area is the same as that of the ambient. For simplicity, we shall assume that the ambient temperature is zero degree so that the calculated temperature is also the rise in temperature. We shall also assume that the motion of the plane persists for a time long enough that a steady state prevails in the contact region.

In problems of frictional heating, one must consider the real area of contact between the slider and the plane rather than the apparent, visible area of contact. For hard surfaces such as metals the real area of contact, A, can be determined by the following approximate formula:[1]

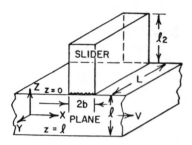 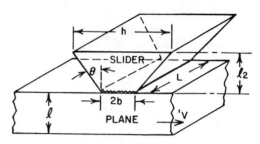

Fig. 1(a): Rectangular slider   Fig. 1(b): Trapezoidal slider
   sliding over a plane   sliding over a plane

$$A = \frac{W}{P_m} . \qquad\qquad\qquad\qquad\qquad\qquad\qquad (1)$$

For elastomeric sliders the real area of contact can be approximately determined by[7]

$$A = A_\infty (1 - e^{-\frac{\beta W}{EA_\infty}}) . \qquad\qquad\qquad\qquad (2)$$

For cross-linked butadiene-nitrile rubbers, Bartenev[7] has shown
that A_∞ is about 90% of the apparent area of contact and the real
area of contact approaches A_∞ at an average pressure of 500 psi.
In most practical applications the pressure in the contact area is
of the above order; therefore, it is reasonable to assume that for
elastomeric sliders the real and apparent area of contacts are
nearly equal. Furthermore, in practice the contact region is a
few mils wide but the length is generally more than a few inches.
Therefore, the temperature is invarient in Y direction and the
calculations can be simplified by assuming that the sliders and
the plane have infinite lengths in Y direction. In cases where the
real area, A, as calculated from Eq. (2) is vastly different from
the apparent area of contact, an equivalent width of the contact
can be defined as follows:

$$2b = A/L. \qquad\qquad\qquad\qquad\qquad\qquad\qquad (3)$$

The following analysis is valid in applications where the width
of the real area of contact is 2b and the length is much larger
than the width.

IV.   DERIVATION OF FORMULAS

The thermal equivalent of the rate at which the energy is dissipated in friction in the contact area is given by:

$$Q_R = \frac{\mu NgV}{2bJ}. \tag{4}$$

The heat Q_R is shared between the slider and the plane. Let a
fraction α of the heat flow into the slider and the remaining
$(1-\alpha)Q_R$ flows into the plane. The fraction α is determined by the
fact that in the contact region the temperature of the slider and
the plane must be equal.

The temperature distribution in the contact region is not uniform. Therefore, it is mathematically difficult to satisfy the
continuity of temperature condition at every point in the contact
region. A common practice[3] is to equate the average temperature
of the slider to the average temperature of the plane in the contact region.

The problem of the rise in temperature due to frictional heating
is equivalent to the calculation of the temperature of the stationary slider due to αQ_R heat inflow and the calculation of the temperature in the moving plane due to $(1-\alpha)Q_R$ heat inflow.

(a) Temperature Distribution in the Plane

A schematic of a moving plane supplied with $(1-\alpha)Q_R$ heat in the
contact region is shown in Fig. 2. The governing differential
equation for the temperature distribution in the plane is:

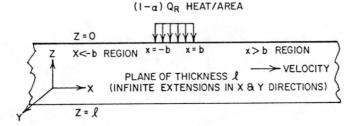

Fig. 2:   A moving plane supplied with a $(1-\alpha)Q_R$ heat/area at the top

$$\nabla^2 T^P = \frac{1}{k_1} \frac{\partial T^P}{\partial t} .$$

$$(5)$$

The two boundary conditions which are needed to solve Eq. (5) can be obtained by assuming that the rate of heat out-flow across the surfaces $Z = 0$ and $Z = \ell$ is directly proportional to the difference in the temperature of the surfaces and the surrroundings. That is:

$$K_1 \frac{\partial T^P}{\partial z} = H_1 T^P \text{ at } Z = 0 ;$$

and

$$K_1 \frac{\partial T^P}{\partial z} = -H_1 T^P \text{ at } z = \ell .$$

$$(6)$$

A Laplace transformation method was used to solve Eq. (5) in a frame of reference moving with the plane. The results are as follows:

$$T_a^P = \frac{4(1-\alpha)}{\rho_1 c_1} \frac{k_1 Q_R}{v^2} \sum_{n=1}^{\infty} \frac{\beta_n}{\sqrt{a}} \left(\frac{1 - e^{-\frac{V}{2k_1}(\sqrt{a} - 1)(b-x)}}{\sqrt{a} - 1} \right.$$

$$\left. + \frac{1 - e^{-\frac{V}{2k_1}(\sqrt{a} + 1)(b+x)}}{\sqrt{a} + 1} \right) \quad (\text{for } -b<x<b) ;$$

$$(7)$$

$$T_b^P = \frac{8(1-\alpha)k_1}{\rho_1 c_1} \frac{Q_R}{v^2} \sum_{n=1}^{\infty} \frac{\beta_n}{\sqrt{a}(\sqrt{a} - 1)} \left(\sinh\left\{ \frac{Vb}{2k_1}(\sqrt{a} - 1) \right\} e^{+\frac{V}{2k_1}(\sqrt{a} - 1)x} \right)$$

$$(8)$$

(for $-\infty<x<-b$); and

$$T_c^P = \frac{8(1-\alpha)k_1}{\rho_1 c_1} \frac{Q_R}{V^2} \sum_{n=1}^{\infty} \frac{\beta_n}{\sqrt{a}\,(\sqrt{a}+1)} \left(\sinh\left\{\frac{Vb}{2k_1}\,(\sqrt{a}+1)\right\} e^{-\frac{V}{2k_1}(\sqrt{a}+1)x} \right)$$

(9)

(for $b<x<\infty$);

where:

$$a = 1 + \frac{4k_1^2\lambda_n^2}{V^2} ;$$

(10)

$$\beta_n = \frac{\lambda_n(\lambda_n\cos\lambda_n z + h_1\sin\lambda_n z)}{\ell(\lambda_n^2 + h_1^2) + 2h_1} ;$$

(11)

λ_n are the n positive roots of $\tan\lambda\ell = \dfrac{2\lambda h_1}{\lambda^2 - h_1^2}$;   (12)

and $h_1 = \dfrac{H_1}{K_1}$.

(13)

(b) Temperature Distribution in the Slider

We shall consider rectangular and trapezoidal sliders as shown in Figs. 3(a) and (b). We shall assume that the sliders are at steady state temperature, that the exposed surfaces of the sliders dissipate heat to the ambient according to Newton's law of cooling, that the holder of the slider is a perfect heat sink so that the temperature at the far end of the slider is ambient. In order to simplify the analysis, we further assume that the temperature at any cross-section perpendicular to the Z axis (Fig. 3) of the

slider is uniform.   The governing heat flow equation in the rectangular slider is given by:

$$\frac{d^2 T^s}{dZ^2} = \frac{H_2}{k_2} \left(\frac{1}{b} + \frac{2}{L}\right) T^s \; . \tag{14}$$

The governing heat flow equation in the trapezoidal slider is
given by:

$$\frac{d}{dZ}\left(Z \frac{dT^s}{dZ}\right) = \frac{H_2}{k_2 \tan\theta}\left(1 + \frac{2Z\tan\theta}{L}\right) T^s \; . \tag{15}$$

The boundary conditions to solve these equations are:

$$-K_2 \left.\frac{\partial T}{\partial Z}\right|_{\text{at contact surface}} = \alpha Q_R ; \tag{16}$$

and

$$T_{\text{at far end}} = \text{Ambient Temperature} = 0. \tag{17}$$

Solutions of Eq. (14) and (15) with (16) and (17) B.Cs. are: for a
rectangular slider:

$$T^s = \frac{\alpha Q_R}{\sqrt{a_2}\left(1 + e^{-2\sqrt{a_2}\ell_2}\right)} \; e^{-\sqrt{a_2}Z}\left[1 - e^{-2\sqrt{a_2}(\ell_2 - Z)}\right] ; \tag{18}$$

where

$$a_2 = \sqrt{\frac{H2}{K_2}\left(\frac{1}{b} + \frac{2}{L}\right)} \; .$$

For a trapezoidal slider:

$$T^S = \frac{2b\ell_2 \log(\frac{h}{h'})}{K_2(h-2b)} \quad \alpha Q_2 + \frac{H_2 T_s \ell_2^2}{K_2(h-2b)^2} \{2b\log(\frac{h}{h'}) + h' - h\} \ . \ (19)$$

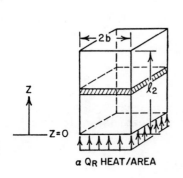

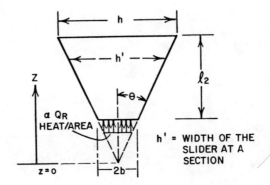

Fig.3(a):   Rectangular Slider   Fig. 3(b):   Trapezoidal Slider

(c) Partition of Heat Between Slider and Plane (α)

The average temperature of the plane in the contact area from Eq. (7) is:

$$T_{ave}^p = \left[\int_{-b}^{b} T_a^p (x) dx\right] \bigg/ 2b$$

$$= \frac{4(1-\alpha)k_1}{\rho_1 c_1 v^2} B' \ Q_R \ ; \tag{20}$$

where:

$$B' = \sum_{n=1}^{\infty} \frac{\beta_n}{\sqrt{a}} \left[\frac{2\sqrt{a}}{a-1} - \frac{2k_1}{Vb(\sqrt{a}-1)^2} \sinh \{\frac{Vb}{2k_1}(\sqrt{a}-1)\} e^{-\frac{Vb}{2k_1}(\sqrt{a}-1)}\right.$$

$$\left. - \frac{2k_1}{Vb(\sqrt{a}+1)^2} \sinh \{\frac{Vb}{2k_1}(\sqrt{a}+1)\} e^{-\frac{Vb}{2k_1}(\sqrt{a}+1)}\right] \ . \tag{21}$$

The temperature of the slider at the contact surface from (18) and (19) is:

$$T_c^s = \frac{\alpha Q_R (1-e^{-2\sqrt{a_2}\ell_2})}{\sqrt{a_2}(1+e^{-2\sqrt{a_2}\ell_2})} \quad \text{(for a rectangular slider)}; \qquad (22)$$

and

$$T_c^s = \frac{\frac{2b}{K_2}\alpha Q_R \log\left(\frac{h}{2b}\right)}{\left[\frac{h-2b}{\ell_2} + \frac{H_2\ell_2}{K_2(h-2b)}\left\{h-2b(1+\log\frac{h}{2b})\right\}\right]} \quad \text{(for a trapezoidal slider)}. \qquad (23)$$

Equating (22) and (23) to (20) we obtain:

$$\alpha = \frac{1}{1 + \frac{\rho_1 C_1 V^2 P}{4\,k_1 B'}} \quad ; \qquad (24)$$

where:

$$P = \frac{1 - e^{-2\sqrt{a_2}\ell_2}}{\sqrt{a_2}(1+e^{-2\sqrt{a_2}\ell_2})} \quad \text{(for a rectangular slider)}; \qquad (25)$$

$$P = \frac{\frac{2b}{K_2}\log\left(\frac{h}{2b}\right)}{\left[\frac{h-2b}{\ell_2} + \frac{H_2\ell_2}{K_2(h-2b)}\left\{h-2b\,(1+\log\frac{h}{2b})\right\}\right]} \quad \text{(for a trapezoidal slider)}. \qquad (26)$$

The knowledge of α from Eq. (24) permits the calculation of the temperature distribution in the plane and the slider from Eqs. (7), (8), (9), (18) and (19). The distribution of the temperature in the contact area is obtained by putting $Z = 0$ in Eq. (7), and the average temperature in the contact region is obtained by Eq. (20), (22) or (23).

V.   COMPARISON WITH EXPERIMENTS

The above results are similar in form to that of Jaeger[3] but their scope of applicability is increased since we have considered real geometries; that is, a plane of finite thickness and rectangular and trapezoidal shaped sliders, and we have accounted for the heat dissipation from the exposed surfaces in the analysis.

This analysis has been applied to study the rise in temperature of an elastomeric slider, trapezoidal in shape, rubbing against a thin layer of chalcogenide material coated over an aluminum substrate. The presence of the substrate does not affect the temperature distribution in the slider and the layer since most of the temperature drop occurs within a few microns depth of the layer as shown in Fig. 8. The temperature was measured by Fisher and Abowitz[8] embedding a thermocouple in the slider at 5 mils from the rubbing surface. The temperature at the rubbing surface was calculated from these measurements using Eq.(27) which can be derived from Eqs.(19) and (23).

$$
T_c^s = \frac{T_e \{ \log(\frac{h}{2b}) / \log(\frac{h}{h'}) \}}{\left[1 - \frac{H_2 \ell_2^{\,2}}{K_2 (h-2b)^2} \{ 2b-h+ \frac{\log(\frac{h}{2b})}{\log(\frac{h}{h'})}(h-h') \} \right]}
\tag{27}
$$

T_e = Measured temperature in the slider at width h'.

The values of the parameters used in the experiments were $\ell = 60 \times 10^{-4}$cm, $K_1 = 4.78 \times 10^{-4}$cal/sec cm °C, $K_2 = 5 \times 10^{-4}$cal/sec cm°C, $H_1 = H_2 = 1 \times 10^{-4}$cal/cm^2 °C sec, h = .31cm, ℓ = .155cm, h' = .025cm, θ = 45°, V = 70cm/sec, ρ_1 = 4.25g/cm^3, C_1 = .084cal/g °C, L = 23cms. The correlation between the predicted temperatures by Eq.(20) and the measured temperatures for various, normal loads, and coefficients of friction is shown in Table 1. The coefficient of friction in the experiment was varied by giving a treatment to the surface of the plane and the corresponding area of contact was estimated by a deformation analysis. Within experimental uncertainty the agreement is very good.

For a fixed width of contact area the rise in temperature increases linearly with the normal load and the coefficient of friction, but the functional dependence of other parameters is quite obscure in Eqs.(20) and (21). Figs. 4 to 9 show the functional dependence of some important parameters under some pre-selected conditions for a trapezoidal slider sliding over a plane. The values of the common parameters used in drawing these figures were the same as those used in the experimental investigation described above.

TABLE I

MEASURED AND PREDICTED RISE IN TEMPERATURE OF A TRAPEZOIDAL SLIDER

				OBSERVED	PREDICTED
NORMAL LOAD g/cm	MEASURED COEFF. OF FRICTION μ	WIDTH OF THE CONTACT AREA (10^-4 cm)	MEASURED RISE IN TEMP. AT 5.0 MILS FROM THE SLIDING SURFACE °C	RISE IN TEMP. AT THE CONTACT SURFACE (CALCU-LATED FROM EQ. 27) °C	RISE IN TEMP. AT THE CONTACT SURFACE (CALCU-LATED FROM EQ. 29) °C
14.8	.91	8	1.5	3.6	5.4
22.7	.90	12	1.5	3.4	6.7
30.5	.91	16.4	3.0	6.4	7.8
40.0	.91	20.6	3.0	6.1	9.1
14.8	2.06	14.4	2.5	5.4	9.1
30.5	2.28	31.4	6.8	12.6	14.2
30.5	1.0	17.5	5.5	11.5	8.3
30.5	1.72	27.2	6.0	11.4	11.8
30.5	.79	14.6	4.7	10.1	7.2
30.5	1.215	19.8	5.5	11.2	8.8

The predicted temperature profile at the surface of the plane at a given instant is shown in Fig. 4. The regions -1<x/b<1, x/b>1 and x/b<-1 represent the surface of the plane in contact with, ahead, and behind the slider, respectively. The temperature propagates a little ahead of the slider (x/b>1 region), and peaks within the contact area; slightly before the trailing edge, and there is an exponential fall of temperature in the region behind the slider.

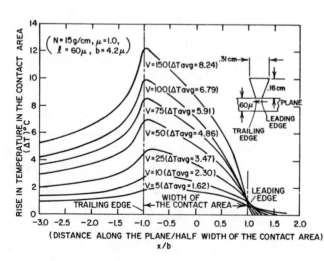

Fig. 4: Temperature profile at the surface of the plane

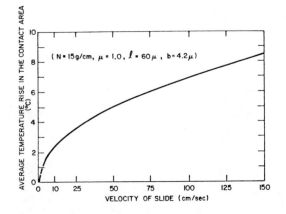

Fig. 5. Average rise in temperature in the contact area as a function of velocity of slide

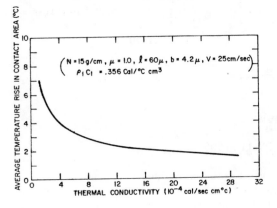

Fig. 6. Variation of average rise in temperature with the thermal conductivity of the plane

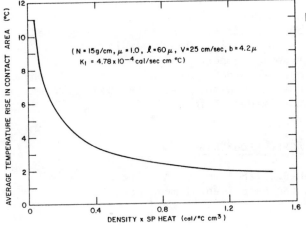

Fig. 7. Variation of the average rise in temperature in the contact area with the product density x specific heat of the plane

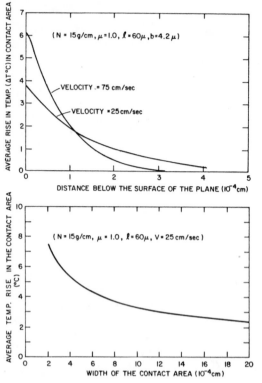

Fig. 8.   Variation of the average rise in temperature within the plane

Fig. 9.   Average rise in temperature in the contact area as a function of the width of the contact area

VI.   CONCLUSIONS

A detailed mathematical analysis has been presented to calculate the rise in temperature due to frictional heating of rectangular and trapezoidal sliders sliding over a plane of finite thickness.   The analysis is applicable to situations where the length of the area of contact is much larger than the width.   The results are expected to hold under a large range of conditions since the heat loss from the exposed surfaces has been accounted for and real geometries are considered.   The analysis has been applied to calculate the rise in temperature of an elastomeric slider, trapezoidal in shape, rubbing against a thin thermal nonconductor solid material coated on a substrate.   A good correlation between theory and experiments is found.

ACKNOWLEDGMENTS

Author wishes to thank D. Fisher and G. Abowitz of Xerox Corporation for providing the experimental measurements used to test this theory.

REFERENCES

1. F.P. Bowden and D. Tabor, The Friction and Lubrication of Solids, Chapters II and III, Oxford University Press, 1958.
2. H. Blok, Proc. Inst. Mech. Engr., 2, 222 (1937).
3. J.C. Jaeger, Journal and Proc. Roy. Soc., N.S.W., 76, 203 (1942).
4. R. Holm, J. Appl. Phys., 19, 361 (1948).
5. F.F. Ling and S.L. Pu, Wear, 7, 23 (1968).
6. P.S. Kounas, A.D. DiMarogonas and G.N. Sandor, Wear, 19, 415 (1972).
7. G.M. Bartenev, V.V. Lavrent'ev and N.A. Konstantinova, Wear, 18, 439 (1971).
8. D. Fisher and G. Abowitz, Private Communication.

DISCUSSION OF PAPER BY G. HARPAVAT

J. J. Bikerman (Case Western Reserve University): The speaker's assumption that the "actual" area of contact was identical with the geometrical area is justified as long as heat transfer is concerned because the resistance of the few gas molecules (between the two solids) to heat conductance is completely negligible compared with the resistance of the slides and the support. But the agreement of the results with the theory does not indicate the absence of adsorption layers.

G. Harpavat: Yes, I would agree with Dr. Bickerman that present the analysis may not indicate an absence of air adsorption layer between the slider and the plane. The presence of an adsorption can be checked by performing an experiment in which the slider is a soft elastomer with a thermal conductivity far better than air. A good agreement between theory and experiments would prove an absence of the adsorption layer.

D. Tabor (Cambridge University): In your analysis you have ignored heat losses by thermal radiation. Presumably, this is not important with polymers, but with metals it could become much more significant.

G. Harpavat: The inclusion of the heat loss by thermal radiation in the analysis makes the differential equation highly nonlinear and impossible to solve. Moreover, the radiation heat loss would be significant only if the rise in temperature is too high, say 200°C or above. Most polymers have a glass transition temperature far below 200°C.

G. Abowitz (Xerox Corporation): A word of caution is in order regarding the experimental measurement of temperatures by fine thermocouples inserted in poor thermal conductors such as elastomers. If the dimension of the thermocouple wire is significant relative to the temperature gradient in which it is inserted, then the meaning of the temperature read by the couple is uncertain. In addition, the metallic wires even if they are very fine may be causing a serious perturbation in the temperatures and heat losses along the wires must be considered as well.

G. Harpavat: Yes, it is a very good point. However, if the thermo- couple is placed away from the steep temperature drop region the measurement error would be small. For elastomers the steep temper- ature drop region is only about 50μ deep. The temperature at the thermocouple end is just a small fraction of the actual temperature in the contact area; therefore, the heat loss from the thermocouple wire would be negligible in most cases.

P. S. Walker (Hospital for Special Surgery): What would be the effect of a liquid surrounding the slider and the plane?

G. Harpavat: The effect of a liquid surrounding the slider and the plane would change the surface heat transfer coefficient for the slider H_2 and the plane H_1. The surface heat transfer coefficients for surfaces immersed in a liquid can be obtained from any standard book on heat transfer. The above analysis can be applied to such applications with proper values of surface heat transfer coefficients.

P. S. Walker: Have you considered the effect of repetitive sliding over the same area of counterface surface?

G. Harpavat: No. We have not considered the effect of repetitive sliding. An exact analysis of the problem of repetitive sliding is much more involved and would require a transient, time varying analysis. However, an approximate estimate of the rise in temperature in such applications can be obtained from our analysis as follows: The analysis is valid for one sliding, and the rise in temperature is directly proportional to the total heat of friction. Therefore, the rise in temperature in N slidings is just N times the rise in temperature as calculated from Eq. 20. This would mean a linear increase in temperature with N. A steady state would be reached eventually. The steady state temperature is such that the convertion heat loss from the heated contact area during the time it is not in contact with the slider is just equal to the heat of friction.

S. Bahadur (Iowa State University): For many reasons, the sliding in conventional friction experiments is performed between a hemispherical indentor and a flat plate. Could you therefore comment on the magnitude of error that would be involved if the results of your pyramidal indentor analysis were to be used for the case of hemispherical indentor?

G. Harpavat: If the hemispherical indentor is a thermal conductor, the error can be quite substantial. Moreover, the geometry of the area of contact is also changed which might affect the temperature distribution. I would expect that for thermally insulator indentors a rough estimate of the rise in temperature could be obtained from the present analysis.

A Laboratory Study of the Friction of Rubber on Ice

E. Southern and R. W. Walker

Malaysian Rubber Producers' Research Association

Welwyn Garden City, England

This paper describes the effect of velocity and temperature on the friction coefficient of both filled and unfilled rubber vulcanizates sliding on smooth ice. It has been shown that the mechanism of the friction of rubber on ice is the same as that on other smooth surfaces under similar low sliding speed conditions and that the maximum friction coefficients are similar. The Williams Landel and Ferry equation is used to superpose curves of the velocity dependence of the friction coefficient at different temperatures to produce a master curve and therefore to demonstrate the viscoelastic nature of the frictional mechanism. The frictional behaviour depends on the condition of the ice track and a tentative explanation for this observation is suggested. The frictional properties of vulcanizates containing various amounts of a reinforcing carbon black filler have been studied. The effect of the carbon black filler on the friction coefficient correlates with changes in the low temperature dynamic properties of the vulcanizate obtained using a torsion pendulum. The effect of contact pressure on the friction coefficient has been investigated using both filled and unfilled vulcanizates

at low sliding velocities. The results
are in accord with a theory which has
been used successfully to describe the
behaviour on other smooth surfaces.
The combined effect of a high contact
pressure of the order of 30 psi
(2 kg cm^{-2}) with sliding speeds up to
40 mph has also received attention and
the results are discussed in terms of
frictional melting of the ice at the
interface.

INTRODUCTION

It is a common experience that the friction of rubbers and of
most other materials on ice is extremely low. Under some circumstances the reasons for the low coefficient of friction may be
ascribed to frictional melting or to pressure melting on the ice.
The friction of rubber on smooth surfaces has been carefully
studied[1,2] but until recently[3,4] little[5-7] scientific work has been
done to investigate the mechanism of the friction of rubber on ice.
This paper describes all the recent work which we have carried out
on the subject and shows that under some circumstances the friction
of rubber on ice may be very high indeed with friction coefficients
well in excess of unity.

EXPERIMENTAL

The apparatus is shown in Figure 1 and consists of circular
turntable containing the ice track and a freely mounted arm holding
the rubber sample. The frictional force generated between the
rubber and ice as the turntable revolves is measured directly by
a force transducer mounted on the arm. The turntable can be rotated
over a wide speed range covering about six decades. The apparatus
excluding motor and gear box was placed in a refrigerator to prevent
frosting of the ice track. The frost tends to form on the coldest
surfaces which are the walls of the refrigerator. The lowest
temperature which could be achieved with this arrangement was
about -30°C but the use of additional cooling coils lowered this
value to -40°C.

The ice was prepared in a similar manner to that used by
Wilkinson[7] and was machined flat in situ using a steel tool mounted
on a carriage. Final polishing of the ice was carried out using a
rubber test sample with a 1 Kgf load and a linear ice velocity of
40 cm s^{-1}. The test sample was a block of rubber 0.7 cm thick and
2.5 cm square with a normal load of 1 Kgf unless stated otherwise.

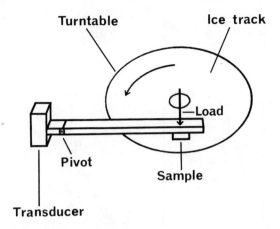

Fig. 1.   Schematic diagram of ice friction machine

Friction coefficients were measured at various speeds within the range 3×10^{-3} to 1 cm s^{-1}.   Higher speeds were not used in order to minimize temperature rises due to frictional heating at the interface between the rubber and the ice.

REPRODUCIBILITY

Initial measurements showed large variations in the friction coefficient from day to day under nominally identical conditions. This lack of reproducibility can be attributed to variability of freshly prepared ice but gradually the results become more consistent if the same ice track is used repeatedly. Once the ice had become stabilized, or conditioned, measurements made at intervals of several weeks were quite reproducible. All the results reported in this paper are for ice surfaces which have been conditioned by repeated testing until reproducible results are obtained.

SPEED AND TEMPERATURE EFFECTS

Measurements of the friction coefficient over a wide range of speeds and temperatures have been made. A typical set of results is shown in Figs. 2 and 3 where it can be seen that increasing the sliding velocity may lead to an increase or a decrease in the friction coefficient depending on the temperature at which the experiment was done. Behaviour of this sort has been reported for the friction of rubber on other smooth surfaces such as glass[1] or polished steel[2]. The individual curves at each temperature may be

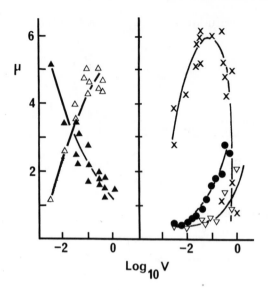

Fig. 2. Friction coefficient as a function of sliding velocity
(cm s^{-1}) at various temperatures for polychloroprene gum.

   ▽ - 1°C,   ● - 5°C,   △ - 15°C,   × - 25°C,   ▲ - 30°C

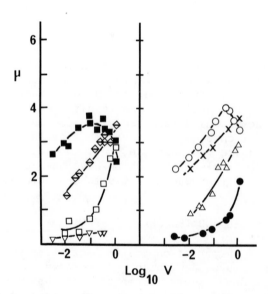

Fig. 3. Friction coefficient as a function of sliding velocity
         (cm s^{-1}) at various temperatures for natural rubber gum.

   ▽ - 1°C,   ● - 5°C,   □ - 10°C,   △ - 15°C
   ⬦ - 20°C   × - 25°C   o - 34°C   ■ - 40°C.

shifted along the abscissa to form a continuous mastercurve as
shown in Figures 4 and 5 choosing the results at one temperature
(-25°C in this case) as a reference. Mastercurves like this have
been obtained for the dynamic properties[8] of rubber in general,
including friction behaviour[1]. The existence of the mastercurve
is taken as good evidence that the process which produces it arises
from the viscoelastic properties of the rubber. The amount, (i.e.
$\log_{10}a_T$) by which each curve is shifted depends on the temperature,
T, at which the measurements were made and the reference temperature,
T_o, according to the Williams, Landel and Ferry equation[9] thus

$$\log_{10}a_T = \frac{A(T-T_o)}{B+(T-T_o)} \qquad (1)$$

where A and B are constants. The shifts, which have been used for
the rubbers tested, are plotted together with the calculated line
from equation 1 in Figure 6 where the excellent agreement is clearly
shown. If the standard reference temperature T_S is selected for
each rubber (i.e. 50°C above the glass transition temperature)
the peak of the master-curve occurs at the same value of $\log_{10}a_T V$
for all rubbers which is additional confirmation of the viscoelastic
behaviour. Although convenient this method of presenting the data
is misleading for practical situations where it is desirable to know
how the friction coefficient varies with speed at a particular
temperature. The data for a temperature of -10°C is plotted in
this way in Figure 7, where it can be seen that natural rubber has
the highest coefficient of friction in the practically important
velocity range (i.e. above 1 mph) whereas polychloroprene has higher
friction at extremely low sliding speeds.

The results obtained on freshly formed ice cannot be shifted
to form mastercurves by the method just described so that it seems
unlikely that the viscoelastic properties of the rubber are responsible for the behaviour. A tentative explanation is that freshly
formed ice is weak structurally and that shearing of the ice surface
is possible. Further experimental evidence which substantiates this
proposal is that the friction coefficient is similar for all rubbers,
as would be expected if the ice is being sheared. The reason for
the change in ice properties is not understood but it may be the
result of a reorientation of the crystal structure. It seems
unlikely that the effect is due to the surface properties only,
since recutting the ice does not affect the ice properties. Melting
of the surface does alter the surface properties and the conditioning
process has to be repeated.

The mastercurves are continuous even near to the peak of the
mastercurve for natural rubber. This contrasts with the behaviour
reported by Grosch[1] for natural rubber sliding on glass when complete
mastercurves near the peak could not be achieved.

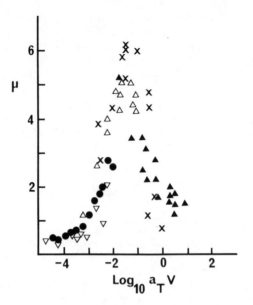

Fig. 4. Master curve for polychloroprene at a reference
temperature T_O = -25°C Data from Figure 2.

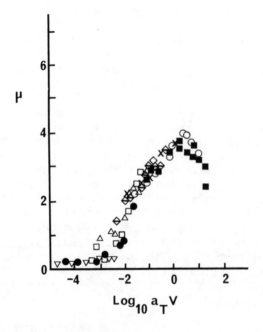

Fig. 5. Master curve for natural rubber at a reference
temperature T_O = -25°C. Data from Figure 3.

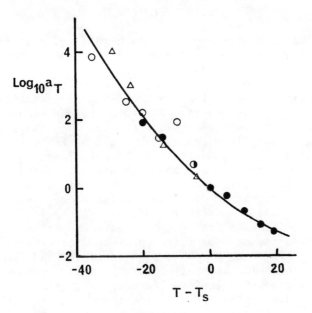

Fig. 6. Horizontal displacements $Log_{10}a_T$ as function of $T-T_S$. Curve is calculated from WLF equation. ● natural rubber (NR), ○ styrene-butadiene rubber (SBR) Δ polychloroprene rubber (CR).

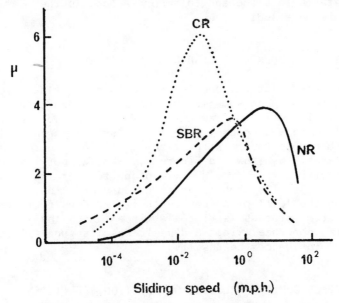

Fig. 7. Master curves of friction coefficient at -10°C of various rubbers showing superiority of natural rubber at practical sliding speeds.

STICK-SLIP BEHAVIOUR

If the friction coefficient is very high (i.e. near to the peak value) stick-slip occurs. The rubber does not slide smoothly but jumps forward in a jerky manner. It has been suggested that this phenomenon occurs only when the friction coefficient is decreasing with increasing velocity thereby producing an instability. Careful observation shows that stick-slip behaviour occurs on ice whenever the friction coefficient is high even when it is increasing with increasing speed. It appears that the adhesion between the rubber and the ice increases with time and this leads to the observed jerky motion. Some improvement can be obtained by using a stiffer force transducer but the rubber sample itself acts as a spring so that the behaviour cannot be completely eliminated. Evidence that the adhesion increases with time is demonstrated by leaving the sample in contact with the ice track for various periods of time and then measuring the maximum force obtained after the track is set in motion. This phenomenon has also been observed for rubber and glass[10]. A possible explanation is that the true contact area between the rubber and the surface increases with time due to the viscous time dependent properties of the rubber. Alternatively sticky substances diffusing to the surface of the rubber may be responsible[10]. On ice it seems that similar considerations are likely to apply. The sliding speed of the sample is equal to the imposed speed of rotation of the ice track only at the instant when the rubber starts to slide and this corresponds to the maximum force which is obtained.

INFLUENCE OF CARBON BLACK FILLER

An important practical consideration is the influence of carbon black fillers and some measurements have been carried out using samples with different amounts of N330 carbon black filler in a polychloroprene compound. The effect of increasing filler content is to reduce the peak friction coefficient on the master-curve considerably as shown in Figure 8. Similar but less dramatic changes are noticeable in the dynamic properties of the rubber as shown in Figure 9. The changes in friction coefficient produced by increasing the level of carbon black filler show further evidence that the friction process is viscoelastic although the lack of complete correspondence with the dynamic results suggests that other factors influence the results.

EFFECT OF CONTACT PRESSURE ON FRICTION COEFFICIENT

Classically the friction coefficient is independent of the normal reaction and hence the superficial contact area but Schallamach[11] has shown that this is not the case for rubber on

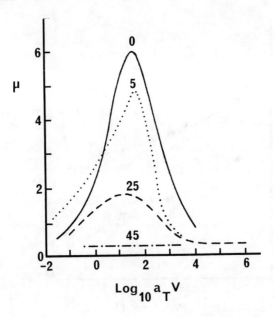

Fig. 8. Master curves for polychloroprene gum. Number refers
to parts of N330 carbon black in 100 parts of gum rubber.

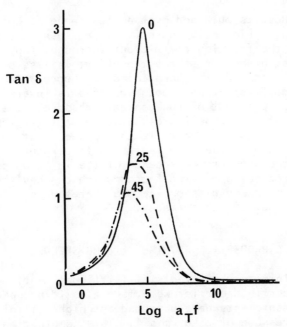

Fig. 9. Torsion pendulum data for polychloroprene gum showing phase
angle, δ, as a function of frequency, f, (ref. 6). Numbers on curves
refer to parts of N330 carbon black in 100 parts of gum rubber.

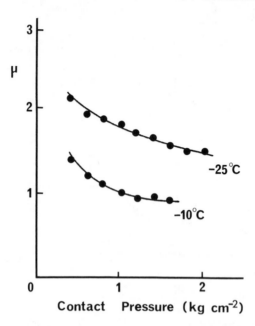

Fig. 10. Variation of friction coefficient with contact
pressure for natural rubber at different temperatures.

smooth surfaces. Results obtained over a wide range of contact
pressures are shown in Figure 10 where it can be seen that as the
pressure increases the friction coefficient decreases rapidly at
low pressures but at higher pressures the effect is less pronounced.
This behaviour is in accord with Schallamach's theory which predicts
that the friction coefficient on smooth surfaces is inversely
proportional to the cube root of the contact pressure. The results
for natural rubber at two temperatures, plotted on logarithmic
scales, are shown in Figure 11 where it can be seen that good
agreement is obtained with the theory at both temperatures. It
would appear that the pressure dependence of the friction coefficient
of rubber on ice produces a scaling factor which changes the ordinate
of the mastercurve. The results on ice are therefore similar to
those obtained on other smooth surfaces.

FRICTIONAL PROPERTIES AT HIGH SLIDING SPEEDS

In order to minimize the effects of frictional heating all
the measurements described earlier were made at sliding speeds below
1 cm s^{-1}. Some measurements have been made at higher sliding speeds
and at a contact pressure of 2 kg cm s^{-2}. This combination covers
the practical range over which tyres operate although the samples
used are unfilled compounds and not therefore tyre compounds. The
results, shown in Figure 12, indicate that the friction coefficient

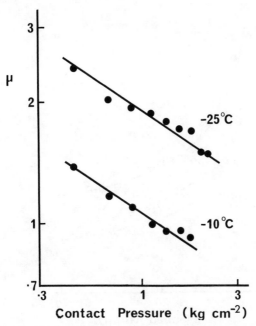

μ

-25°C

-10°C

Contact Pressure (kg cm^{-2})

Fig. 11. Data from Figure 10 plotted on logarithmic scales. Lines have slope of -1/3 in accordance with the theory.

does not change dramatically at sliding speeds over about 10 mph but the friction coefficient is very low - in the region of 0.2 to 0.3. These low results are probably due to frictional heating at the interface between the rubber and ice thereby producing lubrication from melting ice[12]. There is no difference between freshly formed ice and conditioned ice in these tests and this may be taken as evidence that there is melting of the ice thereby preventing conditioning from taking place.

CONCLUSIONS

At low sliding speeds the friction of rubber on ice arises from the viscoelastic properties of the rubber. The behaviour is similar to that obtained on other smooth surfaces and results can be shifted to form mastercurves according to the WLF transform. A rubber containing carbon black filler also exhibits viscoelastic behaviour although the changes in friction coefficient are not completely analogous to changes in dynamic properties. The effect of increasing load or contact pressure is to decrease the friction coefficient in accordance with a theory based on the change in true contact area as a function of contact pressure. At high sliding speeds the friction coefficient falls considerably due to frictional melting of the ice and the effect of conditioning the ice is lost.

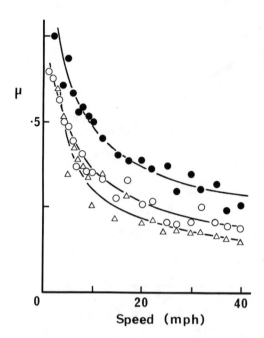

Fig. 12. Speed dependence of the friction coefficient for gum compounds at -15°C and 2 kg/cm² contact pressure. Symbols as in Figure 6.

ACKNOWLEDGEMENT

This work forms part of the Research Programme of the Malaysian Rubber Producers' Research Association.

REFERENCES

1. K. A. Grosch, Proc. Roy. Soc., A274, 21 (1963).
2. K. A. Grosch, thesis, London University (1963).
3. E. Southern and R. W. Walker, Nature Phys. Sci., 237, 142 (1972).
4. W. Gnorich and K. A. Grosch, J. Inst. Rubb. Ind., 6, 192 (1972).
5. F. S. Conant, J. L. Dum, and C. M. Cox, Rubb. Chem. Technol., 22, 863 (1949).
6. P. M. Pfalzner, Canad. J. Res., 28, 468 (1950).
7. C. S. Wilkinson, Ind. Rubb. World, 128, 475, (1953).
8. J. D. Ferry, Viscoelastic Properties of Polymers (second edition), Wiley, (1970).
9. M. L. Williams, R. F. Landel, and J. D. Ferry, J. Amer. Chem. Soc., 77, 3701 (1955).

10. A. Schallamach, Unpublished work.
11. A. Schallamach, Proc. Phys. Soc., <u>65</u>, 657 (1952).
12. F. P. Bowden and D. Tabor, <u>The Friction and Lubrication of Solids</u>, Vol. 2, Ch. 9 (Oxford University Press) (1964).

DISCUSSION OF PAPER BY E. SOUTHERN

J. Harrison (Goodyear Tire and Rubber Co.): Continuing the comparison
of the coefficients of friction on ice between chemically different
types of rubbers, have observations been made which would compare
the effect of more or less methyl group substitutions than natural
rubber possesses, e.g. butadiene versus di-methyl butadiene rubber?

E. Southern: No, we have only studied the commonly used rubbers.
Our results show that natural rubber has a glass transition temperature such that peak friction coefficients are achieved at practical
sliding speeds when the temperature is around 0 to -15°C.

D. Tabor (Cambridge University): I would like to ask a simple
question about frictional heating. Is it possible that if you know
that friction should vary with speed on your master curve, you can
deduce what the interfacial temperature is, and then compare the
friction values based on speed.

E. Southern: We have tried to do this but because of the lack of
precision in the results on the mastercurve it is not possible to
draw useful conclusions. A further complication is the effect of
frictional melting which will cause partial lubrication in the
contact area. We have a simple theory based on Jaeger's solution
of the problem of heat generated during sliding friction but we
have not carried out any experiments yet to verify it.

J. J. Bikerman (Case Western Reserve University): But surely your
natural rubber was vulcanized? If it was, it was polar.

E. Southern: Yes, we used natural rubber vulcanized in the usual
way with sulphur but this does not make the natural rubber as polar
as polychloroprene. The more polar nature of polychloroprene is
easily demonstrated by its lower equilibrium swelling in non-polar
hydrocarbons such as decane.

PART THREE

Characterization and Modification of Polymer Surfaces

Introductory Remarks

Douglas V. Keller, Jr.

Professor of Materials Science

Syracuse University, 409 Link Hall, Syracuse, N. Y.

During the summer of 1962 ten surface chemists and physicists visited Syracuse with the intent of defining a "metal surface" and establishing a method of its characterization. The only agreement that could be achieved during the ensuing three days was on the time to adjoin the meeting. During the following ten years and a few hundred thousand research papers later, agreement on this matter has been achieved and this was probably due to the fact that a standard state was obtained. That is, a pure, uncontaminated, metal-vacuum interface was established and characterized by the very important tools of low energy electron diffraction (LEED), field ion microscopy (FIM) and the like. Once this point was established, adsorbed film kinetics and the surface chemistry of metallic systems could be placed on a more quantitative and orderly basis. The results of these efforts were then applied to metallic adhesion, friction, wear and lubrication which permitted a more quantitative understanding of these to be established.

If we assume that the development of a standard state, i.e. ideal metal-vacuum interface, was in fact the key to the advancements in metallic contact analysis we should also expect that this will also be the case in the polymer, or solid organic, friction or adhesion analysis. The consequence is that polymer surface characterization under various environments becomes the most important issue at hand. The achievement of the standard surface state in polymer or solid organic system will be most difficult due to the relatively weak intermolecular bonding forces and the normal existance of a wide range of impurities within the material itself. Ambient vacuum conditions are required for most of the physics oriented surface characterization techniques, e.g. LEED, FIM, Auger electron spectroscopy and under these conditions the surface can be modified by

239

diffusion processes to an extent that it no longer represents the system under characterization. The change is as irreversible as that of oxide growth on some metals. Argon ion sputtering, which was such a useful tool in inorganic or metal systems in the development of the surface standard will, if applied to organic surfaces, cause untold chemical modifications through molecular cracking and cross-link generation. Again the surface will not be representative of that which we wish to characterize. The answers to these problems will take time as was the case in metal systems; but until that point is achieved we must assume that in all probability the molecular surface film of a polymer is not a simple representative case of the general structure of the polymer. This assumption relative to metals was made prior to 1960 which confounded the analysis of adhesion and friction phenomena so badly that there are still a few people today who look at a bright shiny silver spoon and will not believe that there is a substantial oxide layer present on the metal surface. A case in point will be reported by our group in the very near future in which the adhesion of miron-sized coal dust, a complex fused-ring polymer, and calcite to bulk coal will be reported. Some forms of freshly cleaved coal are essentially hydrophobic yet as surface oxidation takes place over a period of time acid sites are established on the surface and the same coal becomes hydrophylic. The adhesion study indicated that the calcite dust adhesion to freshly cleaved bulk coal is very sensitive to humidity due to the existance of a water film on the calcite, yet the adhesion of coal dust to bulk coal was not influenced by humidity variation.

The lectures of this session demonstrate that the surface characterization of polymers is well underway. The ESCA, electron spectroscopy for chemical analysis, technique is uniquely promising in that it will permit molecular structure analysis of the surface films which is critical for polymeric systems.

The Application of ESCA to Studies of Structure and Bonding in Polymers

D. T. Clark

Department of Chemistry

University of Durham, Durham City, U.K.

The development of ESCA as a major spectroscopic tool for investigating aspects of structure and bonding pertaining to surface, subsurface and bulk polymer systems is described. After an initial discussion of the advantages of the technique, basic instrumentation and types of information available, a review of most of the current areas of applicability to polymers is given. The potential of the technique for the investigations of friction and wear phenomena are considered.

PLENARY LECTURE

THE APPLICATION OF ESCA TO STUDIES
OF STRUCTURE AND BONDING IN POLYMERS

I.   INTRODUCTION

Since solids communicate with the rest of the universe by way
of their surfaces, it is a truism that structure and bonding (in
the chemical sense) of the surface of solids is of fundamental
importance in any detailed discussion at the molecular level of
friction and wear in polymers. Despite the obvious importance of
the nature of the surface and immediate sub-surface of polymer
systems, there are relatively few techniques currently available
for routinely delineating the aspects of structure and bonding
which are of crucial importance in determining many of the physical,
chemical and mechanical properties of polymers or polymer coated
surfaces. This is more particularly the case when the horizon is
broadened to encompass not only 'academic' studies under relatively
idealized conditions but real situations corresponding to working
conditions. For example, it may be important to study friction
and wear of polymer coatings subject to atmospheric conditions such
that ageing, weathering, etc. are of importance. Techniques capable
of handling such widely differing situations non-destructively (as
far as the samples are concerned) have to this date been conspicuous
by their absence.

Over the past few years, we have been applying the relatively
newly developed technique of Electron Spectroscopy for Chemical
Applications (ESCA) to studies of structure and bonding across a
broad front encompassing organic, inorganic and polymeric systems[1].
These studies together with complementary theoretical analysis have
demonstrated that ESCA is an extremely powerful tool for investigations of structure and bonding with an information content
per spectrum unsurpassed by any other spectroscopic technique. Of

particular interest as far as this symposium is concerned, is the
application of ESCA to polymer studies which has revealed the great
potential of the technique in this area. Although not dealing
specifically with problems of polymer friction and wear, these studies
indicate that ESCA is potentially a very powerful technique for such
studies and this can most readily be appreciated from a summary of the
current 'state of the art'. The applications outlined below derive
almost exclusively from my own research program at Durham. I make
no apologies for this bias since it may readily be justified. In
such a rapidly expanding field and in an article of this size, it is
not a feasible proposition to include a comprehensive literature
survey. Rather the objective should be twofold. Firstly, to
delineate the areas of applicability of the technique and hence
give the uninitiated some idea of the sort of problem to which ESCA
might readily be applied particularly in the field of polymer
friction and wear. Secondly, to indicate some of the likely 'growth
areas' over the next few years. The application of ESCA to structure
and bonding in polymers has largely been pioneered at Durham, and
therefore my own material is readily to hand for the preparation of
slides, etc. and is illustrative in the areas of applicability of
ESCA which have so far been delineated.

It is clear from the literature classification of friction
(usually classified as a branch of Physics or of Mechanical
Engineering) and wear (for metallic systems often considered to be
part of Metallurgy) that an ACS Symposium on polymer friction and
wear encompasses an enormous range of traditional disciplines.
This fact in itself is particularly encouraging since history teaches
us that in such situations the interaction of ideas from workers of
widely differing backgrounds often leads to spectacular advances.
One drawback, however, of a meeting of specialists from differing
backgrounds is the obvious possibility of lack of communication.
To obviate this, however, before outlining the current areas of
research in ESCA applied to polymers a brief outline of ESCA as a
technique will be given.

II.   FUNDAMENTALS OF ESCA

1.   Introduction

In common with most other spectroscopic methods, ESCA is a
technique originally developed by physicists and now gradually being
taken over by chemists to be developed to its full potential as a
tool for investigating structure and bonding. The technique has
largely been developed by Professor Kai Siegbahn[2] and his
collaborators at the University of Uppsala over the past twenty years
or so, and much of the early work has been extensively documented in
the 'first ESCA book'[2a]. It is only within the last decade, however,

that the potential of the technique has been revealed with the development of spectrometers of sufficient resolution and sensitivity. The field has been opened to chemists with the advent of commercially produced instruments (at the last count there were ten instrument manufacturers). In addition to the aesthetically pleasing designation as 'Electron Spectroscopy for Chemical Applications'* originally coined by Siegbahn, the technique is also variously known as X-ray Photoelectron Spectroscopy (XPS), High Energy Photoelectron Spectroscopy (HEPS), Induced Electron Emission Spectroscopy (IEES), and Photoelectron Spectroscopy of the Inner Shell (PESIS).

The principal advantages of the technique may be summarized as follows.

(i) The sample may be solid, liquid or gas (it is as easy to study a high molecular weight polymer as it is to study a gas) and the technique is essentially non-destructive. One noticeable exception to this generalization is poly(thiocarbonyl fluoride) which depolymerizes rather rapidly under X-irradiation.

(ii) The sample requirement is modest, in favorable cases 1 mg. of solid, 0.1 μl of liquid or 0.5 cc. of gas (at STP). These sample requirements are based on the minimum readily handled with conventional techniques. As will become apparent, the sensitivity of the technique is such that a fraction of a monolayer coverage may be detected.

(iii) The technique has high sensitivity and is independent of the spin properties of any nucleus and is applicable in principle to any element of the periodic table. H and He are exceptions being the only elements for which the core levels are also the valence levels.

(iv) The information it gives is directly related to the electronic structure of a molecule, and the theoretical interpretation is relatively straightforward.

(v) Information can be obtained on both the core and valence energy levels of molecules.

These particular advantages of ESCA as a technique make it eminently suitable for the study of polymers.

2. Properties of Core Orbitals

A clearer understanding of ESCA as a technique is obtained with

* Originally designated '... for Chemical Analysis'

some knowledge of the properties of core orbitals. The material
presented below provides a convenient introduction and is particularly
apposite since it was from a research program involving non-empirical
quantum mechanical calculations of cross sections through potential
energy surfaces for simple reactions that this author proceeded to
an experimental interest in ESCA as a technique.

Traditionally, chemists have discussed the electronic structure
of molecules, dealing only with the valence electrons and neglecting
inner shell or core electrons. The reason for this being:

(i)   core electrons are not explicitly involved in bonding (although
most of the total energy of a molecule resides in the core electrons);

(ii) it is only in the past ten years that sufficient computing
capability has become available to allow non-empirical quantum
mechanical calculations on molecules in which the core electrons
are explicitly considered.

It has become clear, however, that although core electrons are not
involved in bonding, the core energy levels of a molecule encode a
considerable amount of information concerning structure and bonding.
This is illustrated by examples of work carried out at Durham over
the past few years.

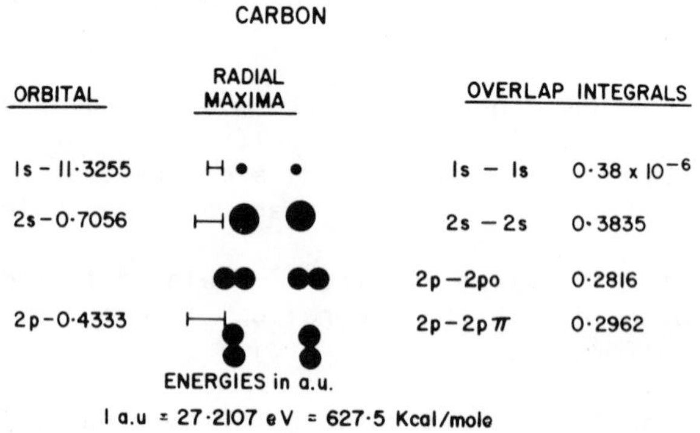

CARBON

ORBITAL	RADIAL MAXIMA	OVERLAP INTEGRALS	
1s – 11·3255		1s – 1s	$0·38 \times 10^{-6}$
2s – 0·7056		2s – 2s	0·3835
		2p – 2pσ	0·2816
2p – 0·4333		2p – 2pπ	0·2962

ENERGIES in a.u.

1 a.u = 27·2107 eV = 627·5 Kcal/mole

Fig. 1.   Orbital energy diagram for carbon

Fig. 1 shows the orbital energies, total energy and radial
maxima for the carbon atom calculated non-empirically in a Gaussian
basis set. Considering first the orbital energies, it is clear
that the 1s (core) level is very much lower in energy than the 2s
and 2p (valence) levels. From the radial maxima, the 1s orbital is
confined to a region in the immediate vicinity of the nucleus whereas
the valence orbitals are much 'larger'. Since the core orbital is
essentially localized around the nucleus, overlaps with orbitals
on adjacent atoms are negligible. Shown in Fig. 1 are overlap

integrals between orbitals on two adjacent carbon atoms with a bond
length of 1.39Å. The negligible value for the overlap integral
involving the core orbitals is one reason why they are not involved
in bonding.

It has tacitly been assumed by chemists in the past that in
discussing the transformation of one molecule into another the
energies of the inner shell or core electrons could be taken as
constant and effectively ignored. A particularly interesting
transformation which illustrates that this is not the case is the
transformation of cyclopropyl to allyl cation which occurs in a
disrotary fashion. The relevant energy levels are shown in Fig. 2.
As far as the carbon atoms are concerned, there are fairly drastic
charge migrations involved in this transformation. Thus, C_1 which
carries a substantial positive charge in cyclopropyl cation becomes
C_2 with a substantial negative charge in allyl cation. As a result,
the C_{1s} orbital energy changes from -11.7122 a.u. to -11.5613 a.u.

The change in energy of this particular core level is in fact
almost three times the total energy change in the transformation.†
The almost degenerate pair of C_{1s} orbitals for C_2 and C_3 in
cyclopropyl cation change in energy by 0.044 a.u. in the
transformation to allyl cation. Inspection of the charge
distributions and core energy levels reveals that a more negative
energy (i.e. increased binding energy††) is associated with an
increased <u>positive</u> charge on an atom. The charge on a given atom
is determined by the valence electron distribution, and the core
levels of a molecule reflect this. Clearly, although the core levels
are not involved in bonding, they are a sensitive function of the
electronic environment about a given atom.

† It should be remembered that the total energy of a molecule may
be expressed as

$$E = \sum_{r=a}^{k} \varepsilon_r - \sum_{pairs\ rs} (2J_{rs} - K_{rs}) + V_{nn},$$

where the ε_r are the occupied orbital energies, J_{rs} and K_{rs} are
coulomb and exchange repulsion integrals and V_{nn} is the nuclear
repulsion energy.

†† Binding energy is defined here as the energy required to remove
the electron in a given orbital to infinity and may be equated to
the -ve of the orbital energy (Koopmans' Theorem). The
approximations involved in this are discussed elsewhere.

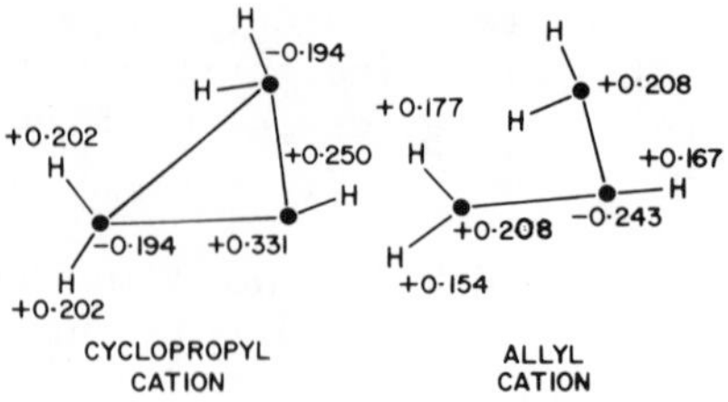

	Cyclopropyl cation	Allyl cation
CORE LEVELS	−11·5801	−11·5613
	−11·5806	−11·6237
	−11·7122	−11·6238
TOTAL ENERGIES	−115·0885	−115·1447

ΔE = −0·0562 a.u. = 35·27 Kcal/mole

Fig. 2. Charge distributions and orbital energies for cyclopropyl and allyl cations.

The different electronic environment about C_1 and $C_2(C_3)$ in cyclopropyl cation is therefore reflected in the 'shift' in C_{1s} binding energies of ~4.1 eV.

Core electrons are localized in space close to a nucleus and this is reflected in the fact that the binding energies are characteristic of a given element. This is emphasized on consideration of the approximate core binding energies for first and second row elements given in Fig. 3. Clearly on the basis of their core binding energies, it is an easy matter to distinguish say sulphur from chlorine.

To summarize:

Core orbitals are essentially localized on atoms, their energies are characteristic for a given element and are sensitive to the electronic environment of an atom. Thus, for a given core level of a given element whilst the absolute binding energy for that level is characteristic for the element, differences in electronic environment of a given atom in a molecule give rise to a small range of binding energies (i.e. 'shifts' in binding energies are apparent) often characteristic of a particular structural feature.

	Li	Be	B	C	N	O	F	Ne
1s	55	111	188	284	399	532	686	867

	Na	Mg	Al	Si	P	S	Cl	Ar
1s	1072	1305	1560	1839	2149	2472	2823	3203
2s	63	89	118	149	189	229	270	320
$2p_{1/2}$			74	100	136	165	202	247
	31	52						
$2p_{3/2}$			73	99	135	164	200	245

Fig. 3.   Approximate core binding energies for first and second row elements (in eV).

3.   The ESCA Experiment

ESCA involves the measurement of binding energies of electrons ejected by interactions of a molecule with a monoenergetic beam of soft X-rays.   For a variety of reasons, the most commonly employed X-ray sources are $AlK\alpha_{1,2}$ and $MgK\alpha_{1,2}$ with corresponding photon energies of 1486.6 eV and 1253.7 eV respectively.   In principle all electrons, from the core to the valence levels can be studied and in this respect the technique differs from u.v. photoelectron spectroscopy (UPS) in which only the lower energy valence levels can be studied.   The basic processes involved in ESCA are shown in Fig. 4.

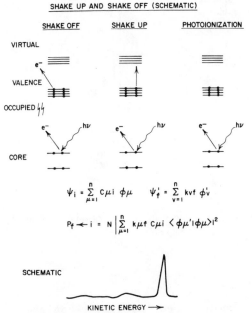

$$\psi_i = \sum_{\mu=1}^{n} c_{\mu i}\, \phi_\mu \qquad \psi_f' = \sum_{v=1}^{n} k_{vf}\, \phi_v'$$

$$P_{f \leftarrow i} = N \left| \sum_{\mu=1}^{n} k_{\mu f}\, c_{\mu i}\, \langle \phi_{\mu'} | \phi_\mu \rangle \right|^2$$

Fig. 4.   Schematic illustrating photoionization, shake up and shake off processes.

The removal of a core electron (which is almost completely shielding as far as the valence electrons are concerned) is accompanied by reorganization of the valence electrons in response to the effective increase in nuclear charge. This perturbation gives rise to a finite probability for photoionization to be accompanied by simultaneous excitation of a valence electron from an occupied to an unoccupied level (shake up) or ionization of a valence electron (shake off). These processes giving rise to satellites to the low kinetic energy side of the main photo-ionization peak, follow monopole selection rules and may well be of considerable importance in the future in elucidating particular aspects of structure and bonding in polymer systems.

De-excitation of the hole state can occur (Fig. 5) via both fluorescence and Auger processes, for elements of low atom number the latter in fact being the more probable.

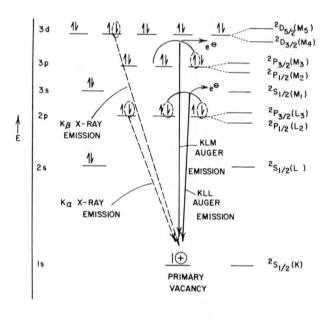

Fig. 5.   Fluorescence and Auger processes for de-excitation of core hole states.

It should be perhaps emphasized that the lifetimes of the hole states involved in ESCA are typically in the range $10^{-12} - 10^{-17}$ secs. The basic experimental setup for ESCA is shown in Fig. 6.

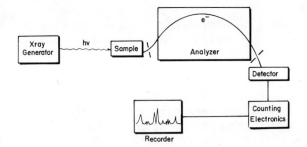

Fig. 6.   Schematic of basic experimental setup for ESCA.

The components are:

X-ray generator. The most commonly used X-ray sources are $MgK\alpha_{1,2}$
and $AlK\alpha_{1,2}$ with photon energies (and linewidths) of 1253.7 eV
(~0.7 eV) and 1486.6 eV (~1.0 eV) respectively. Harder X-ray
sources, e.g. $CrK\alpha_1$ 5414.7 eV and $CuK\alpha_1$ 8047.8 eV have much larger
inherent linewidths and thus give lower resolution. If this is
not important, however, the use of sources with widely differing
photon energies can be very useful for studying escape depth
dependence (see later). In many cases as will become apparent,
the inherent width of the excitating radiation provides the dominant
contribution to overall linewidth.

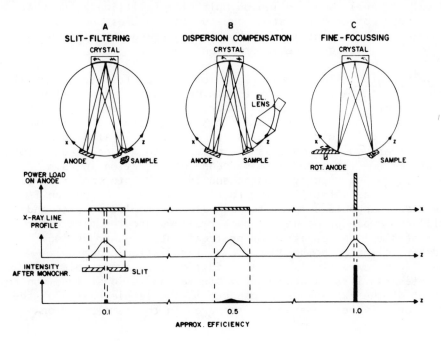

Fig. 7.   Basic schemes for X-ray monochromatization.

For AlK$\alpha_{1,2}$ radiation, the lattice spacing for quartz is such that monochromatization may be achieved by dispersion from suitably bent quartz discs. The three variants which have been proposed are shown in Fig. 7[3]. The first two namely slit filtering and dispersion compensation have been commercially exploited. The most recently developed, however, is the elegant fine focussing X-ray monochromatization scheme of Siegbahn and coworkers[3]. As will become apparent, composite linewidths for individual C$_{1s}$ levels obtained with an unmonochromatized MgK$\alpha_{1,2}$ source are typically > 1.0 eV[+] whereas with a monochromatized X-ray source linewidths can typically be ~0.5 eV with concomitant improvement in resolution. A further advantage occurring from X-ray monochromatization is the removal of satellites (K$\alpha_{3,4}$ Kβ etc.) and bremsstrahlen (white radiation) leading to cleaner cut spectra with much improved signal/background.

With MgK$\alpha_{1,2}$ and AlK$\alpha_{1,2}$ photon sources, there is sufficient energy to study the 1s and valence levels of first row elements, the 2s,2p and valence levels of second row elements and so on. There is no particular virtue in studying the most tightly bound core levels of a given element (e.g. say the 1s level of gold), since this may well have a very large natural linewidth (see later) and the required higher energy photon source inevitably has a larger linewidth. Information concerning the valence electron distribution is encoded in all core levels (the information content may differ from level to level, see section on multiplet splittings), so that the core level selected for study depends on the cross section for photoionization, inherent linewidth, binding energy, photon energy and photon linewidth and the information required.

Sample region. The sample region of the spectrometer is usually separated from the X-ray tube by a thin metal window (typically 3/10 thou Al or better, Be) which ensures that electrons from the gun, used to excite the characteristic X irradiation, do not enter the sample region. Polymer samples are conveniently studied as films or powders mounted on a sample probe which may be taken into the spectrometer from atmosphere by means of insertion locks. Samples may thus be readily mounted and inserted into the spectrometer greatly facilitating routine analyses. Provision is usually made to enable samples to be heated or cooled in situ and an ancillary sample preparation chamber allows greater flexibility in terms of sample preparation or pretreatment (e.g. argon ion bombardment, electron bombardment, u.v. irradiation, chemical treatment (e.g. oxidation), etc.). Addition of a quadrupole mass spectrometer facilitates degradation studies and allows close control to be kept of the extraneous atmosphere in the sample region of the spectrometer. Typical operating pressures in the sample region

[+] Defined as Full Width at Half Maximum (FWHM)

would be < 10^{-8} Torr. The sample area irradiated might typically
be 5 mm x 5 mm.

   Analyzer. The analyzer must typically have a[†] resolution of
something approaching one part in 10^4. In the precise energy analysis
of the photoelectrons, therefore, two types of analyzers have mainly
been used (Fig. 8).

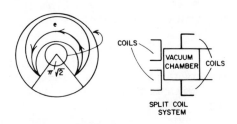

MAGNETIC

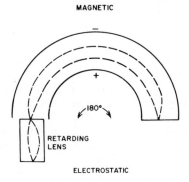

ELECTROSTATIC

Fig. 8.   Schematic of magnetic and electrostatic analyzers
          used in ESCA spectrometers.

   (i) Magnetic. These are largely of the double focussing type
and are generally made from brass or aluminum typically with a 30 cm.
radius. Double focussing is provided by an inhomogeneous magnetic
field produced by a set of four cylindrical coils placed about the
electron trajectory. The chief advantage of a magnetic analyzer
is relative ease of construction, the major disadvantage, the
requirement to eliminate stray magnetic fields by employing
necessarily bulky Helmholtz coils.

[†] The main barrier to the development of the technique has been the
    design of analyzers of sufficiently high resolution and luminosity.

(ii) Electrostatic. Most are based on the hemispherical double focussing design first described by Purcell as long ago as 1938. By employing a retarding field on the electrons before they enter the analyzer, the dimensions of the latter may be considerably reduced and μ metal may also be used for screening. Although more difficult to construct, therefore, spectrometers employing electrostatic analyzers can be made more compact. In fact, the majority of the commercial instruments available are all of this basic design. Typical dimensions would be 25 cms. mean diameter for the hemispheres which are usually constructed of stainless steel or gold-plated aluminum or glass.

(iii) Detector. The minute electron currents involved means that detection is via counting and most spectrometers employ channel electron multipliers. With most designs of double focussing analyzers, their focal plane properties may be exploited by incorporating multi-channel detectors which can give spectacular increases in the rate of data acquisition, and this together with the development of X-ray monochromators is the most important development in commercial instruments in the short term. The output from the multiplier is then amplified and fed to counting electronics.

(iv) Scan and Readout. There are basically two ways of generating a spectrum either continuous or step scan. In the continuous mode of operation, the field (either electric or magnetic) is increased continuously and the signal from the detector is continuously monitored by a rate meter. If the signal to background and overall count rates are sufficiently high, this is the routine way to obtain a spectrum which may be plotted out directly on an XY recorder. Alternatively, the field may be incremented in small steps and at each setting either a fixed number of counts may be timed or a count can be made for a fixed length of time. Where signal to background is poor then this is the method of choice. It is advisable to have both wide and narrow scan facilities available, the former for carrying out preliminary searches and the latter for detailed study.

4. Processes Involved in ESCA

The fundamental processes involved in ESCA are:

(i) Photoionization: $A + h\nu_1 \longrightarrow (A^+)^* + e_1^-$

(ii) Electronic relaxation by either

(a) X-ray emission: $(A^+)^* \longrightarrow A^+ + h\nu_2$

or (b) Auger process: $(A^+)* \longrightarrow A^{++} + e_2^-$

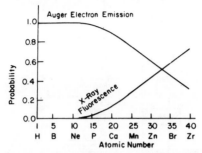

PROBABILITY OF AUGER ELECTRON EMISSION AND X-RAY
FLUORESCENCE AS FUNCTION OF ATOMIC NUMBER

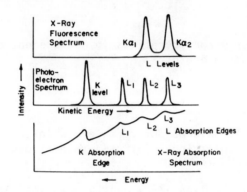

Fig. 9.   Schematic of Auger and fluorescence probabilities
and simulated X-ray fluorescence X-ray absorption
and photoelectron spectra for the same elements.

The relative probabilities for electronic relaxation by X-ray
emission and the Auger process depend on the atomic number of the
element concerned and this is outlined in Fig. 9. Since the
spectrometer is set up to detect and measure the energies of
electrons expelled from the sample, both the photoelectron and
Auger spectra may be studied.

For photo ejection of a core electron in a solid sample of an
insulator (in electrical contact with the spectrometer) the energy
considerations in the measurements of the electron binding energies
are shown in Fig. 10. The reference is the Fermi Level and if
the work function of the spectrometer is known then the absolute
binding energy of a given level may be calculated. This will
differ from the absolute binding energy defined with respect to the
vacuum level by approximately the work function of the sample.

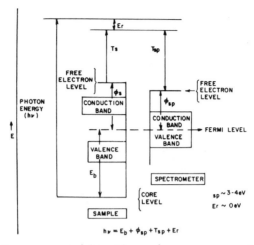

Fig. 10.   Energy considerations in measurement of electron
           binding energies.

5. Linewidths

The measured linewidths for core levels (after taking into
account spin orbit splittings if these are not resolved) may be
expressed as

$$(\Delta E_m)^2 \;=\; (\Delta E_x)^2 \;+\; (\Delta E_s)^2 \;+\; (\Delta E_{C_1})^2,$$

where: ΔE_m is the measured width at half height, the so called
           full width at half maximum (FWHM),
         ΔE_x is the FWHM of the photon source,
         ΔE_s is the contribution to the FWHM due to the spectrometer
           (i.e. analyzer),
         ΔE_{C_1} is the natural width of the core level under invest-
           igation. (For solids, this includes solid state
           effects not directly associated with the lifetime
           of the hole state but rather with slightly differing
           binding energy due to differences in lattice environment.)

It has previously been pointed out that $MgK\alpha_{1,2}$ and $AlK\alpha_{1,2}$
are the most useful photon source from the standpoint of keeping
the contribution of ΔE_x to the total linewidth small. With well
designed magnetic or electrostatic analyzers, the contribution ΔE_s
can be reduced to negligible proportions so that the major limiting
factors in terms of resolution are photon linewidths (which may be
reduced by monochromatization) and the inherent width of the level
itself. (For solids in which longer range interactions are important,
e.g. ionic lattices or hydrogen bonded covalent solids, solid state
effects can contribute to the overall linewidths.)

Some examples of natural linewidths (ΔE_{C_1}) derived from X-ray spectroscopic studies are given in Table 1. The uncertainty principle in the form

$$\Delta E \cdot \Delta t \geq \frac{h}{4\pi}$$

shows that for a hole state lifetime of 6.6×10^{-16} sec. the linewidth, i.e. uncertainty in the energy of the state is ~ 1 eV.

Table 1

APPROXIMATE NATURAL WIDTHS OF SOME CORE LEVELS (in eV)

Level		Atom							
		S	A	Ti	Mn	Cu	Mo	Ag	Au
	1s	0.35	0.5	0.8	1.05	1.5	5.0	7.5	54
	$2p_{3/2}$	0.10		0.25	0.35	0.5	1.7	2.2	4.4
Radiative widths	1s	0.04	0.07	0.2	0.33	0.65	3.6	6.0	50
Fluorescence yields	1s	0.1	0.14	0.22	0.31	0.43	0.72	0.8	0.93

L.G. Parratt, Rev. Mod. Phys., 31, 616 (1959)

It is evident from Table 1 that there are large variations in natural linewidths both for different levels of the same element and for the same levels of different elements. These reflect differences in lifetimes of the hole state. From Fig. 9, it is clear that the lifetime is a composite of radiative (fluorescence) and non-radiative (Auger) contributions, the importance of the former increasing with atomic number (approximately as Z^4). This is clearly shown in Table 1. This emphasizes the fact that there is no particular virtue in studying the innermost core level as has been previously pointed out. For gold for example, the 1s level has a halfwidth of ~54 eV so that even if a monochromatic X-ray source with the requisite photon energy were available any subtle chemical shift effects we might wish to investigate would be swamped.

Typically, the lifetimes of the core hole states involved in ESCA are in the range 10^{-14} ~ 10^{-17} secs. emphasizing the extremely short time scales involved in ESCA compared with molecular vibration and the process may fairly be called sudden with respect to the nuclear (but not electric) motions.

6. Simple Examples Illustrating Points Discussed In (1-5)

A typical wide scale ESCA spectrum (i.e. plot of number of electrons of given kinetic energy arriving at the detector in unit time) is shown in Fig. 11 for a PTFE sample.

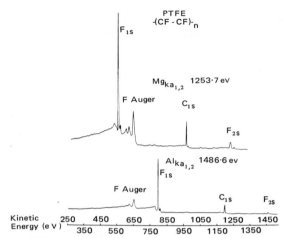

Fig. 11. Wide scan ESCA spectra for PTFE samples studied
          with MgKα₁,₂ and AlKα₁,₂ radiation.

The 'sharp' photoionization peaks due to F_{1s} and C_{1s} levels
are readily identified by their characteristic binding energies.
The group of three rather broader peaks whose KE are independent
of the photon source are readily identified as Auger peaks arising
from de-excitation of the F_{1s} hole states. As will become apparent,
one of the most important attributes of ESCA as a technique is the
possibility of providing information relating to surface, subsurface
or essentially bulk properties, and this depends on differences in
escape depths for photoemitted (or Auger) electrons corresponding to
different kinetic energies. For the most commonly used soft X-ray
sources, the penetration of the incident radiation for a solid sample
is typically in excess of 1000Å (dependent on the angle of incidence
etc.). However, the number of photoemitted electrons contributing
to the elastic peak (i.e. corresponding to no energy loss) is
determined by the mean free path of the electron in the solid.
Thus, in general the ESCA spectrum of a given core level consists of
well resolved peaks corresponding to electrons escaping without
undergoing energy losses, superimposed on a background tailing to
lower kinetic energy arising from inelastically scattered electrons.
(This is clearly evident in Fig. 11. With an unmonochromatized
X-ray source as in this example, there is of course a general
contribution to the background arising from the bremsstrahlen.)

    In the applications to be discussed below, we may assume that
the X-ray beam is essentially unattenuated over the range of surface
thickness from which the photoelectrons emerge. The intensity of
electrons of a given energy observed in a homogeneous material may
be expressed as

$$dI \ = \ F\alpha N k e^{-x/\Lambda}dx,$$

where F is the X-ray flux, α is the cross section for photoionization
in a given shell of a given atom for a given X-ray energy, N is the
number of atoms in volume element, k is a spectrometer factor for the
fraction of electrons that will be detected and depends on geometric
factors and on counting efficiency, Λ is the electron mean free path
and depends on the KE of the electron and the nature of the material
that the photoelectron must travel through.   The situations which
are of common occurrence as far as this and the related paper in
this symposium are concerned is that of a single homogeneous
component or of a surface coating of thickness d on a homogeneous
base.   This is illustrated in Fig. 12.   The intensity for a film
A of thickness d may be expressed as

$$I^A = I_\alpha^A (1 - e^{-d/\Lambda_A}),$$

whereas for the film B (considered infinitely thick) on the surface

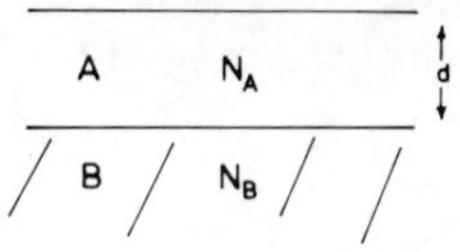

$$I_\alpha = \int_0^{d} F_\alpha N K e^{-x/\Lambda} \, dx = F_\alpha N K \Lambda$$

$$I^A = \int_0^{d} F_\alpha N_A K e^{-x/\Lambda_A} dx = F_\alpha N_A K \Lambda_A (1 - e^{-d/\Lambda_A})$$

$$= I_\alpha{}^A (1 - e^{-d/\Lambda_A})$$

$$I^B = \int_{d}^{d} F_\alpha N_B K e^{-(x-d)/\Lambda_B} e^{-d/\Lambda_A} \, dx$$

$$= F_\alpha N_B K \Lambda_B e^{-d/\Lambda_A}$$

If $N_A \times N_B$ $\Lambda_A \approx \Lambda_B$

$$I^B = I_\alpha e^{-d/\Lambda_A}$$

Fig. 12.   Intensities of elastic peaks as a function
            of film thickness

of which A is located, the intensity is given as $I^B = I_\alpha e^{-d/\Lambda_B}$
(with $N_A = N_B$).   Fig. 13 shows data pertaining to the escape depth
dependence on kinetic energy for electrons, derived from Auger
and ESCA studies on films of known thickness[4].   The striking
feature clearly evident in Fig. 13 is that the experimental data
for solids, in which the contributions of various scattering
processes might be expected to differ, fit closely onto a 'universal'
curve of escape depth versus kinetic energy.   The importance of
this will become apparent in the ESCA studies relating to surface
fluorination.   In general, however, it is clear that for photon
energies of 1253.7 eV or 1486.6 eV the mean free paths for photoemitted electrons should be < 20Å.

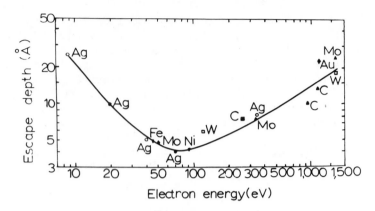

Fig. 13.   Escape depth dependence on kinetic energy for electrons.

     To put matters in perspective, it is worth noting that for
escape depths of 5Å, 10Å and 15Å, 90% of the signal intensity from
a homogeneous sample derives from the topmost 11.5Å, 23.0Å and
34.5Å respectively.

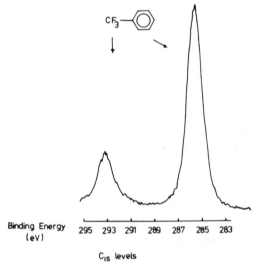

Fig. 14. C_{1s} levels for benzotrifluoride.

     Fig. 14 shows the C_{1s} levels for benzotrifluoride and
illustrates the large shifts in core levels which can occur.   The
electronegative fluorines withdraw electron density from the carbon
of the trifluoromethyl group with concomitant large increase in
C_{1s} core binding energy.   As will become apparent, a peak at ~294 eV
binding energy may be used to 'fingerprint' -CF_3 groups.   As a
simple example again illustrating how ESCA can provide direct
information on electron distributions and also illustrating the
phenomena of shake up processes, Fig. 15 shows the N_{1s} and C_{1s} core

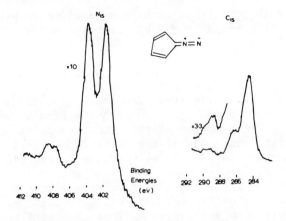

Fig. 15. N_{1s} and C_{1s} levels for diazocyclopentadiene
showing shake up structure.

levels of diazocyclopentadiene. The large difference in electron
density about the two nitrogens is clearly evidenced by the shift
of 2 eV between the two peaks. Clearly evident to the low kinetic

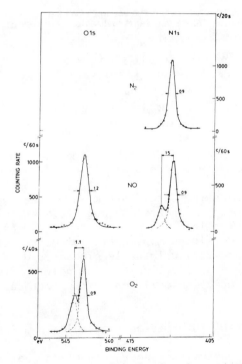

Fig. 16. O_{1s} and N_{1s} core levels for O_2 and NO showing
multiplet splittings.

energy side of both core level peaks are satellites which are in
fact identified as $\pi \longrightarrow \pi^*$ shake up transitions for the relevant
hole states.

So far, we have indicated the importance in ESCA measurements
of inter alia - the absolute binding energies and shifts in core
levels, relative peak intensities, and their dependence on mean free
path, kinetic energy, photoionization cross sections, etc. A further
possible source of information arises from paramagnetic species due
to multiplet splittings. One of the first demonstrations of this
due to Siegbahn and coworkers is illustrated in Fig. 16 for the
paramagnetic NO and O_2 molecules (2b). The theory in the cases of
core ionization from S levels is particularly simple as outlined in
Fig. 17. In the case of NO, the single unpaired electron is
delocalized on nitrogen and oxygen so that the magnitude of the
multiplet splitting of the O_{1s} and N_{1s} core levels will depend upon
the unpaired spin densities at the two atoms. The pronounced

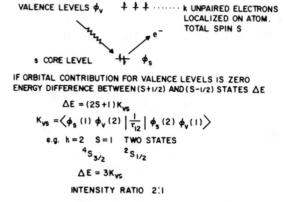

Fig. 17. Multiplet splittings arising from photoionization
from S type core orbitals

satellite for nitrogen indicates that most of the unpaired spin
density is on nitrogen and hence the magnitude of multiplet
splittings can provide information comparable to that from ESR
studies. The best developed application of multiplet effects is in
transition metal chemistry originally pioneered by Fadley and
coworkers[5]. Possible applications in the polymer field might be
the study of free radicals localized at the surfaces of polymers.

7. Basic Types of Information Available From ESCA Studies

Both the core and valence energy levels of polymers may be
studied and the basic types of information available are summarized
in Table 2.

Table 2

INFORMATION FROM ESCA

<table>
<tr><td>

Core Electrons

(1) Binding energy characteristic
of a given level of a given
element, therefore, useful
for analysis. Different KE
dependencies for different
core levels provides means
for analytical depth
profiles.

(2) Absolute BE may be charact-
eristic of particular
structural features, (e.g.
CF_3, C=O, $-NH_2$ etc.).
'Shifts' can be related
to electron distribution.

</td><td>

Valence Electrons

(1) Can study valence energy
levels of insulators.
Densities of states for
conduction bands of metals.

(2) Studies of differential
changes in cross section
with photon energy provides
information on symmetries
of orbitals (σ, π etc.).

</td></tr>
</table>

(3) Multiplet Splittings

For paramagnetic species observation of
multiplet splittings provides information
on spin states of atoms or ions and
distribution of unpaired electrons.

(4) Shake up and shake off satellites.
Information on excited states of hole
states.

The application of ESCA to studies of Structure and Bonding in polymers is discussed in Section III; at this stage, however, it is worth emphasizing that no detailed investigations of either multiplet effects or shake up phenomena in this field have been published, although several situations where such studies may well be of importance can be envisaged. Most studies to date can, therefore, be categorized under headings (1) and (2).

III. APPLICATIONS OF ESCA TO STUDIES OF STRUCTURE AND BONDING IN POLYMERS

1. Introduction

The great advantage of ESCA as a technique, in being able to study in principle the core and valence levels of any element, (regardless of nuclear properties such as magnetic or electric quadrupole moments), coupled with the low sample requirements, and the ability to study involatile insoluble solids, is nowhere more apposite than in the study of polymers. We have noted in previous sections the particular feature of ESCA in being in principle, capable of providing information concerning the surface and immediate subsurface of solid samples as well as that pertaining to the bulk. One obvious question of relevance to any detailed investigation of friction and wear is whether the surface is typical of the bulk. ESCA provides a convenient means of delineating differences in structure and bonding between the surface and bulk of a sample. This particular feature of the technique is crucial to an under-standing of surface treatments of polymers such as oxidation, fluorination, argon ion bombardment, etc., which will be discussed in the next paper. As a simple example, however, of how ESCA may routinely be used to monitor surface composition mention might be made of pressed films of PTFE. Fig. 11 previously discussed, pertains to a thin film of PTFE pressed from the powder between sheets of 'clean' aluminum foil at the minimum temperature necessary for plastic flow using a hand press. At much higher temperatures ~ 320°C, films are produced which visually, bulk chemically and by transmission infrared and multiple attenuated total reflectance measurements, appear to be the same as those produced at lower temperatures. The ESCA spectra of such films, however, are very revealing (Fig. 18). It is clear from the appearance of peaks associated with the core levels of both oxygen and aluminum that in the high temperature pressing process a contaminant surface layer of alumina Al_2O_3 is deposited at the surface. The thickness of the layer is almost certainly < 10Å. Although this is a trivial example, it can readily be appreciated that if surfaces are mechanically prepared, the possibility of contamination is always present.

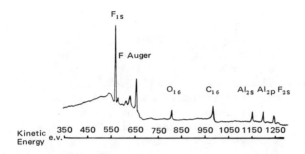

Fig. 18.   Wide scan ESCA spectra for PTFE sample pressed
           at high temperature between Al foil.

Before discussing particular applications, it is perhaps worth
digressing to outline our general philosophy in applying ESCA to
studies of polymers. The general guidelines adopted follow directly
from our previous development of ESCA as a tool for studying
structure and bonding in organic systems in general and halocarbon
systems in particular[6]. Namely, to start by studying simple
well characterized systems to build up banks of data (on relative
peak intensities, absolute binding energies and shifts) from which
trends may be discerned and comparison drawn with simple monomers.
Then develop a theoretical framework to quantify the results, and
this provides a strong basis for studying more complex systems.

As a logical first step, we therefore consider studies of simple
well characterized homopolymer and copolymer systems. These
'homogeneous' samples then form a valuable yardstick for gauging the
utility of ESCA for studies pertaining to information characteristic
of the bulk.

The applications of ESCA to structure and bonding in polymer
systems discussed below largely pertains to fluoropolymer systems.
The reasons for this are threefold. Firstly, our research interests
are centered around the halocarbon field (both monomeric and
polymeric species) and hence samples and the expertise in preparing
them are readily to hand. Even without an underlying research
interest in these systems however, there are two further reasons why
an initial research program into the application of ESCA to polymers,
should concentrate initially on fluoropolymer systems. Thus, the
large shift in (e.g. C_{1s}) core levels induced by fluorine gives the
most favorable situation for delineating the likely areas of
applicability of ESCA in this field. Finally, fluoropolymers are
amongst the technologically most important systems of interest and
are often difficult to study by other spectroscopic techniques.

In applying the technique to the investigation of simple polymer systems there are several distinct aspects about which one would hope to gain information. Firstly, the gross chemical composition of the polymers. This would include determination of elemental composition and in the case of copolymers the percentage incorporation of comonomers. Secondly, information concerning the gross structure, for example, for copolymers the block and/or alternating or random nature of the linkages. Thirdly, finer detail of structure such as structural isomerism, the nature of end groups, branching sites, etc. Finally, deductions made from ESCA measurements, of charge distributions and nature of the valence bands for polymers. It should be emphasized that the data obtainable from ESCA is rather coarser in detail than that often obtainable by more conventional techniques (I.R., N.M.R., inelastic neutron scattering, etc.) where information regarding for example, conformational aspects of polymer structure, may often be inferred and which are in principle not amenable to direct study by ESCA. ESCA should be regarded as a powerful technique providing information complementary to that from other branches of spectroscopy, but with unique advantages which mean that for many studies of polymeric systems it may well be the most important. In particular aspects of polymer chemistry, such as dynamic studies of thermal or photochemical degradation, and in studies of polymeric films produced at surfaces by chemical reaction (e.g. fluorination, oxidation, etc.) the information derived from ESCA studies is not obtainable by other techniques. The application of ESCA to such problems is considered briefly in this article, however, more extensive discussion of particular aspects are given in the paper on surface fluorination.

2.   Sample Preparation

Before discussing the results, it is of some relevance to consider the ways in which polymer samples may be prepared for examination by ESCA. When the polymer is available as a powder, it is often convenient to study it as such by applying the powder to double sided scotch tape mounted on the sample probe. The pitfalls to beware of in this approach are that no extraneous signals are observed from the sample backing and also that no chemical reaction occurs between sample and substrate. The incomplete coverage and uneven topography of samples prepared in this way generally lead to lower signal/noise ratios than polymers studied directly as films. The most generally useful methods of preparing polymer films for ESCA studies may be classified as follows.

(i)   <u>From Solution</u>. If the polymer is sufficiently soluble, then thin polymer films may be deposited directly on a gold backing (ready for mounting on the sample probe), by conventional dip or

bar coating. Since ESCA is such a surface sensitive technique, it
is important to use clean apparatus and pure solvents containing no
non-volatile residues (e.g. antioxidants, etc.) which would
segregate at the surface on evaporation of the solvent. With
readily oxidized systems or with systems with sites capable of
hydrogen bonding with extraneous water, it is imperative to maintain
a suitable inert atmosphere during the slow evaporation of solvent.
Solvent entrainment can also be a problem, and indeed the technique
lends itself well to studying diffusional problems in polymers.

(ii) From Pressing or Extrusion. Because of problems of
contamination in solvent casting films, it is convenient to study
most polymers in the form of pressed or extruded films mounted on
a suitable backing (e.g. gold). For elastomers, of course, it is
often possible to 'melt' a small amount of the sample and allow it
to spread in the form of a thin film on the tip of a sampling probe
or to slide a thin film from a larger sample. In preparing samples
from powders, it is often convenient to press films between sheets
of clean aluminum foil at an appropriate temperature and pressure.
There are two precautions to be taken in doing this: (a) the
temperature and pressures used should be such that no decomposition
or adhesion of surface contamination occurs, (b) since typically
only the top ~50Å of the sample is studied by ESCA, it is important
to avoid chemical reaction at the surface during preparation. Thus,
pressing polyethylene films in air at the minimum temperature and
pressure necessary, results in considerable surface oxidation. This
may be obviated by pressing in an inert atmosphere (e.g. N_2 or Ar).
Surface contamination, e.g. hydrocarbon etc., arising for example
by inadvertent handling during processing can most readily be
removed by careful treatment with an appropriate solvent.

(iii) Polymerizations In Situ. ESCA is particularly suited
to dynamic studies and would be useful for monitoring polymerizations
carried out in situ at the sample probe by e.g. irradiation ($h\nu$ or
e^-) or from impinging beams of precursors generated by e.g. pyrolysis
or irradiation.

3.   Energy Referencing

Of their very nature, most polymers of interest are extremely
good insulators and as such in studying samples as thin films there
is only a fortuitous possibility that during the ESCA experiment
sufficient charge carriers will be available such that the sample is
in electrical contact with the spectrometer. Sample charging will,
therefore, occur resulting in a shift of the energy scale, and some
form of referencing back to the Fermi Level is, therefore, necessary.

Referencing of the energy scale is most readily accomplished by
depositing a thin coating of a suitable reference (e.g. C_{1s} signal
from a hydrocarbon $4f_{7/2}$ levels of gold with binding energies of
285 eV and 84 eV respectively) and monitoring the core levels of
the sample and reference (cf. Ref. 1).   To put matters in perspective
however, it should be emphasized that with most commercial
spectrometers (based on unmonochromatized X-ray sources and retarding
lens system, double focussing hemispherical electrostatic analyzers
and associated slit systems), sample charging is not too serious
a problem and at worst involves only a few eV correction to the
energy scale.   With slitless designs and a monochromatized X-ray
source as conventionally applied in dispersion compensation,
sample charging can, however, reach serious proportions involving
shifts in the energy scale of tens of eV for insulating samples.
The problem can be alleviated to some extent by use of low energy
electron flood guns, but this introduces other complications.

4.   Preliminary Considerations

        An interesting observation is that for homopolymers, even in
the presence of overall sample charging, with suitable preparation
of sample, linewidths for given core levels are comparable with those
for monomers, cf. Fig. 19.   On this basis, in principle the range

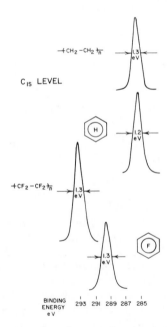

Fig. 19. C_{1s} levels for homopolymers and typical monomers
illustrating linewidth similarities.

of applicability of ESCA in studying homopolymers is essentially
the same as in studying simple monomers. With some preparations of
thin films, by pressing or more usually with samples prepared by
slicing from a larger sample or melted directly onto a probe tip,
the less regular nature of the surface is often manifest in slight
increase in linewidths arising from non-uniform distribution of
charge on the sample. This is often of no consequence since it
merely affects the overall resolution, however, it may be obviated
in many cases by modifying the way in which the film is produced.
A simple example illustrating this rather effectively arises from
our studies of copolymers of ethylene and tetrafluoroethylene (to
be discussed in more detail later on).

   The first spectra recorded with these samples seemed to
indicate that the degree of fluorinated monomer incorporation was
not as great as expected. Thus, in the carbon-1s spectra the peaks
at low binding energy due to CH_2 groups were always bigger than the
higher energy bands due to $\overline{CF_2}$ groups, even for samples where the
degree of incorporation of tetrafluoroethylene was reliably known
(from conventional carbon and fluorine elemental analysis) to be
greater than 50 mole %. This observation is clearly indicative
of surface contamination of the samples by hydrocarbon and serves
to emphasize the fact that the ESCA technique is primarily concerned
with the topmost ~50Å of the sample. The surface contamination
could conceivably arise in one or more ways which will not concern
us here. When the surfaces of the films were cleaned with
methylene chloride (by lightly rubbing the surface with a tissue
dampened with solvent), the contaminants were removed and the
elemental compositions of the cleaned surfaces could be shown (see
later) to be the same as those measured by conventional combustion
analysis of bulk samples. The effect of this surface cleaning
on the carbon-1s spectrum of a copolymer sample is illustrated in
Fig. 20, spectrum (a) shows the spectrum of the polymer prior to
cleaning and spectrum (b) that of the same sample after it had been
cleaned with methylene chloride. Two effects are apparent from
inspection of the spectra; firstly,the peaks due to hydrocarbon
contamination of the surface which occur at low binding energy have
been removed by the cleaning process, and secondly both the peaks in
the spectrum of the cleaned sample are narrower than those in the
original untreated material. This second observation can be
explained on the assumption that the surface contamination is
unevenly distributed resulting in an uneven distribution of
charges at the sample surface and a consequent line broadening,
such an uneven distribution of contaminants could arise from
handling of the sample.

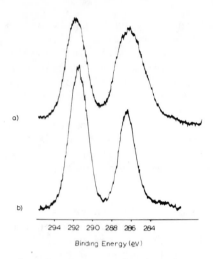

294 292 290 288 286 284

Binding Energy (eV)

Fig. 20. C_{1s} levels for copolymer of ethylene-tetrafluoroethylene illustrating the effect of removal of
surface contamination.

5.   Homopolymers[7]

The simplest systems to start with are the high m.wt. homopolymers of the fluoroethylenes for which complications due to
branching, end groups, etc., are minimal. In this section, we
therefore outline the results from studies on polyethylene (high
density), polyvinyl fluoride, polyvinylidene fluoride, polyvinylene
fluoride, polytrifluoroethylene and polytetrafluoroethylene.
Spectra for the C_{1s} spectra of these polymers are shown in Figs.
21-23.

Typically, measurements of all the core levels of a polymer
take ~30 mins. whereas under the conditions employed (pressure in
sample chamber typically ~5 x 10^{-7} Torr) hydrocarbon build up on
the surface only becomes appreciable after several hours. The
technique employed for calibration of the energy scale, therefore,
is to measure the core levels of a given polymer immediately on
introduction of the sample into the spectrometer when hydrocarbon
contamination is unimportant. After several hours in the sample
chamber, further spectra may then be recorded and the appearance
of an extra peak, (clearly evident in the spectra in Figs. 21-23)
in the C_{1s} region at 285.0 eV binding energy may then be used to
reference the energy scale.

The C_{1s} spectrum of polyvinyl fluoride shows two partially
resolved peaks of equal area corresponding to $\underline{C}HF$ and $\underline{C}H_2$ carbons
whilst for polyvinylene fluoride in addition to the main peak
corresponding to $\underline{C}HF$ carbons and hydrocarbon calibration peak, there

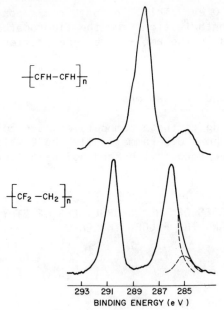

Fig. 21. C$_{1s}$ levels for polyethylene and polyvinyl fluoride.

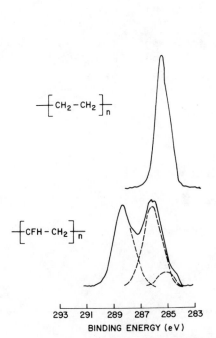

Fig. 22. C$_{1s}$ levels for poly-
          vinylene fluoride and
          polyvinylidene fluoride.

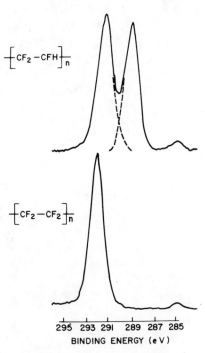

Fig. 23. C$_{1s}$ levels for
          polytrifluoroethylene
          and PTFE.

is a weak peak at 292.0 eV. This, in fact, corresponds to $\underline{C}F_2$ type carbon arising from contamination from the fluorocarbon soap $(H(CF_2)_8COO^-NH_4^+)$ used in the emulsion polymerization. For polyvinylidene fluoride, well resolved peaks of equal area corresponding to $\underline{C}F_2$ and $\underline{C}H_2$ carbons are evident, and for poly-trifluoroethylene, partially resolved peaks of equal area corresponding to $\underline{C}F_2$ and $\underline{C}FH$ carbons.

The data pertaining to these homopolymers are collected in Table 3, and Fig. 24 shows diagrammatically the C_{1s} levels including data for polytrifluorochloroethylene and polyhexafluoropropene.

Table 3

BINDING ENERGIES OF THE HOMOPOLYMERS OF ETHYLENE
AND THE FLUOROETHYLENES

			C_{1s}	$\Delta(C_{1s})$	F_{1s}	$\Delta(F_{1s})$
Ia	$\{CH_2-CH_2\}_n$		285·0	(0)	-	-
Ib	$\{CH_2-CH_2\}_n$		285·0	(0)	-	-
II	$\{CFH-CH_2\}_n$	$-\underline{C}FH-$	288·0	3·0	689·3	(0)
		$-\underline{C}H_2-$	285·9	0·9	-	-
III	$\{CFH-CFH\}_n$		288·4	3·4	689·3	0·0
IV	$\{CF_2-CH_2\}_n$	$-\underline{C}F_2-$	290·8	5·8	689·6	0·3
		$-\underline{C}H_2-$	286·3	1·3	-	-
V	$\{CF_2-CFH\}_n$	$-\underline{C}F_2-$	291·6	6·6	690·1	0·8
		$-\underline{C}FH-$	289·3	4·3	690·1	0·8
VI	$\{CF_2-CF_2\}_n$		292·2	7·2	690·2	0·9

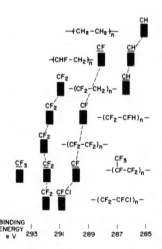

Fig. 24. Substituent effect in C_{1s} levels of homopolymers.

The assignment of peaks arising from $\underline{C}F_3$, $\underline{C}F_2$, $\underline{C}F$ and $\underline{C}H$ structural units is readily apparent. By taking appropriate pairs of polymers, it is possible to investigate both primary and secondary effects of replacing hydrogen, Figs. 25 and 26. The average primary (2.9 eV) and secondary (0.7 eV) substituent effects are in excellent agreement with those found for monomer systems and their relative constancy emphasizes the characteristic nature of substituent effects.

POLYMER PAIRS	SHIFT IN C_{1s} BINDING ENERGY ON REPLACING H BY F eV
$(\underline{C}HFCH_2)_n$, $(\underline{C}H_2CH_2)_n$	3.0
$(\underline{C}F_2CH_2)_n$, $(\underline{C}HFCH_2)_n$	2.8
$(CHF\underline{C}HF)_n$, $(CHF\underline{C}H_2)_n$	2.5
$(CF_2\underline{C}HF)_n$, $(CF_2\underline{C}H_2)_n$	3.0
$(CF_2\underline{C}F_2)_n$, $(CF_2\underline{C}HF)_n$	2.9
$(\underline{C}F_2CHF)_n$, $(\underline{C}HFCHF)_n$	3.2
AVERAGE	2.9

Fig. 25. Primary substituent effects in homopolymers (C_{1s} levels)

POLYMER PAIRS	SHIFT IN C_{1s} BINDING ENERGY ON REPLACING H BY F (Per Substituent) eV
$(CHF\underline{C}H_2)_n$, $(CH_2\underline{C}H_2)_n$	0.9
$(\underline{C}HFCHF)_n$, $(\underline{C}HFCH_2)_n$	0.4
$(CF_2\underline{C}H_2)_n$, $(CH_2\underline{C}H_2)_n$	0.7
$(\underline{C}F_2CHF)_n$, $(\underline{C}F_2CH_2)_n$	0.8
$(\underline{C}F_2CF_2)_n$, $(\underline{C}F_2CHF)_n$	0.6
$(CF_2\underline{C}HF)_n$, $(CHF\underline{C}HF)_n$	0.9
AVERAGE	0.7

Fig. 26. Secondary substituent effects in homopolymers (C_{1s} levels).

The rapid fall off in effect of the fluorine substituent provides a crude but immediate manifestation of the σ inductive effect exerted by fluorine. As the degree of fluorine substitution increases in going through the series, the fluorine substituents to some extent compete for the sigma electron drift from the carbon and hydrogen atoms and this is clearly evidence by the increase in binding energy for the F_{1s} levels. It is gratifying to note that the shifts in core binding energies are qualitatively in agreement with what might

be termed the organic chemists intuitive ideas concerning charge
distributions and this will be quantified in a later section.
In studying fluoropolymer systems in general in addition to the
F_{1s} and C_{1s} core levels and their associated Auger transitions,
it may be noted that the F_{2s} levels at the bottom of the valence
band are essentially core like so that even with a single photon
source the escape depth dependences on kinetic energy span a large
range.   For $MgK\alpha_{1,2}$ KE's for F_{1s}, C_{1s} and F_{2s} are ~560 eV,
~960 eV and ~1220 eV respectively whilst the Auger processes
for de-excitation of the F_{1s} and C_{1s} hole states are in the
kinetic energy range 610-660 and 240-270 eV respectively.   This
being the case it is a relatively easy matter to use ESCA to 'depth
profile' samples to investigate their homogeneity.   Considering the
data for the homopolymers of the fluorinated ethylenes, mention has
already been made of the fact that the C_{1s} spectra consist of two
peaks of equal area.   Since the KE energies for the photoemitted
electrons are virtually identical, the sampling depths must be the
same and the results, therefore, suggest a homogeneous sample.   This
may be confirmed by comparison with data from the F_{1s} level where
the sampling depth will be different.   Table 4 shows the relative
carbon-1s to fluoride-1s peak areas for the four samples (with
our particular instrumental arrangement).

Table 4

Polymer	Carbon-1s peak area:Fluorine-1s peak area	Number of carbon atoms: Number of fluorine atoms	Relative* carbon-1s peak area:Fluorine-1s peak area
$(CH_2CHF)_n$	1:1	2:1	1:2.0
$(CH_2CF_2)_n$	1:2.08	1:1	1:2.1
$(CHFCF_2)_n$	1:3.15	2:3	1:2.1
$(CF_2CF_2)_n$	1:3.97	1:2	1:2.0

*
With probable error limits of ±0.05

The constancy of the ratio demonstrates unambiguously the
uniformity of the samples since the escape depth dependencies are
significantly different (mean free paths F_{1s} ~7Å (KE ~564 eV),
C_{1s} ~10Å (KE ~965 eV).   The determination of the intensity ratios
of core levels for model homogeneous systems is an important
precursor in discussing copolymer systems.

6.   Theoretical Models for Quantitative Discussion of Results

At this stage, a brief digression is necessary to consider the theoretical interpretation of results for polymer systems. Detailed consideration of the theoretical models for quantitatively discussing both absolute binding energies and shifts have been given elsewhere and a minimum amount of background will hence be given here[1,5]. The photoionization process itself, at least for systems containing first and second row atoms, can in general be quantitatively described within the Hartree Fock formalism although this is not true for the accompanying shake up and shake off processes for which detailed considerations of electron correlation are necessary. Detailed non-empirical LCAO-MO-SCF calculations can, therefore, provide a quantitative discussion of experimental ESCA data, (other than for shake up and shake off phenomena), however, such computations are only feasible on relatively simple systems. In attempting to quantify theoretically the results for polymer systems, the great complexity of even model systems renders a rigorous approach impossible. We have, therefore, developed a less sophisticated but still theoretically valid approach based on an expansion for the expression for the Fock eigenvalues in the zero differential overlap approximation. By grouping terms which are essentially independent of the local electron density, it may be shown that the binding energy of a given core level of an atom in a molecule is related to the overall charge distribution as equation (1).

$$E_i = E_i^o + kq_i + \sum_{j \neq i} \frac{q_j}{r_{ij}} . \tag{1}$$

The so called 'charge potential model' originally developed by Siegbahn et al. (2b) relates the core binding energy E_i to the charge on atom i and the potential provided by the other charges within the molecule. E_i^o is a reference level and k represents approximately the one center coulomb repulsion integral between a core and valence electron on atom i. In dealing with complex organic molecules, analysis of experimental data in terms of the charge potential model and all valence electron SCF-MO calculations in the CNDO/2 formalism (formally an approximation to a non-empirical treatment) has proved highly rewarding. From studies of series of related molecules, values of k and E_i^o may be established for a given core level of a given atom. The values of k depend on the bonding situation as is evident from Fig. 27 where data pertaining to C_{1s} levels are displayed with the average being ~25 eV/unit charge.

The central role of the charge potential model in quantifying experimental data on complex molecules is shown in Fig. 28. Starting on the LHS if we have geometries and appropriate charge distributions, (e.g. from CNDO/2 SCF-MO calculations) we may use

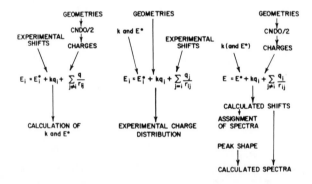

C$_{1S}$ LEVELS

SERIES	k
(ring with H)—X	24·6
$\underline{C}Cl_3$-X, $\underline{C}Cl_2$ H – X	28·7
$\underline{C}H_3$ $\underline{C}OX$	25·0
AROMATICS PERHYDRO PERFLUORO	25.0
SIX MEMBERED RING HETEROCYCLES PER H PER F PER Cl	22·4
FLUOROBENZENES	23·5
FIVE MEMBERED RING HETEROCYCLES	·25.4

Fig. 27.   Values of k in the charge potential model for the
            C$_{1S}$ levels of series of organic compounds.

Fig. 28.   The central role of the charge potential model in
            quantitative discussion of core binding energies.

experimental shifts on series of model compounds to obtain values
of k and E° for a given level of a given element. On the RHS, if
k and E° values are available, then theoretical charge distributions
may be used for the assignment of spectra and if peak shapes and
widths have been established then theoretically calculated spectra
may be simulated. Last but not least if appropriate values of k
and E° and geometries are available it is possible to invert the
charge potential model to obtain 'experimental' charge distributions
and the application to polymer systems is of particular importance
and will be discussed more fully later on.

The short range nature of substituent effects on core binding
energies in saturated systems suggests that it may be feasible to

quantitatively discuss the results for polymers in terms of
calculations on simplified model systems which contain the
essential structural features and accommodate all short range
interactions. The success of the charge potential model, coupled
with CNDO-SCF-MO calculations of charge distributions in
quantitative discussion of data for simple monomers mentioned
earlier, suggests that this is a feasible approach, since the
model accounts quite nicely for the short range nature of
substituent effects.

As our model for polyethylene and polyvinylene fluoride, we
have therefore taken our representative model unit as being the
monomer linked to other monomer units and then appropriate end
groups. In this way, substituent effects over three carbon atoms
are taken into account. (The calculations show that such long
range effects are negligible so that the model incorporates all
of the important short range interactions.) All valence electron
SCF-MO calculations have then been carried out with appropriate
values for k's and E°'s used to calculated absolute binding
energies for the core levels of the representative structural unit
of polymer. The adequacy of the theoretical treatment for polyethylene and polyvinylene fluoride is apparent from Fig. 29.
The small discrepancy in the case of PTFE undoubtedly arises from
the fact that computer limitations dictated that a smaller model
system than optimum be studied. In fact, recent upgrading in
computer hardware has allowed us to investigate this, and for the
larger model the results are in quantitative agreement with
experiment.

Polymer Models C_{1s}
                      Binding Energy

CH_3 CH_2 CH_2 $\underline{CH_2}$ $\underline{CH_2}$ CH_2 CH_2 CH_3
Calc. 284·9
Observed 285·0

CH_2F CHF CHF $\underline{CHF}$ $\underline{CHF}$ CHF CHF CH_2F
Calc.   288·6
Observed   288·4

CF_3 CF_2 $\underline{CF_2}$ $\underline{CF_2}$ CF_2 CF_3
Calc. 292·6
Observed 292·0

Fig. 29.   Absolute binding energies calculated for model systems
of polyethylene, polyvinylene fluoride and PTFE.

For the unsymmetrical monomers: polyvinyl fluoride, polyvinylidene fluoride and polytrifluoroethylene, the possibility arises
of structural isomerism by way of head-to-tail and/or head-to-head
addition. In fact, ^{19}F studies show that structural isomerism does

occur in both of the first two and indeed provides information on
tacticity which is not available by ESCA studies. Since the C_{1s}
spectra of the three polymers are relatively well resolved, it is
evident that information regarding structural isomerism if available
must be encoded in the lineshapes and/or linewidths.

Theoretical calculations on suitable models incorporating the
relevant structural features viz. head–to–tail and head–to–head
linkages give the results shown in Fig. 30. Considering firstly

$(CHF-CH_2)_n$ C_{1s}

Regular

CFH_2 CH_2 CFH CH_2 $\underline{C}FH$ $\underline{C}H_2$ CFH CH_2CFH CH_3

Calc.   2881 285·4

Observed   288·0 285·9

Irregular

CFH_2 CH_2 CH_2 CFH $\underline{C}FH$ $\underline{C}H_2CH_2CFH$ CFH CH_3

Calc.   2880 2856

(CF_2-CH_2)

Regular

CF_2 HCH_2 CF_2 CH_2 $\underline{C}F_2$ $\underline{C}H_2$ CF_2 CH_2 CF_2 CH_3

Calc. 291·0 286·1

Observed 290·8 286·3

Calc. 291·0 286·3

CF_2 HCH_2CH_2 CF_2 $\underline{C}F_2$ $\underline{C}H_2$ CH_2CF_2 CF_2 CH_3

Irregular

$(CF_2-CFH)_n$

Regular

CF_2H CFH CF_2 CFH $\underline{C}F_2$ $\underline{C}FH$ CF_2 CFH CF_2CFH_2

Calc.   291·8 289·3

Observed   291·6 289·3

Irregular

CF_2H CFH CFH CF_2 $\underline{C}F_2$ $\underline{C}FH$ CFH $CF_2CF_2CFH_2$

Calc.   291·7 289·4

Fig. 30.   Absolute binding energies calculated for model
           systems of polyvinyl fluoride, polyvinylidene fluoride
           and polytrifluoroethylene.

polyvinyl fluoride, the calculated binding energies for both types
of structural arrangements are the same within experimental error
and are in fact in excellent agreement with the observed values.
It seems clear, therefore, that for this particular polymer ESCA is
unable to provide information on structural isomerism along the
chain. (This contrasts with the situation to be described later
concerning nitroso rubbers.) The same considerations apply to both
polyvinylidene fluoride and polytrifluoroethylene, again the
theoretical models are in excellent agreement with experiment.

7.   Copolymers[8,9]

   A.   <u>Introduction</u>. It is evident from the previous section
that substituent effects on core binding energies in polymers can
be understood both qualitatively and quantitatively on the same
basis as those for simple monomeric systems. We are now in a
position, therefore, to proceed to more complex systems such as
copolymers. The first feature of interest is the determination of
compositions viz. percentage comonomer incorporations. We illustrate
the applicability of ESCA in this field by reference to our work on
Viton and Kel-F type copolymers. Further features of interest
arise in studying copolymers of ethylene and tetrafluoroethylene
namely the gross structure of the polymers in terms of block,
alternating or random features. This leads to a brief discussion
of ESCA for studying domain structure in block copolymers. The
use of ESCA for studying finer details of structure such as
structural isomerism is exemplified by studies of nitroso rubbers.

   B.   <u>ESCA Studies of Copolymer Compositions</u>.

<u>Viton Type</u>   <u>Kel-F Type</u>

$$CF_3 \qquad\qquad CF_3$$
$$|\qquad\qquad\quad |$$
$$(CF\text{-}CF_2)_n, \quad (CF\text{-}CF_2)_m(CF_2\text{-}CH_2)_n \; ; \quad (CF_2\text{-}CFCl)_n, \quad (CF_2CFCl)_m(CF_2CH_2)_n.$$

   Before discussing the results, it is interesting to consider
how ESCA may be used as a non-destructive technique, with virtually
no sample pre-preparation necessary for routine identification of
polymers. As a simple example, we suspected that one of the Kel-F
samples in our possession was incorrectly labelled. Two rapid ESCA
experiments, taking ~20 mins each, in fact identified the polymer
as a Viton type polymer. The carbon-1s spectra are shown in Fig.
31. The spectrum of the Viton sample exhibits the characteristic
high binding energy peak of the $\underline{C}F_3$ carbon at 294.3 eV and a
pronounced shoulder on the $\underline{C}F_2$ peak at 292 eV attributable to the

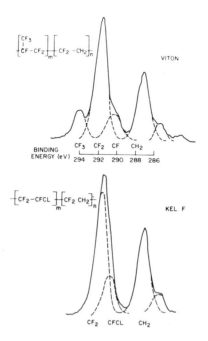

Fig. 31. C_{1s} levels for Viton and Kel-F polymers.

CF carbon at 290.4 eV. As a double check, the Viton sample also showed no levels attributable to chlorine.

Viton Polymers. The C_{1s} levels for the parent polyhexafluoro-propene and the 30/70 and 40/60 copolymers with vinylidene fluoride are shown in Fig. 32. The binding energies are tabulated in Table 5 and may again be understood in terms of simple substituent effects. In obtaining the binding energies of the CF₂ and CF carbon-1s levels a simple deconvolution is necessary. In any estimation of copolymer compositions, however, it is obviously desirable to avoid even such a minor complication. The procedure, therefore, is to measure the area of the CF₃ peak, the total area of the (CF₂ + CF) peak and the area of the CH₂ peak. The degree of incorporation of hexafluoro-propene (HFP) in the copolymers may then be calculated from the percentage of the total area due to C_{1s} levels represented by each peak:

(i) The mole percent incorporation of HFP must be three times the area of the peak due to CF₃, on the basis of the stoichiometry of the HFP unit.

(ii) The area of the peak due to CF₂ and CF (A) is made up of half the total C_{1s} peak area due to vinylene fluoride (1/2 VF₂) and two thirds the total C_{1s} peak area due to HFP, i.e. A = 1/2 VF₂ + 2/3 HFP and 100 = VF₂ + HFP, by definition, hence % HFP = 6(A-50).

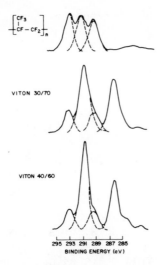

Fig. 32. C_{1s} levels for Viton polymers.

TABLE 5

Binding energies of polyhexafluoropropene and the Vitons

			C_{1s}	$\Delta(C_{1s})$	F_{1s}	$\Delta(F_{1s})$
VIIa	40/60 Viton	CF_3	293·3	8·3	690·2	0·9
		CF_2	291·1	6·1	690·2	0·9
		CF	289·4	4·4	690·2	0·9
		CH_2	286·6	1·6	-	-
VIIb	30/70 Viton	CF_3	293·4	8·4	689·9	0·6
		CF_2	290·9	5·9	689·9	0·6
		CF	289·3	4·3	689·9	0·6
		CH_2	284·6	1·4	-	-
VIII	$\begin{array}{c} CF_3 \\ \overline{\text{-(-CF-CF}_2\text{-)-}_n} \end{array}$	CF_3	293·7	8·7	690·2	0·9
		CF_2	291·8	6·8	690·2	0·9
		CF	289·8	4·8	690·2	0·9

(iii)   The area of the peak due to CH_2 is half the total area due to CF_2 and hence the mole percent incorporation of HFP is given by the expression

100-(2 x % Area due to CH_2).

Using these three methods to determine the degrees of incorporation of HFP gives an internal check on the reliability of the method, the results are as follows:

% HFP Incorporation

	Method of Calculation		
	(i)	(ii)	(iii)
Sample 40/60	39	42	40
Sample 30/70	33	30	32

The internal consistency is good to within 3% and the values obtained are in good agreement with those quoted for the copolymers investigated.

Kel-F Polymers. The C_{1s} spectrum of the parent polychlorotri-fluoroethylene consists of a single, broad peak with a flattened top corresponding to overlapping lines from CF_2 and $CFCl$ carbons, Fig. 33. The absolute binding energies and shifts, Table 6, can again be neatly rationalized in terms of simple substituent effects. The C_{1s} spectra for the two copolymers differ considerably, the most noticeable feature being the high proportion of $-CH_2$ units in the 30/70 copolymer coupled with the drastically reduced linewidth for the composite line at high binding energy.

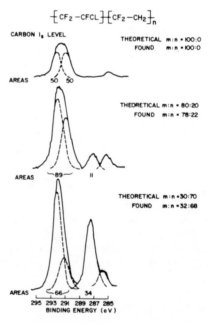

Fig. 33. C_{1s} levels for Kel-F polymers.

TABLE 6

Binding energies of the Kel-F polymers

			C_{1s}	$\Delta(C_{1s})$	F_{1s}	$\Delta(F_{1s})$	$Cl_{2p_{3/2}}$	Cl_{2s}
IXa	80/20 Kel-F	$-CF_2-$	291.7	6.7	690.3	1.0	-	-
		CFCl	290.5	5.5	690.3	1.0	201.4	272.8
		CH_2	286.8	1.8	-	-	-	-
IXb	30/70 Kel-F	CF_2	291.5	6.5	690.5	1.2	-	-
		CFCl	290.6	5.6	690.5	1.2	201.6	272.5
		CH_2	286.9	1.9	-	-	-	-
X	$\{CF_2-CFCl\}$	CF_2	291.9	6.9	690.8	1.5	-	-
		CFCl	290.8	5.8	690.8	1.5	201.1	272.2

In estimating the compositions of these copolymers it is again desirable to avoid reliance on deconvoluted peak areas. Two methods are available: measurement of the total ($-CF_2-$) + (CFCl) peak area, and measurement of the (CH_2) peak area. Since the total CF_2 content of each polymer is 50 mole %, the difference between the percentage of the total C_{1s} peak area attributable to (CF_2 + CFCl) and 50 gives the percentage of the total C_{1s} area due to CFCl, twice this figure gives the amount of chlorotrifluoroethylene units in the polymer.

For the 80/20 and 30/70 copolymers this gives 78% and 32% respectively both being within 2% of the values of 80% and 30% based on elemental analysis. If the areas of the CH_2 peaks are used, the proportions of chlorotrifluoroethylene so obtained are again 78% and 32%. Thus, both methods give exactly the same composition within 2% of the quoted values.

C.   ESCA Studies of Structural Details. Copolymers of ethylene/ tetrafluoroethylene. So far we have shown how a consideration of the detailed structure of the C_{1s} core levels of certain copolymers may be used to obtain information on composition. A further example which also illustrates the utility of ESCA for providing structural data is provided by studies of copolymers of ethylene and tetra- fluoroethylene. Fig. 34 shows the C_{1s} and F_{1s} levels for a series of samples of varying bulk composition. From the ESCA data, the copolymer compositions may be calculated in two independent ways. Firstly, from the relative ratios of the high to low binding energy peaks in the C_{1s} spectrum attributable to CF_2 and CH_2 type environ- ments respectively. Secondly, from the overall C_{1s}/F_{1s} intensity ratios taken in conjunction with data obtained from the study of the homopolymers previously discussed. The results are tabulated in Table 7.

The agreement between the two sets of ESCA data is striking and demonstrates the uniformity (within the outermost $\sim$50Å) of the copolymer since the escape depth dependencies for the F_{1s} and C_{1s} levels are significantly different as previously mentioned. This is indicative of a largely alternating structure for the copolymers

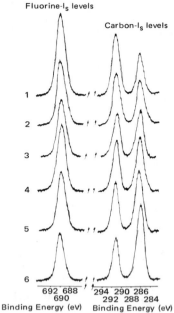

Fig. 34. F_{1s} and C_{1s} levels for a series of copolymers of ethylene-tetrafluoroethylene

TABLE 7

Sample	Composn monomer mixture mole % C_2F_4	Copolymer composition (mole % C_2F_4)				
		Predicted monomer reactivity ratio	Calc. from C analysis	Calc. from F analysis	Calc. area ratios of C_{1s} peak: F_{1s} peak	Calc. from C_{1s} (CH_2 peak): C_{1s} (CF_2) peak)
1	94	63	61	61	63	62
2	80	53	52	54	52	52
3	65·5	50	49	48	57	46
4	64	50	47	45	44	45
5	35	45	41	40	42	40
6	15	36	-	-	32	31

since the PTFE domains of a block copolymer would be expected (on the basis of their lower free energy) to predominate at the surface. In this hypothetical case, estimates of TFE incorporation based on either ESCA method would be unlikely to agree and both procedures would give values higher than those computed from elemental analysis of bulk samples. Comparison of the ESCA data with classical techniques corresponding to bulk analyses shown in Table 7 reveals overall good agreement and also demonstrates that for these systems ESCA is competitive with microanalytical techniques in terms of accuracy and reproducibility with the added advantage of being nondestructive and much faster.

Having satisfactorily established the compositions of the copolymer samples, the next question which arises is that of their structure.   Thus, although preliminary observations mentioned above indicate the unlikelihood of essentially block structure, we should like to establish whether the monomer units are linked in an alternating, block or random manner and to what extent we can quantify an analysis of the distributions of the various structural possibilities.

The previous studies of the homopolymers of fluorinated ethylenes reveals that structural information is most readily derived from absolute binding energies and shifts in C_{1s} core levels.   The shifts can be understood both qualitatively (in terms of simple substituent effects) and quantitatively, (in terms of the charge potential model and SCF-MO computed charge distribution for appropriate models), and lead to a clear cut distinction between the two extremes of essentially block or alternating structure. Thus, the carbon-1s binding energies to be expected for block sequences of ethylene and tetrafluoroethylene are the same as those measured for the respective homopolymers, viz. 285.0 eV and 292.2 eV. (There will be appropriate minor contributions to the predicted overall spectra arising from $-CF_2-$ and $-CH_2$ groups at the junction of the block sequences.)   By contrast, for an alternating structure, the C_{1s} binding energies are expected to be 286.3 eV and 291.0 eV, on the basis of (i) the experimentally observed values for polyvinylidene fluoride, (ii) predicted values based on substituent effects, and (iii) theoretical calculations for the model compound shown below.

$$HCF_2CH_2CH_2CF_2\underline{CF_2}-\underline{CH_2}CH_2CF_2CF_2CH_3$$

291.0, 286.3

A clear distinction should, therefore, be evident if either of these features predominate both in terms of the predicted shifts (7.2 eV for the block and 4.7 eV for the alternating cases respectively) and absolute binding energies for the two major components.

A consideration of the shifts between the two component peaks in the C_{1s} spectra, Fig. 34, illustrates this quite strikingly. The measured shifts in all cases are ~5 eV, which establishes without further refinement of the data that the predominant structural feature is alternation.   However, a close examination of the spectra reveals two features of importance to a complete interpretation of the data.   In the first place, the total linewidths (FWHM) are commonly greater than 2 eV compared with typical linewidths for individual core levels in homopolymers of

~1.3 ± 0.1 eV. Secondly, the peak shapes are distinctly
asymmetric; indeed, the observed shape varies with copolymer
composition. Both these observations indicate that the
experimentally observed spectra are the envelopes of numbers of
overlapping peaks arising from different molecular environments.
To obtain detailed structural information it is, therefore, necessary
to deconvolute the experimental spectra. Meaningful deconvolutions
of spectra requires a careful systematic approach which makes use
of all the relevant information (chemical and physical) concerning
both the sample and the instrumental measurements[1]. Although
in principle there are an infinite number of ways of fitting a
spectral envelope with a combination of curves, it is often (but
not invariably) possible to find a unique solution which fits all
the physical and chemical data. This has previously been
demonstrated in the deconvolution of complex lineshapes for simple
molecules. In our approach to the deconvolution of these spectra,
the information derived from our earlier studies of homopolymers is
invaluable. Thus, both linewidth and lineshape for unique
carbon-1s levels are closely defined under a given set of
experimental conditions. The shape is to a very good approximation
Gaussian with a slight tail to the low kinetic energy side due to
inelastically scattered electrons, and the full width at half maximum
(FWHM) lies in the range 1.2 to 1.4 eV. The spread in linewidths
is almost certainly attributable to differences in surface topography
giving rise to non-uniform sample charging characteristics.

    Calculation of expected carbon-1s binding energies for the
various structural features which a priori might be considered as
possible components of the overall structure of these copolymers is
also indispensable to the complete interpretation of these spectra.
The absolute values and shifts for pure block and pure alternating
sequences have been discussed earlier; the predicted binding
energies for irregular structural features resulting from one or two
monomer units in a block of the other comonomer are shown below
together with the structures of the model compounds for which the
calculations were carried out. These model compounds accommodate
all the important short range effects which it is necessary to
include for a quantitative description of a polymer incorporating
the particular structural feature.

I $CH_3CH_2CH_2CH_2\underline{C}F_2CF_2\underline{C}H_2\underline{C}H_2\underline{C}H_2CH_3$

290.5, 285.6, 285.4, 285.2.

II $CH_3CH_2CH_2\underline{C}F_2\underline{C}F_2CF_2\underline{C}H_2CH_2CH_3$

291.0, 291.2, 285.9, 285.6

III $\quad CF_3CF_2CF_2CF_2\underline{C}H_2\underline{C}H_2\underline{C}F_2\underline{C}F_2\underline{C}F_2CF_3$

287.0, 291.8, 292.1, 292.1 .

IV $\quad H\underline{C}F_2CF_2\underline{C}F_2\underline{C}H_2\underline{C}H_2\underline{C}H_2\underline{C}H_2\underline{C}F_2\underline{C}F_2CF_2H$

286.3, 286.1, 291.4, 291.3 .

The procedure for deconvolution of experimental spectra is, therefore, as follows:

(i)   Starting with the smallest of the two carbon-1s peaks, the best fit possible is made with a single standard line (Gaussian, FWHM 1.4 eV).

(ii)   A second standard line of the same intensity as that in (i) is then fitted under the bigger peak, and positioned so that it is as close to the first peak as possible, consistent with fitting the experimental envelope. These two bands then represent the alternating part of the structure and their binding energies and intensities are listed in columns b and e in Table 8. It is gratifying to note that the average shift in binding energy between these two peaks is 4.7 eV in excellent agreement with the predicted value as are the absolute binding energies.

(iii)   Next two standard lines are fitted to account for the block parts of the copolymer structure at binding energies of approximately 292.2 eV (polytetrafluoro-ethylene) and 285.0 eV (polyethylene) columns a and g respectively in Table 8.

TABLE 8

DECONVOLUTION OF CARBON - 1s SPECTRA

Binding energies (area ratios)

a	b	c	d	e	f	g
292·1(25	291·2(31)	290·4(5)	287·8(3)	286·3(30)	285·7(5)	285·0(1)
292·1(12)	291·0(33)	290·2(7)	287·6(4)	286·3(32)	285·7(6)	285·0(6)
292·3(4)	291·0(37)	290·1(5)	287·7(5)	286·4(36)	285·7(9)	285·0(4)
292·4(6)	291·1(35)	290·4(4)	287·7(4)	286·3(35)	285·7(9)	285·0(7)
292·2(2)	291·0(35)	289·4(3)	287·8(4)	286·4(34)	285·6(17)	285·0(5)
292·1(3)	291·1(27)	289·5(1)	287·3(3)	286·4(26)	285·7(30)	285·0(10)

(iv)   Finally, as many extra standard lines as are required
        to obtain a good fit to the spectral envelope are added
        with both line positions and intensities being treated
        as variables.

In practice, three further peaks of low intensity are required to
obtain a quantitative fit to the overall spectral envelope. It
may be significant that an excellent fit to the overall envelope
may be obtained with the addition of just two further peaks
(columns c,f) except for a minor discrepancy for the valley between
the two major components for which a small peak (~3-4% of the
total) is required. The results for these detailed deconvolutions
are shown in Table 8 and a typical example is shown in Figure 35.

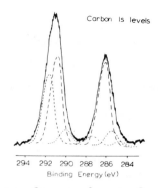

Carbon is levels

294 292 290 288 286 284
Binding Energy (eV)

Fig. 35. C_{1s} spectra for copolymer of ethylene and tetra-
            fluoroethylene showing the deconvolution into
            component peaks.

     The interpretation of binding energies and intensities arising
from the deconvolution in general terms is straightforward. The
major peaks centered at ~291.1 eV (b) and ~286.3 eV (e) arise from
the alternating component to the structure. Whilst those at
~292 eV (a) and ~285.0 eV (g) are accounted for by essentially
block runs of tetrafluoroethylene and ethylene respectively. The
model compounds (I-IV) suggest that peaks c and f arise from an
isolated tetrafluoroethylene unit in an ethylene sequence and an
isolated ethylene unit in a tetrafluoroethylene sequence respectively.

     The remaining component is that tabulated in column 3, from its
binding energy this might qualitatively be expected to arise from an
isolated ethylene unit in a polytetrafluoroethylene sequence;
however, the predicted binding energy for model compound III (viz.
287.0 eV) is not particularly good agreement with the value derived
by deconvolution viz. an average value of 287.7 eV. This does,
however, correspond very closely to the binding energy expected for
carbon-1s levels of carbonyl $>C=O$ groups (cf. CH_3COCH_3 288.0 eV).
A detailed experimental analysis of the kinetic energy region

appropriate to O_{1s} core levels reveals a peak of low overall
intensity with binding energy corresponding to oxygen in a carbonyl
environment (binding energy 533.4 eV). The relative intensities
for the C_{1s} peaks (d) and corresponding O_{1s} peaks coupled with a
knowledge of the escape depth dependence for electrons of differing
kinetic energies shows that the small extent of oxidation is
localized within the immediate surface (~5Å) of the samples.
This almost certainly arises during the production of the polymer
films.

The raw data in terms of the separation between the two major
components of the C_{1s} spectra reveals the essentially alternating
nature of these copolymers without recourse to more refined
analysis. Since, however, kinetic data in terms of monomer
reactivity ratios are available for these systems, it is of interest
to consider a more refined analysis in terms of the detailed
deconvolutions, Table 8.

In analyzing in detail the experimental data, we have therefore
computed the most probable monomer sequences from reactivity ratios
and starting monomer compositions. The simplest distributions to
calculate for direct comparison with experimental data and
computations on model systems are pentad sequences. The dominant
short range interactions are thereby accommodated for the central
triad allowing a direct comparison to be drawn between theory and
experiment. The calculated sequence distributions are given in
Table 9, where the copolymer compositions used in the computations
were taken to be those determined by ESCA. This data in conjunction
with model calculations I-IV allows a direct comparison to be drawn
with the experimental data and this is illustrated in Table 10. In
constructing this table, the core levels corresponding to the central
triads of each pentad sequence have been assigned to the six distinct
peaks obtained in the deconvolution (peaks d arising from surface
oxidation of methylene groups have been excluded from the analysis
for this purpose). With the aid of previous studies on homopolymers
and model calculations I-IV, this assignment is relatively straightforward and representative examples are given in Table 11 for the
more important contributing sequences.

There is a remarkable measure of agreement between the two
sets of data in Table 10 which is most encouraging. Thus, the
sequence distributions computed from the reactivity ratios show
the marked degree of alternation inferred qualitatively from the
ESCA data. As a broad generalization, the ESCA data would suggest
that when the monomer ratio of tetrafluoroethylene (TFE) to ethylene
is greater than unity the 'excess' TFE tends to prefer a block
arrangements whereas if the ratio is less than one there is a greater
tendency for a random distribution of the small percentage of TFE
monomers amongst the predominantly ethylene sequences. Thus, for
samples 1 and 5 where the excess of appropriate comonomers is

TABLE 9

Sample No.	1	2	3	4	5	6
Sequence			Probability			
AAAAA	1·8	–	–	–	–	–
BAAAA AAAAB	5·2	·3	–	–	–	–
BAAAB	3·8	·9	·3	·3	–	–
ABAAA AAABA	12·3	2·0	·5	·4	–	–
ABAAB BAABA	17·8	11·3	6·3	5·9	1·6	·3
BBAAA AAABB	·1	–	–	–	–	–
BBAAB BAABB	·2	·4	·5	·5	·4	·3
ABABA	21·0	34·9	33·9	32·9	24·6	9·5
BBABA ABABB	0·4	2·5	5·5	5·8	13·9	16·9
BBABB	–	–	·2	·3	1·9	7·5
AABAA	6·2	1·0	·2	·2	–	–
BABAA AABAB	17·8	11·3	6·3	5·9	1·6	·3
BABAB	12·8	32·2	39·8	40·5	35·4	21·1
ABBAA AABBA	·3	·4	·4	·4	·3	·1
ABBAB BABBA	·4	2·6	5·5	5·8	13·9	16·9
BBBAA AABBB	–	–	–	–	·1	·1
BBBAB BABBB	–	–	·4	·5	3·9	15·0
ABBBA	–	–	·2	·2	1·4	3·4
ABBBB BBBBA	–	–	–	–	·8	6·0
BBBBB	–	–	–	–	·1	2·6

TABLE 10

PEAK DESIGNATION

Sample		a	b	c	e	f	g
1	Experimental	25	31	5	30	5	1
	Theory	26	37	-	37	-	-
2	Experimental	12	33	7	32	6	6
	Theory	8	43	1	45	1	-
3,4	Experimental	5	37	5	37	9	4
	Theory	4	46	2	46	3	1
5	Experimental	2	35	3	34	17	5
	Theory	1	39	6	42	10	3
6	Experimental	3	27	1	26	30	10
	Theory	-	24	10	31	23	12

<u>Experimental</u> from deconvoluted ESCA spectra Table 8

<u>Theory</u> computed from pentad probabilities, assignments as in
Table 9

TABLE 11

ASSIGNMENT OF TRIADS FOR MOST PROBABLE PENTAD SEQUENCES FOR C_{1s} CORE LEVELS

Sequence

<u>ABAA</u>A $-CF_2CF_2(CH_2CH_2CF_2CF_2CF_2CF_2)CF_2CF_2-$
              e e b a a a

<u>ABAA</u>B $-CF_2CF_2(CH_2CH_2CF_2CF_2CF_2CF_2)CH_2CH_2-$
              e e b a a b

<u>ABAB</u>A $-CF_2CF_2(CH_2CH_2CF_2CF_2CH_2CH_2)CF_2CF_2-$
              e e b b e e

<u>BBAB</u>A $-CH_2CH_2(CH_2CH_2CF_2CF_2CH_2CH_2)CF_2CF_2-$
              f e c b e e

<u>BABA</u>A $-CH_2CH_2(CF_2CF_2CH_2CH_2CF_2CF_2)CF_2CF_2-$
              b b e e b a

<u>BABA</u>B $-CH_2CH_2(CF_2CF_2CH_2CH_2CF_2CF_2)CH_2CH_2-$
              b b e e b b

<u>ABBA</u>B $-CF_2CF_2(CH_2CH_2CH_2CH_2CF_2CF_2)CH_2CH_2-$
              e f f e c b

<u>BBBA</u>B $-CH_2CH_2(CH_2CH_2CH_2CH_2CF_2CF_2)CH_2CH_2-$
              g g f e c c

approximately the same the contributions from peak a for sample 1 (representing TFE units linked together) is much larger than for peak g in sample 5 (representing ethylene units linked together). This conclusion is strongly supported by the calculated pentad probabilities shown in Table 9.

These results are qualitatively in good agreement with the physical and more general spectroscopic properties of tetrafluoro-ethylene-ethylene copolymers. Improvements in resolution and signal/background which will become available with spectrometers employing monochromatized X-ray sources and multiple collector assemblies will considerably extend the scope of ESCA investigations of such copolymer systems. Indeed, the results presented here would suggest that it may well be feasible to determine monomer reactivity ratios directly from ESCA data.

D.   Domain Structure in Block Copolymers.   The differing sampling depths for photoemitted electrons from core levels corresponding to differing kinetic energies suggests that ESCA may be a very useful tool for studying domain structure of block copolymers[10].

As a simple example, we discuss here results for a 70/30 block copolymer of dimethylsiloxane and styrene which illustrates the great potential of ESCA in this area.

From studies of simple model systems, overall intensity ratios for homogeneous samples may be established for the principle core levels pertinent to the copolymer system (viz. I_α values may be established for O_{1s}, C_{1s}, Si_{2s}, S_{2p} core levels). For the block copolymer, comparison of the intensity ratios shows that essentially only the polydimethylsiloxane component is observed. This is an important result for two reasons. Firstly, it demonstrates that ESCA is potentially a very valuable technique for identifying the domain structure at the surfaces of block copolymers. This is of considerable importance if the friction and wear characteristics of such systems are being investigated. Secondly, since on the ESCA scale the sample appears to be homogeneously polydimethylsiloxane, we may use the approximate knowledge of escape depth dependencies to put a lower limit on the dimension of the domains. A reasonable estimate in this particular case would be >50Å. (This is based on an estimate of the mean free paths, and taking 98% of the infinity intensity values.)

E.   ESCA Studies of Structural Isomerism.   Nitroso Rubbers. We have thus far illustrated how raw ESCA data may be used to obtain information on copolymer compositions, gross structural features and domain structure. A more complex example illustrating how ESCA may be used to study structural isomerism is provided.

Nitroso Rubbers

CF$_3$
|
[NO] [CF$_2$-CFX]   X = F, C$_1$, H.

The C$_{1s}$ levels for these three polymers are shown in Fig. 36.
The assignment of peaks is straightforward and again the shifts are
understandable in terms of simple substituent effects, Fig. 37.
The 1:2 area ratio for CF$_3$ with respect to CF$_2$ carbons for the
copolymer involving tetrafluoroethylene together with the relevant
binding energies demonstrates the 1:1 alternating nature of this
copolymer. Similar arguments apply to the other polymers.
Having demonstrated the 1:1 alternating nature of these copolymers,
it is of interest to investigate the possibility of detecting
structural isomerism in these systems. It is clear, for example,
that for the C$_{1s}$ levels of the polymer involving trifluoroethylene,
although the three peaks are of equal area, there are substantial
differences in linewidth that are not apparent in the spectrum for
the polymer formed from the symmetrical olefin, tetrafluoroethylene.
To investigate the possibility that the lineshapes may encode
information concerning structural isomerism, therefore, calculations
have been performed on model systems. It is clear from tables of
binding energies that the binding energy of the CF$_3$ carbons are
virtually the same in all three polymers. Charge potential
calculations on model systems, Fig. 38, also reproduce this feature.

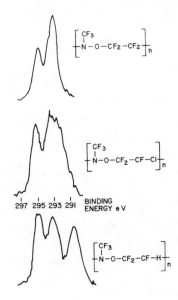

Fig. 36.   C$_{1s}$ levels for nitroso rubbers.

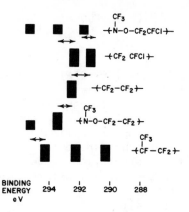

Fig. 37. Schematic of C_{1s} energy levels for some copolymers.

POLYMER MODELS

Ei – Ei° FOR CF$_3$ FROM CNDO CHARGES AND POTENTIAL MODEL

$$
\begin{array}{c}
CF_3 \\
| \\
R_1-N-O-R_2
\end{array}
$$

Ei – Ei°

CF_3CF_2 –	10·2	$-CF_2\,CF_3$
CF_2H-CF_2 –	10·2	$-CFHCF_3$
CF_3-CFCl –	10·1	$-CF_3$
CF_3 –	10·2	$-CFClCF_3$

Fig. 38. Charge potential calculations on model systems for nitroso rubbers showing constancy of binding energy for CF_3 carbon-1s levels.

POLYMER MODELS

COPOLYMER			SHIFTS IN RELATIVE B.E. (eV)	
			CALCULATED	OBSERVED
CF_3 NO and $CF_2=CF_2$	$CF_3\dotplus CF_2-N-O-CF_2\dotplus CF_3$ (with CF_3 branch)	$\Delta\ \underline{C}F_3-\underline{C}F_2$ (N)	1.7	1.8
		$\Delta\ CF_3-CF_2$ (O)	1.5	1.8
CF_3NO and $CF_2=CFCl$	$CF_3\dotplus N-O-CFCl\ \dotplus CF_3$ (with CF_3 branch)	$\Delta\ CF_3-\underline{C}FCl$	3.4	3.1
	$CF_3\dotplus CFCl-N-O\ \dotplus CF_3$ (with CF_3 branch)	$\Delta\ CF_3-CFCl$	3.6	3.1
CF_3 NO and $CF_2=CFH$	$CF_3\dotplus CFH-N-O-CF_2\dotplus CF_2H$ (with CF_3 branch)	$\Delta\ \underline{C}F_3-\underline{C}FH$	4.0	4.4
		$\Delta\ \underline{C}F_3-\underline{C}F_2$	1.4	1.9
	$CF_2H\dotplus CF_2-N-O-CFH\dotplus CF_3$ (with CF_3 branch)	$\Delta\ \underline{C}F_3-\underline{C}FH$	4.3	4.4
		$\Delta\ \underline{C}F_3-\underline{C}F_2$	2.1	1.9

Fig. 39. Computed shifts in C_{1s} levels for model systems for nitroso rubbers.

The models for discussing structural isomerism are shown in
Fig. 39 together with relative binding energies with respect to CF_3.
Considering firstly the copolymer of tetrafluoroethylene, the model
calculations suggest that CF_2 attached to oxygen or nitrogen have
closely similar binding energies thus accounting for the fact that
the composite CF_2 peak is only slightly broadened compared to that
for the CF_3 carbons.

The results for the trifluoroethylene copolymer suggests that
structural isomerism should indeed be detectable by ESCA. In view
of this, deconvolutions may be attempted for the trifluoro- and
chlorotrifluoro-ethylene copolymers using five individual peaks with
lineshape and linewidth corresponding to that for the CF_3 C_{1s} levels.
The five peaks in each case corresponding to four CF_2 and CFX carbons,
(in two pairs corresponding to the two distinct modes of bonding)
and only one for the CF_3 carbon.

No unique deconvolution is possible in the case of chlorotrifluoroethylene copolymers, however, if two sets of two lines are
used, with each member of each pair of equal intensity, a range of
good deconvolutions are found, the extremes of the range being
shown in Fig. 40. The range of the relative areas of the two pairs
is between 4:1 and 2:1 in the two cases. The average shifts in C_{1s}
binding energies between CF_3 and $CFCl$ within this range are 3.0 eV
and 3.8 eV compared with calculated values of 3.4 eV and 3.6 eV.

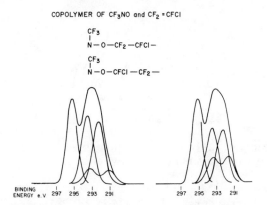

COPOLYMER OF CF$_3$NO and CF$_2$ =CFCl

CF$_3$
|
N—O—CF$_2$—CFCl—

CF$_3$
|
N—O—CFCl—CF$_2$—

BINDING ENERGY e.V   297   295   293   291   297   295   293   291

Fig. 40.   Deconvolution of C_{1s} envelope for nitroso rubber
involving chlorotrifluoroethylene.

The measured ratios of 4:1 to 2:1 indicate a 20-33% contribution
of structure

CF$_3$
|
(-NO-CFCl-CF$_2$)

in the polymer and this agrees with other evidence. The deconvolution for the polymer involving trifluoroethylene is less complicated and the results are shown in Fig. 41. The experimental shifts in C_{1s} binding energies between $\underline{C}F_3$ and $\underline{C}F_2$ are 1.6 and 2.2 eV and for the shifts between CF_2 and $C\overline{F}H$ 4.1 and 4.9 eV. The corresponding theoretical values from the model calculations are in good agreement ($\underline{C}F_3$ and $\underline{C}F_2$, 1.4 and 2.1 eV; $\underline{C}F_3$ and $\underline{C}FH$, 4.0 and 4.3 eV). This assignment would, therefore, appear to be perfectly reasonable. The area ratio between the two sets of pairs is 1:1, indicating a 50% contribution from

$$CF_3$$
$$|$$
$$(N-O-CFH-CF_2)$$

which is in excellent agreement with other studies. For these particular copolymers, therefore, in addition to establishing their elemental composition and comonomer incorporations it is possible to say something about structural isomerism.

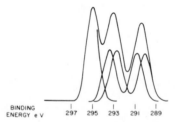

COPOLYMER OF CF_3 NO AND CF_2 =CFH

Fig. 41. Deconvolution of C_{1s} envelope for nitroso rubber involving trifluoroethylene.

8. Structural Elucidation of Polymers

With the background theoretical and experimental studies detailed in the previous sections, it is now possible to study structure and bonding in polymer systems which have proved intractable by more conventional techniques. As an interesting example of this, we may consider the identification of a polymer produced in substantial yield as a by-product in the fluoride ion initiated reaction of hexafluorobut-2-yne with fluorinated heterocyclic molecules. The reaction produces substantial amounts of an

insoluble off white polymer, with the carbon-1s spectrum shown in
Fig. 42.   The hydrocarbon peaks at ~285 eV arise from solvent
and/or the Scotch tape backing.   (It is difficult to remove the
last traces of solvent used in the reaction and it proved impossible
to press a film of the sample.) The spectrum for the polymer is

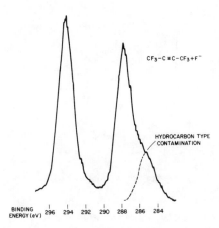

$CF_3-C\equiv C-CF_3+F^-$

HYDROCARBON TYPE
CONTAMINATION

BINDING
ENERGY (eV)   296   294   292   290   288   286   284

Fig. 42. C_{1s} spectrum of polymer produced as by-product in
nucleophilic substitution of perfluoroheteroaromatics
with fluoride ion and perfluorobut-2-yne.

therefore extremely simple with just two peaks of approximately
equal areas at 287.7 eV and 294.1 eV.   From our studies of
substituent effects (and since the polymer contains only C and F),
the peak at high binding energy may be unambiguously assigned to
$-\underline{C}F_3$ groups.   The assignment of the other peak is not as
straightforward; however, the most likely structure for the polymer
formed from fluoride catalyzed polymerization would be a polyene
of the form:

This, however, still leaves the question of the white colour of the
polymer open.   As confirmation of this structure a double check
may be performed; firstly, a theoretical calculation of the
relative binding energies of the two types of carbon, and, secondly
experimental studies of model compounds.

Considering firstly the calculations on model systems, it is of
interest to investigate the likely stereochemistry of such a polyene.
Considering the prototype diene for which the trans-configuration
expected on chemical grounds has been assumed, it is clear from
Fig. 43 that the energy minimum corresponds quite closely to the

ethylene units being at right angles to each other. On this basis,
the extended polymer is predicted to have a spiral structure and
the off white colour of the material is also accounted for.

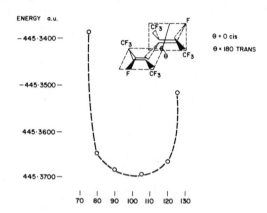

Fig. 43. The energy minimization curve (with respect to the
          dihedral angle between adjacent ethylenic units) for
          the diene polymer model.

As a better model, the corresponding triene may be used for
the calculation of binding energies with dihedral angles as found
for the diene, i.e. θ ≈ 90°. Considering the central unit as
characteristic of the polymer chain, the calculated relative binding
energies for the CF_3 and the central $\geqslant$ C=C $\leqslant$ carbons are 293.9 and
287.8 eV respectively, in excellent agreement with the experimental
values of 294.1 eV and 287.7 eV (Fig. 44).

MODEL FOR CHARGE POTENTIAL CALCULATIONS

E TOTAL θ=90 −559·4023
        θ=105 −559·3974

E_0 = 284·6 eV   k = 25

ATOM	BINDING ENERGY (eV)		EXPT.[l]
	CALC.	EXPT.[l]	
CF_3 $\underset{\diagup}{\overset{\diagdown}{C}}$ =	287.8	287.7	287.9
$\underline{C}F_3$ $\underset{\diagup}{\overset{\diagdown}{C}}$ =	293.9	294.1	294.2

Fig. 44. Comparison of the theoretical and experimental
          carbon-1s binding energies for the polyene and models.

The C_{1S} spectrum for the substituted pyridine model compound
Fig. 45 yields the data given in Fig. 44. This molecule contains
both CF_3 groups and the polyene structure proposed for the polymer
(u.v. spectral data confirms that in this system also the double
bonds are twisted with respect to the ring). The assignment of the
spectrum for this molecule (Fig. 45) is made easier by comparison
with the substituted vinyl pyridine, Fig. 46. The two peaks of
equal intensity correspond to the CF_3 carbons and the three carbons
of the diene portion (the terminal carbon is at higher binding
energy), and energy separation for this model are in excellent
agreement with the model calculations and experimental data for
the polymer thus completing the identification.

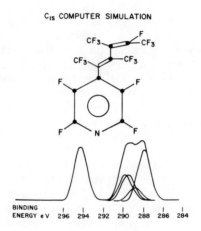

Fig. 45. C_{1S} levels and deconvolution for model diene system.

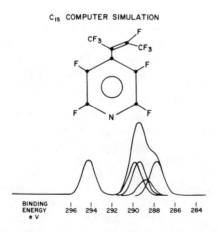

Fig. 46. C_{1S} levels and deconvolution for model monoene system.

9.   Surface Treatment of Polymers

The emphasis in the discussion so far has been predominantly on the application of ESCA to aspects of structure and bonding in polymer systems in the particular case where the surface is closely representative of the bulk. Situations often arise, however, where this is not likely to be the case. Chemical treatment, for example, must be initiated at the surface of a sample and differences between the surface, immediate subsurface and bulk might then be expected. Similarly, friction and wear processes may modify the surface with respect to the bulk. In the case of chemical treatment, the reaction in the bulk may be diffusion controlled and a technique which is capable of providing a profile of structure and bonding as a function of depth is therefore particularly valuable. Two chemical reactions of considerable technological importance are oxidation and fluorination. The former because of its importance in understanding polymer degradation and the latter because of the desirable mechanical, chemical and physical properties arising from surface fluorination.   In the following paper, we discuss in some detail the application of ESCA to such studies in which the unique advantages of ESCA are clearly apparent. In this section, we discuss two simple examples illustrating the great power of ESCA in differentiating surface from bulk properties.

A. Casing. A particular interesting surface treatment is that designated as Crosslinking by Activated Species of Inert Gases, developed by Hansen and Schonhorn[11]. By exposing polymer samples to activated species of inert gases, produced by electrodeless discharge, it has been shown that dramatic improvements in adhesive bonding may be brought about. The evidence presented suggested that the improvements arose from extensive crosslinking and it was inferred that this occurred at the surface.

We have made preliminary studies of the application of ESCA to this process choosing for our study an ethylene/tetrafluoroethylene copolymer. Samples were irradiated with a low energy (2KV) beam of argon ions with a beam current of 5μA for successive periods of 5 seconds and the C_{1s} and F_{1s} core levels monitored. To enable a reasonably complete study in a short time, resolution was sacrificed to sensitivity and the measurement of the core levels of a sample typically took ~2 mins. The results, Fig. 47, are quite striking and it is clear that argon ion bombardment causes extensive rearrangement in the outermost 50Å or so of the sample. The main features are as follows. The F_{1s} signal decreases whilst the total C_{1s} signal remains approximately constant in intensity. The original copolymer, with its largely alternating structure of ethylene and tetrafluoroethylene units, exhibits initially the characteristic doublet structure in the C_{1s} region. On successive argon ion treatment, the high binding energy peak corresponding to tetrafluoro-ethylene units progressively decreases in intensity and is replaced

by a region of intermediate binding energy. This is illustrated
in Fig. 48. More detailed studies of both these and other core
levels will be required to allow a complete elucidation of the
structure as a function of depth for this system. However, even
with the results from Figs. 47, 48 a few general conclusions may be
drawn. We will consider briefly at this stage, however, the
possible mechanism for the process.

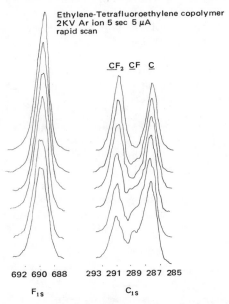

Fig. 47. C_{1s} spectra for copolymer of ethylene and tetrafluoroethylene subjected to argon ion treatment.

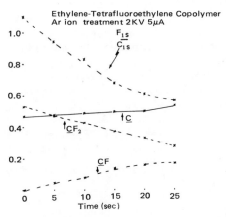

Fig. 48.   Relative intensities of core levels as a function
of irradiation time.

From the previously published evidence a plausible mechanism for CASING in the particular case of polyethylene would be as follows. Production of radicals by initial argon ion bombardment (i).

$$M^* \ + \ -CH_2-CH_2- \ \longrightarrow \ -CH_2-\overset{\bullet}{CH}- \ + \ H\bullet \ + \ M \qquad (i)$$

(M^* is electronically excited species of Ar)

The hydrogen atoms produced in (i) since they are in a fairly rigid matrix, may abstract a hydrogen from a CH_2 group in close proximity. Two situations seem probable. Firstly, abstraction from the CH_2 group next to a radical center, (which would be energetically particularly favorable because of the reduced C-H bond strength), leads to an unsaturated system (ii) viz.

$$-CH_2-\overset{\bullet}{CH}- \ + \ H\bullet \ \longrightarrow \ -CH=CH- \ + \ H_2 \qquad (ii)$$

(It is, of course, entirely feasible that collision of M^* with the polymer chain could lead to direct elimination of molecular hydrogen. The crosslinking, however, shows that although this could be a contributing pathway to the formation of unsaturated sites, radicals must be involved.) Abstraction of a hydrogen from an adjacent chain would lead to radical combination and hence crosslinking (iii).

$$\begin{matrix} -CH_2-\overset{\bullet}{CH}- \\ -CH_2-\overset{\bullet}{CH}- \end{matrix} \qquad \longrightarrow \qquad \begin{matrix} -CH_2-CH- \\ | \\ -CH_2-CH- \end{matrix} \qquad (iii)$$

For a fully fluorinated system (e.g. PTFE), the corresponding process to (i) might be expected to be energetically less favorable because of the greater bond strength of the C-F as compared to the C-H bond and also because of the low bond dissociation energy of molecular fluorine. Indeed, it has been noted that PTFE films require extended treatment to improve the adhesive properties[11].

In the case of a largely alternating copolymer of ethylene and tetrafluoroethylene, the initial process analogous to (i) might, therefore, be expected to be (iv).

$$M^* \ + \ -CF_2-CF_2-CH_2-CH_2- \ \longrightarrow \ -CF_2-CF_2-\overset{\bullet}{CH}-CH_2- \ + \ H\bullet \ + \ M \qquad (iv)$$

The high bond strength for HF suggests that (v) might be an efficient process,

$$-CF_2-CF_2-\overset{\bullet}{CH}-CH_2 \ + \ H\bullet \ \longrightarrow \ -CF_2CF=CH-CH_2- \ + \ HF \qquad (v)$$

although the possibility of direct molecular elimination of HF
cannot be discounted. Hydrogen atom abstraction from an adjacent
chain corresponding to (vi) and (vii) could then lead to crosslinked
products (viii) and (ix).

$$-CF_2-CF_2-CH_2-CH_2- \; + \; H\cdot \; \longrightarrow \; -CF_2\overset{\bullet}{C}F-CH_2-CH_2- \; + \; HF \qquad (vi)$$

$$-CF_2CF_2-\overset{\bullet}{C}H-CH_2- \; + \; H_2 \qquad (vii)$$

$$
\begin{array}{ccc}
-CF_2-CF_2-\overset{\bullet}{C}H-CH_2- & & -CF_2-CF_2-CH-CH_2- \\
 & \longrightarrow & \qquad\qquad\quad | \qquad\qquad\qquad (viii) \\
-CF_2-CF_2-\overset{\bullet}{C}H-CH_2- & & -CF_2-CF_2-CH-CH_2-
\end{array}
$$

$$
\begin{array}{ccc}
-CF_2-CF_2-\overset{\bullet}{C}H-CH_2- & & -CF_2-CF_2-CH-CH_2- \\
 & \longrightarrow & \qquad\qquad\quad | \qquad\qquad\qquad (ix) \\
-CF_2-\overset{\bullet}{C}F-CH_2-CH_2- & & -CF_2-CF-CH_2-CH_2-
\end{array}
$$

The ESCA data clearly show the decrease in F_{1s} signal
corresponding to loss of fluorine; in the process $\underline{C}F_2$ sites being
converted to $\underline{C}F$ type. This would suggest that (v) and (ix) are
the major product steps. The experiment lends itself to dynamic
studies, in which any volatiles produced could be analyzed by mass
spectrometry in parallel with the ESCA experiments, and such studies
will be reported in due course.

   B. Structure and Bonding in the Surface Layers of Fluorographite. In a symposium devoted to advances in polymer friction
and wear, it is not perhaps inappropriate to include one example of
how ESCA can provide data of immediate and obvious interest in this
field. The lubricating properties of layer lattice type materials
are well known and comparatively well understood[12]. In the particular
case of graphite, the favorable frictional properties have been
attributed to the relatively weak interactions between the sheets
due to the large lattice spacing of ~3.4Å[12]. In 1934, Ruff and
co-workers showed the graphite could be smoothly fluorinated by
the action of fluorine in the temperature range 420-460°C and
obtained a grey material with bulk composition $CF_{0.92}$. Since that
time numerous investigations have been carried out (particularly
by Margrave[14] and Watanabe[15] and their co-workers), into the
fluorination of graphite under a wide variety of conditions and
fluorinated graphite in the bulk composition range $CF_{0.25} - CF_{1.2}$
have been well authenticated. The material of nominal composition
$CF_{1.0}$ often referred to as poly(carbon monofluoride) is white with
highly desirable lubricating properties. Extensive studies have
been made of its friction and wear life properties and applications

range from such diverse uses as pressure and heat resistant
lubricant, to its use in high energy primary cells. For certain
applications, films were found to have wear lives greater than
either graphite or molybdenum disulphide, and the coefficient of
friction was lower than the former and about the same as the latter.
The useful limiting upper temperature is ~400°C[16]. Largely
as a result of the work of Margrave and co-workers, a considerable
amount of thermodynamic data has been accumulated on the system,
and it has also been the subject of several X-ray investigations.
Although as recent as ten years ago one standard text[17] considered
the structure to be a layer type with fluorines located such that
the lattice spacing between the carbon planes increased to 8.17Å,
more recent data suggested a network structure in which each carbon
was covalently bonded to three other carbons and a fluorine, the
carbons being arranged in puckered inter-connecting six-membered
rings with lattice spacing of 5.84A between the average planes,
Fig. 49[18].

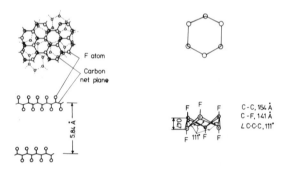

Fig. 49. Structure of crystalline graphite fluoride[18].

        The reaction of fluorine with graphite is undoubtedly
initiated at the surface and is followed by diffusion into the bulk.
In such circumstances, it would be surprising if reaction produced a
homogeneous material progressing from surface to subsurface to bulk.
Although the recent X-ray data, therefore, pertaining as it does to
the bulk undoubtedly gives a correct overview of the structure, it
is clear that the surface may well be drastically different and if
so have important ramifications as far as the interpretation of
friction and wear characteristics are concerned. The surface
sensitivity of ESCA is, therefore, particularly valuable for
investigating this in some detail. Fig. 50 shows the C_{1s} and F_{1s}
spectra for commercially available samples† of fluorographite of
composition C_1F_1. For comparison, data are also presented for
samples of nominal composition $C_1F_{0.8}$ and $C_1F_{1.2}$. The C_{1s} spectra

---

† We are indebted to Ozark-Mahoning Co. Special Chemical Division,
  Tulsa, Oklahoma, for these and other samples which will be
  discussed in detail elsewhere.

are immediately revealing. In each case the peak of low intensity at 285.0 eV is identified as arising from hydrocarbon contamination at the surface almost certainly arising from the packaging of the samples. The main peak centered at ~290.3 eV by comparison with model compounds previously studied[6] is consistent with tertiary -C-F type environments in a perfluoro system. This would tend to support, therefore, the major features of the structure proposed on the basis of diffraction studies, Fig. 49. The F_{1s} levels of binding energy 690.3 eV are also entirely consistent with covalently bond -C-F groups and rules out the possibility of a structure based on a loosely bound intercalate. The most significant feature displayed in Fig. 50, however, is the progressive increase in intensity of the shoulder centered around 292.4 eV to the high binding energy of the main C_{1s} peaks. This feature, together with the relative overall intensities of the C_{1s} with respect to the F_{1s} levels, strongly suggests that the shoulder arises from a differing structure at the surface of the samples. It is interesting to note that the C_{1s} binding energy is almost exactly the same as that for PTFE and we might speculate that the unusual chemical and mechanical properties of fluorographite may well arise from this surface feature. Clearly, further discussion must await more definitive studies and these are currently in hand. By studying the direct fluorination of different crystallographic faces of single crystal graphite, ESCA should shed considerable light on the surface structure of fluorographites particularly since for the well ordered system the possibility of carrying out electron channelling experiments exists (cf. ref. 5).

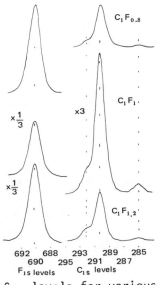

Fig. 50. F_{1s} and C_{1s} levels for various fluorographites.

10.   Charge Distribution in Polymers

In Section 6, a brief outline was given of the application of the charge potential model to the theoretical interpretation of ESCA data for polymers, by computing electron densities for model systems. If values of the parameters k and $E°$ (Fig. 28) are established for all the relevant core levels of a system, it is possible to invert the charge potential model to obtain experimental charge distributions. This procedure has obvious application in the calculation of charge distributions in molecules that are of such a size that conventional molecular orbital calculations are impracticable. A crude idea of the charge distribution is very often useful as a rule of thumb in discussing the chemistry of complex systems and also for preliminary assignment of other spectroscopic data. The simplicity of the procedure is illustrated in the particular case of CCl_3CF_3 in Figs. 51, 52. From studies of closely related systems, values of k and $E°$ may be established for each core level. Knowing the geometry and the measured core binding energies, the charge distributions may readily be obtained by solution of the system of linear simultaneous equations, Fig. 52. In fact, there is one more equation than unknowns

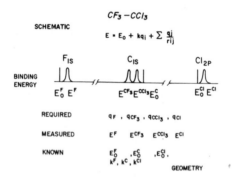

Fig. 51. Experimental charge distributions from inversion of charge potential model.

$$F_{1s} \quad E^F = E^F + k^F q_F + \frac{2q_F}{r_{F-F}} + \frac{3q_{Cl}}{r_{C-F}} + \frac{q_{CF_3}}{r_{F-CF_3}} + \frac{q_{CCl_3}}{r_{F-CCl_3}}$$

$$C_{1s} \quad E^C_{CF_3} = E^C_0 + kq_{CF_3} + \frac{3q_F}{r_{C-F}} + \frac{3q_{Cl}}{r_{C-Cl}} + \frac{q_{CCl_3}}{r_{C-C}}$$

$$Cl_{2p} \quad E^{Cl} = E^{Cl}_0 + kq_{Cl} + \frac{3q_F}{r_{Cl-F}} + \frac{2q_{Cl}}{r_{Cl-Cl}} + \frac{q_{CF_3}}{r_{Cl-CF_3}} + \frac{q_{CCl_3}}{r_{Cl-cCl_3}}$$

$$\text{ALSO} \quad 3q_F + 3q_{Cl} + q_{CF_3} + q_{CCl_3} = 0$$

Fig. 52. Linear simultaneous equations to be solved to obtain experimental charge distributions.

because of the condition of overall electroneutrality. This can in fact be used to advantage to study sample charging[19].

As a practical example of this procedure, Fig. 53 shows the experimental charge distribution in a complex fluorocarbon system which we have recently studied. This typically represents the practicable upper limit for SCF-MO calculations since even with a large computer (IBM 360/67) the cpu time required is 20 mins. which at current commercial rates represents an investment of several hundred dollars. By contrast, the ESCA experiment and the trivial computer operations necessary to solve the equations can be accomplished in about the same time. For more complex systems, therefore, the determination of charge distributions from ESCA data is a quicker and more economical possibility than direct calculation. For polymer systems, this approach is particularly valuable, and to illustrate the great potential in this area, Fig. 54 shows experimental charge distributions determined for polyvinylene fluoride and polytrifluoroethylene. These show rather effectively the large net migration of electron densities from carbon and hydrogen to fluorine.

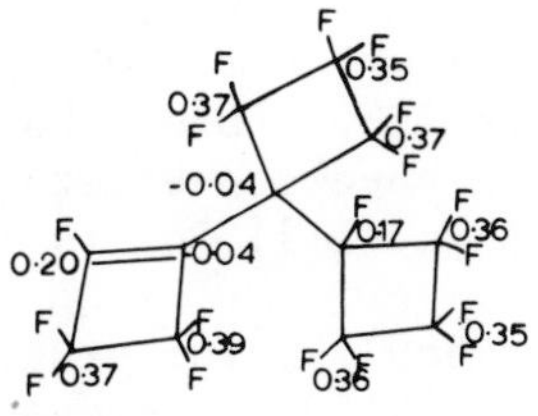

$$q_{expt^l} = 0.007 + 1.004 \; q_{CNDO}$$

Fig. 53.   Experimental charge distributions compared with direct theoretical computations for a typical complex organic molecule.

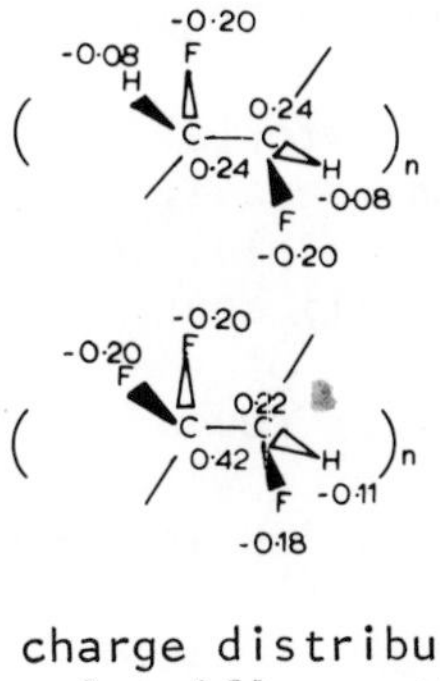

Fig. 54.   Experimental charge distributions for polyvinylene fluoride and polytrifluoroethylene.

11. Valence Bands of Polymers

In the previous sections, emphasis has been placed on the study of the core levels of polymer systems. Information concerning structure and bonding has then been largely inferred from shifts in core binding energies which reflect differences in valence electron distributions. Of obvious interest is the direct investigation of the valence levels of polymers; the derived information being relevant to the detailed interpretation in particular of the electrical properties of samples.

The valence energy levels of simple molecules have been extensively studied in the gas phase by low energy photoelectron spectroscopy, cf. Ref. 20. The inherent widths of the exciting radiations which are most commonly used, He(I) and He(II) (photon energies ~21 eV and ~40 eV respectively) are such that in favorable cases vibrational progressions may be resolved which considerably aids assignment. Although the development of ultraviolet photoelectron spectroscopy, (UPS) as the technique has come to be known, has primarily been in the hands of chemists, the application to the study of the valence bands of solids has primarily been the province of the physicist (the technique often being referred to as u.v. photoemission) and has dealt mainly with metals studied as evaporated films under UHV conditions[21]. It is only within the past few years, therefore, that interests and differences in instrumentation have converged and the UPS study of the valence bands of polymers is still in its infancy.

In the case of simple molecules, the study of the valence energy levels by ESCA has two distinct disadvantages compared with the corresponding UPS measurements. (It should be emphasized, however, that comparison between the two is very valuable since differential changes in cross sections with photon energy are useful for assigning the symmetries of occupied orbitals). Firstly, cross sections for photoionization are generally lower than for the longer wavelength photon sources used in UPS, and secondly the resolution is much poorer, (viz. photon linewidths He(I) ~ 5meV, $MgK\alpha_{1,2}$ ~ 800 meV).

In studying involatile materials such as polymers, however, these disadvantages are considerably offset. Thus, since there are so many vibrational modes possible, resolution becomes less of a problem since even with a He(I) source only broad unresolved bands would be obtained. Inspection of the escape depth versus kinetic energy curve discussed earlier reveals three major problems with using a low energy photon source which do not arise in ESCA examination of valence energy levels. Firstly, with a low energy photon source not all of the valence energy levels may be studied, only the higher occupied levels. Secondly, since the kinetic energy range for electrons will typically be in the range 0-21 eV

(<u>He</u>I), 0-40 eV (<u>He</u>II) it is clear that this is a region of rapidly
varying escape depth. Surface contamination is, therefore, very
critical, much less so for an X-ray source. Thirdly, in the
absence of contamination there are still difficulties of interpreting
the data because of marked differences in escape depths which do not
arise in ESCA since the escape depth dependence is virtually constant
across the valence band.

For these reasons, it is very convenient to study the valence
bands of polymers by ESCA. Typical of the data which may be obtained
is that shown in Fig. 55 for PTFE, polyvinylidene fluoride, and
polyethylene. This data, together with theoretical calculations,
leads to an assignment as follows. The large peak at highest
binding energy clearly evident in the fluorinated polymers arises
from molecular orbitals essentially F_{2s} in character. The prominent
peak at lowest binding energy for PTFE which is also clearly evident
in polyvinylidene fluoride is assigned to M.O.'s corresponding
essentially to fluorine 2p lone pairs. The shoulder at lower
binding energy in polyvinylidene fluoride which has its counterpart
in polyethylene may then be assigned to carbon hydrogen bonding
orbitals (essentially $C_{2p} H_{1s}$). The assignments for the remaining

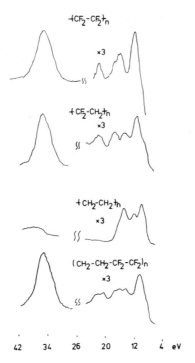

Fig. 55. Valence bands for PTFE, polyvinylidene fluoride,
polyethylene and copolymer of ethylene and
tetrafluoroethylene.

peaks are then essentially C-F and C-C (2s) bonding orbitals. With an appropriate correction, for work function (~ 5 eV), the binding energies correspond quite nicely to those obtained from UPS studies of simple systems containing the essential structural features.

For comparison purposes, the valence band for the ethylene-tetrafluoroethylene of composition close to 50-50 is also included. This bears a striking resemblance to that for polyvinylidene fluoride as one might expect on the basis of its largely alternating structure discussed earlier.

From the foregoing discussion, it should be clear that ESCA has a bright future in unravelling problems of structure and bonding in polymers in general and should in particular find considerable application in investigating those aspects of relevance to friction and wear.

ACKNOWLEDGEMENTS

In developing the research program sketched out in this lecture, I would like to acknowledge the help and encouragement of my colleagues Professor W. K. R. Musgrave and Dr. W. J. Feast. However none of the work would have been possible without the enthusiasm, skill and perseverance of my research workers, D. Kilcast, D. B. Adams, I. Scanlan, I. Ritchie and J. Peeling. Last but not least, I would like to express my gratitude to Mrs. E. McGauley for painstakingly transcribing my (often illegible) lecture notes into a publishable manuscript.

REFERENCES

1. D. T. Clark, "Chemical Aspects of ESCA", pp. 373-507, Electron
    Emission Spectroscopy, Ed. W. Dekeyser, D. Reidel Publishing
    Co., Dordrecht, Holland (1973).
2. (a) K. Siegbahn et al., Nova Acta R. Soc. Sci., Uppsala Ser.
        IV, 20 (1967);
    (b) K. Siegbahn et al., ESCA Applied to Free Molecules ,
        North Holland Publishing Co., Amsterdam (1969).
3. U. Gelius, E. Basilier, S. Svensson, T. Bergmark and
    K. Siegbahn, J. Electron Spectroscopy, 2, 405 (1974).
4. J. C. Tracey, "Auger Electron Spectroscopy for Surface Analysis",
    pp. 295-372, Electron Emission Spectroscopy, Ed. W. Dekeyser,
    D. Reidel Publishing Co., Dordrecht, Holland (1973).
5. C. S. Fadley, "Theoretical Aspects of X-ray Photoelectron
    Spectroscopy", pp. 151-224, ibid., idem.
6. Cf. Ref. 1 and D. T. Clark, D. B. Adams and D. Kilcast, J. Chem.
    Soc. Disc. Faraday Soc., 54, 182 (1972).
7. D. T. Clark, D. Kilcast, W. J. Feast and W. K. R. Musgrave,
    J. Pol. Sci., Polymer Chemistry Edn., 11, 389 (1973).
8. D. T. Clark, D. Kilcast, W. J. Feast and W. K. R. Musgrave,
    J. Pol. Sci. Al, 10, 1637 (1972).
9. D. T. Clark, W. J. Feast, I. Ritchie, W. K. R. Musgrave,
    M. Modena and M. Ragazzini, J. Pol. Sci., Polymer Chemistry Edn.,
    in press (1974).
10. H. Kawai, T. Soen, T. Inove, T. Ono and T. Uchide, Memoirs
    of Faculty of Engineering, Kyoto University XXXIII, Pt. 4
    (1971).
11. R. H. Hansen and H. Schonhorn, Polymer Letters, 4, 203 (1966).
12. F. P. Bowden and D. Tabor, The Friction and Lubrication of
    Solids , Part II, Oxford (1964).
13. O. Ruff, O. Bretschneider and F. Elert, Z. Anorg. Allg. Chem.,
    217, 1, (1934).
14. Cf. (a) J. L. Wood, R. B. Badachhape, R. J. Lagow and
        J. L. Margrave, J. Phys. Chem., 73, 3139 (1969);
    (b) R. J. Lagow, R. B. Badachhape, P. Ficalora, J. L. Wood,
        and J. L. Margrave, Synthesis in Inorganic and Metal
        Organic Chemistry, 2(2), 145 (1972).
15. N. Watanabe, Y. Koyama, A. Shibuya and K. Kuman, Memoirs of
    Faculty of Engineering, Kyoto University, 33, 15 (1971).
16. R. L. Fusaro and H. E. Sliney, Chem. Abs., 71, 5115s.
17. C. F. Wells, Structural Inorganic Chemistry , 3rd Edn.,
    Oxford (1964).
18. N. Watanabe and M. Takashima, Abstract p. 19, 7th International
    Symposium on Fluorine Chemistry, Santa Cruz, California
    (1973).

19. D. T. Clark, D. B. Adams and D. Kilcast, Chem. Phys. Letters,
     13, 439 (1972).
20. D. W. Turner, C. Baker, A. D. Baker and C. R. Brundle,
     Molecular Photoelectron Spectroscopy , Wiley, New York
     (1970).
21. W. E. Spicer, "Optical Density of States Ultraviolet Photo-
     electric Spectroscopy", Proceedings - Electron Density of
     States - Ed. T. H. Bennett, National Bureau of Standards
     Special Publication 323, 139-158 (1971).
     U.S. Government Printing Office, Washington, D. C.

DISCUSSION OF PAPER BY D. T. CLARK

R. F. Roberts (Bell Laboratories): Dr. Clark, since all of the data which you have described are concerned with polymers, which are electrical insulators, I wonder what sort of technique you have employed to correct for sample charging effects during the ESCA measurements?

D. T. Clark: The most convenient method we have found is to use the hydrocarbon peak at 285 ev which builds up over many hours in the spectrometer. The technique is fully described in the paper.

Participant: In your slide, you showed the universal curve of escape depth versus kinetic energy, the escape depth for electron is 10-15Å. I believe that for other materials, e.g. stearic acid, the escape depth could be 100Å. Could you elaborate on this?

D. T. Clark: There is some confusion in the literature over escape depths and sampling depths. If we define the electron mean free path or escape depth as that distance in which $1/e^{th}$ of the electrons will not have suffered any energy loss then we may note the following. For escape depths of 5, 10, 15 and 20Å, 95% of the signal intensity in a homogeneous material derives from the topmost ~15, 30, 45 or 60Å respectively. We might loosely refer to a sampling depth of 60Å say when in actual fact the escape depth might be say 20Å. The data presented in the generalized escape depth vs kinetic energy curve are reasonably well authenticated. Although the elegant experiments of Siegbahn et al. on α-iodo stearic acid indicated the surface sensitivity of ESCA, they cannot be regarded as definitive with regards to quantitative delineation of escape depths. The only data available pertaining to carbon is that indicated in the generalized curve derived from studies of graphite. These have been used as a starting point for our analysis of surface fluorination. The most significant feature of our results is that allowing the relevant escape depths to float between wide limits still brackets the "refined" values for F_{1s}, C_{1s}, and F_{2s} levels in fluoropolymer systems, close to the generalized curve.

D. Tabor (Cambridge University): Can you use ESCA to determine the bonding between the individual planes within the bulk of a graphite crystal? If so, can you show up whether interlayer bonds (π-bonds) are relaxed at the free surface?

D. T. Clark: I don't think it would be possible to study directly the interaction between the planes in graphite by ESCA; however, I can envisage that it might be possible to do so indirectly. If good theoretical calculations were available for example for a model graphite plane, it might be possible to infer something about the interaction between planes by comparison of theory and experiment as far as the band structure for the real and model system was concerned.   In this connection, angular dependence and studies at different photon energies would be particularly useful.

The Atomic Nature
of Polymer–Metal Interactions
in Adhesion, Friction, and Wear

Donald H. Buckley and William A. Brainard

National Aeronautics and Space Administration

Lewis Research Center, Cleveland, Ohio 44135

The polymers PTFE (polytetrafluoroethylene) and polyimide were studied contacting various metals in adhesion and
sliding friction experiments. Field ion
microscopy and Auger emission spectroscopy,
were used to examine the nature of the
polymer-metal interactions. Strong adhesion
of polymers to all metals in both the
clean and oxidized states was observed.
Adhesive bonding was sufficiently strong
with the cohesively weaker metals such as
aluminum that metal transferred to the
polymers. Adhesion coefficients measured
approach those for clean metals in contact
and field ion microscopy indicates that
the polymer to metal bonds are chemical in
nature. The field ion microscope also
indicates that polymer fragments transferred
to cohesively strong metals such as tungsten
and that these fragments are highly oriented.
Auger emission spectroscopy indicates that
a single pass of PTFE across a metal surface
is sufficient to generate a transfer film.
Electron induced desorption of the PTFE
from the metal surface indicates that
bonding to the metal is via the carbon atom
with little to no fluorine to metal interaction.

315

INTRODUCTION

In lubrication systems polymers are frequently used in sliding or rolling contact with metal surfaces. It is of interest therefore to study the fundamental surface interactions between these polymers and metals and the manner in which those interactions influence adhesion, friction, and wear.

Of solids the polymers used, PTFE (polytetrafluoroethylene) and polyimide are two structures with inherently good self-lubricating and transfer characteristics. Unlike polymers such as nylon they do not depend upon the presence of adsorbates such as water vapor or other environmental constituents for their low adhesion, friction and wear properties[1-3].

Another type of polymer-metal interaction that has been observed is associated with the formation of polymeric surface films from the interaction of monomers or (lower molecular weight species) with metal surfaces in sliding contact. Frictional energy provides the activation energy for polymerization. This phenomenon was first observed in electrical contacts[4,5]. And more recently it has been observed in friction and wear studies[6-7].

The objectives of this paper are twofold, first, to examine the adhesion, friction and wear behavior of the solid polymers PTFE and polyimide with various metals and second, to gain insight into the role of the friction process in the generation of polymer surface films. Analytical surface tools including field ion microscopy and Auger emission spectroscopy were used to examine the atomic nature of polymer-metal surface interactions.

MATERIALS

The solid polymer bodies of PTFE and polyimide (obtained from E. I. Dupont Co.) used in these studies were research grade materials. The polymer forming gaseous monomer vinyl chloride, was 99.95 percent in purity as was the ethyl chloride.

The metals used were iron, tungsten, gold, and aluminum. The tungsten, gold, and aluminum were all 99.999 percent and the iron was 99.99+ percent in purity.

EXPERIMENTAL RESULTS

Field Ion Microscopy

The field ion microscope (FIM) is a powerful tool for studying the adhesion process. A combination of high magnification and

resolution of 2 to 3 Å permits the adhesion process to be studied
in atomic detail. The principals and operation of the FIM are
described by Müller[8], who also demonstrated that adhesion could be
studied with the FIM[9].

A schematic of the FIM used as an adhesion apparatus is shown
in Fig. 1. A specially constructed contact device can be moved into
place under the field ion tip by the compression of bellows. With
suitable voltage applied to the electromagnets, the beam can be made
to move upward and contact the field ion tip. The load applied
during contact is determined by the change in output voltage from a
solar cell. PTFE or polyimide is mounted on the surface of a thin
(0.002 cm) gold foil. When contact is made, the foil deflects downward cutting off some of the light illuminating a solar cell surface
thus changing the output voltage (see Ref. 10 for details). When the
beam is moved downward away from the tip, the adhesion between the
tip and the PTFE or polyimide pull the foil upward past the original
zero load point until separation occurs. Calibration of the loads
applied to the foil, to cause given voltage changes, was done with
an electronic balance with a sensitivity of <1 µg.

A series of PTFE-tungsten contacts were made at loads between
20 to 30 µg and the force of adhesion measured. Figure 2 is a
micrograph taken prior to contact of a tungsten tip. Figure 3 is
a photomicrograph taken following contact with PTFE for a few seconds.
Many extra image points are apparent on the post-contact micrograph
particularly on the (110) plane. Adsorbed or adhered atoms can be
observed because the geometry of the extra atoms on the surface of
the flat creates points of localized field enhancement resulting in
increased probability of ionization. Thus, clusters visible on the
(110) plane are fragments of PTFE which remain adhered to the tungsten
surface after separation has occurred. The other bright image spots
also represent PTFE on the metal surface but their cluster-like
nature cannot be resolved. The fragments of PTFE have the appearance
of the end of a PTFE chain which is normal to the (110) plane. The
fact that the fragments are stable at the very high electric field
required for helium ion imaging implies that the bonding between the
PTFE and tungsten is very strong, otherwise field desorption of the
adhered PTFE would occur.

To obtain a measure of the bonding between the PTFE and tungsten,
the forces of adhesion were measured for a number of contacts over
varying periods of contact time. The results are summarized in
Fig. 4. For short contact times, the forces of adhesion were
immeasurably small. After 2 minutes, however, the force of adhesion
increased markedly. At contact times of 4 to 6 minutes, adhesion
coefficients approaching those for clean metals in contact were
obtained.

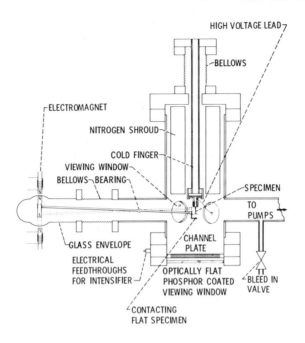

Fig. 1. Schematic of field ion microscope.

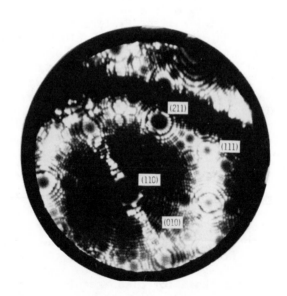

Fig. 2. FIM image of tungsten tip prior to contact (18.0 kV,
helium image gas, liquid nitrogen cooling).

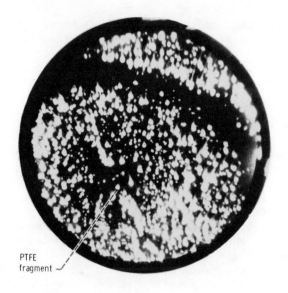

PTFE
fragment

Fig. 3.   FIM image of tungsten tip following contact with
          PTFE (16.5 kV, helium image gas, liquid helium cooling).

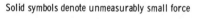

Solid symbols denote unmeasurably small force

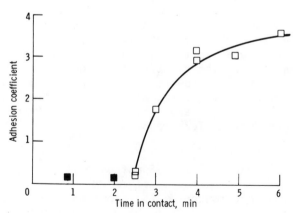

Fig. 4.   Adhesion coefficients for PTFE-tungsten contacts.

It was observed that negligible adhesion force was obtained when a second contact was made with a previously contacted tip indicating the adhesive polymer to metal bond is stronger than the cohesive polymer bond. Polymer radicals can be expected to occur as a result of the breaking of chains, by the chemical interaction between polymer and metal. Thus, for PTFE contacting a clean tungsten surface, the possibility of reactive valence states of carbon atoms in PTFE bonding to tungsten exists.

A heavily loaded tungsten-PTFE (approximately three times more load ~1 mg) contact gave the rather surprising result that extensive deformation of the tungsten occurred. The deformation extended far into the bulk of the material as shown in Fig. 5, which was taken after extensive field evaporation (an effective increase in tip radius of approximately 50 percent). Generally, during heavy loaded contacts the tip will bend so that the image is often displaced on the screen; however, there is no displacement of (110) in the micrograph of Fig. 5. The tip, when pressed into the PTFE, may be prevented from bending and thus absorbing the stress further up the shank by the very large area of contact (relative to metal-metal contacts).

Mechanical contacts with a polyimide polymer contacting tungsten tips were made in vacuum of 10^{-9} torr with both light and heavy loads. At light loads the results obtained were analogous to those obtained with PTFE. Random distribution of bright spots were visible indicative of polymer fragments adhering to the tungsten. Spots (polymer fragments) were particularly heavily clustered on the (110) surface as was observed with PTFE.

To determine if load would significantly influence the polyimide-tungsten contacts, an experiment was conducted with polyimide contacting tungsten using a load approximately three times greater than that used for light load contacts. Figure 6 is the image of the same tungsten tip after a heavily loaded polyimide contact. The results show a series of elongated spots all radially projected outward from a dark nonimaging area on the right side of the micrograph. As was also observed with PTFE, it appears that the polyimide polymer fragments have been oriented by the radial stresses projecting outward from the center of contact, where the center of contact is the dark region to the right in the micrograph.

Auger Emission Spectroscopy

Auger emission spectroscopy (AES) provides a technique for determining the chemical composition of a surface for elements heavier than helium with a high degree of sensitivity (i.e., 1/100th of a monolayer). AES was used in conjunction with a vacuum friction apparatus to provide an instantaneous chemical analysis of a metal

Fig. 5. Tungsten after heavy load PTFE contact.

Fig. 6. Tungsten after polyimide contact (9.25 kV, helium
         image gas, liquid helium cooling).

surface during both static and dynamic contact with PTFE. The
experimental apparatus is shown in Fig. 7 and further details on the
technique and equipment are available in the literature[12].

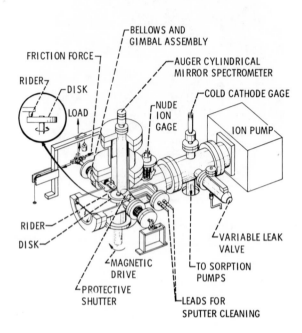

Fig. 7. Friction apparatus with Auger spectrometer.

Static Contact. Transfer of PTFE to atomically clean metals
by static contact was observed for all metals used. These included
iron, tungsten, aluminum, and gold. Figure 8 shows three Auger
spectrograms for a gold surface before and after adhesive contact.
The presence of carbon and fluorine in Fig. 8(b) indicate the transfer
of PTFE to the gold surface.

The possibility that the transfer of PTFE to metal might be
adversely affected by the presence of an oxide film on the metal
was investigated by two methods. In the first method, high-purity
oxygen was admitted to the chamber after the disk surface had been
sputter cleaned with the oxygen being chemisorbed on the surface
to monolayer coverage. Static contact was then initiated, and again
transfer of PTFE was observed. Thus, the presence of a monolayer of
chemisorbed oxygen does not prevent the transfer observed.

The second method involved the use of a preoxidized aluminum
disk. It is known that the natural oxide layer on aluminum is many
layers thick. Removing the normally present adsorbed carbon dioxide
and carbon monoxide by a short sputtering (20 min) exposed the "clean"

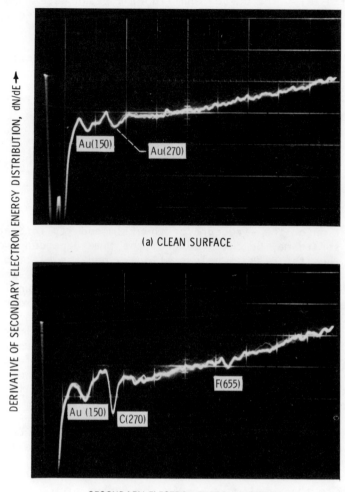

DERIVATIVE OF SECONDARY ELECTRON ENERGY DISTRIBUTION, dN/dE →

Au(150)

Au(270)

(a) CLEAN SURFACE

Au (150)

C(270)

F(655)

SECONDARY ELECTRON ENERGY, VOLTS →

(b) AFTER 500-GRAM-LOAD STATIC CONTACT FOR 1 MINUTE WITH PTFE

Fig. 8. Auger spectrogram of gold surface

aluminum oxide layer. Static contact was again initiated, and again PTFE was found on the surface. Thus, PTFE transfers to the oxide of aluminum as well as to the clean metal, which implies that the chemical activity of the substrate was not an important factor in the transfer observed in these static contact experiments.

Sliding Contact. Sliding contact was initiated on an atomically clean tungsten disk. The velocity was 1.0 centimeter per second, and the load was 100 grams. A transfer film of PTFE was generated on the disk during the first revolution. The Auger spectrogram is

shown in Fig. 9(a). The carbon and fluorine peaks were much larger
than those in the static contact experiments and indicated the
presence of larger amounts of PTFE on the surface.

The film appeared uniform across the track, as indicated by
the constance of the peaks when the deflection plates in the electron
gun moved the beam across the track. The film also appeared uniform
and continuous along the circumference of the track, as indicated by
the constancy of the peaks throughout the first revolution of the
disk.

Notice that the tungsten peak is still visible in the Auger
spectrum. This indicates that the film was only a few monolayers
thick, since AES is sensitive only to the first few layers on the
surface.

Information on the structure of the film and its interaction
with the substrate may be obtained from the time dependence of the
Auger peaks when the disk is held stationary and the electron beam
impinges on one spot of the surface. In Fig. 9, the two spectra
were taken 60 seconds apart. It can be seen that the fluorine peak
has decreased while the carbon and tungsten peaks have grown. The
incident 2000-eV electrons used to excite the Auger transitions have
severed the carbon-fluorine bonds in the PTFE, and the fluorine has
desorbed from the surface. With the departure of the fluorine, Auger
electrons from the carbon and tungsten can leave the surface, enter
the analyzer, and cause growth of these peaks. Exposure of the
surface to the electron beam for about 1 minute resulted in complete
disappearance of the fluorine peak (Fig. 2(b)).

In Ref. 13 the cross section value for electron induced fluorine
desorption was calculated from the time dependence of the fluorine
Auger peak as 5×10^{-18} cm^2, implying no interactions exist between
the fluorine in the PTFE and the metal substrate. This indicates
that bonding to the metal must be via the carbon atom.

Fig. 10 shows the curve of friction coefficient versus number
of passes of the disk for PTFE sliding on atomically clean tungsten
and aluminum. The value of 0.08 obtained for PTFE on tungsten is
consistent with values usually reported for PTFE sliding on metals
in air. The friction for PTFE on aluminum, however, rose drastically
from 0.08 at the start to over 0.5 in less than one complete
revolution. Severe "machining" of the aluminum disk occurred and
metal cut from the weak track could be seen at the rider-disk contact
zone, as well as chips of aluminum covering the surface. The severe
scoring of the aluminum occurred both in the presence or absence of
an oxide film.

DERIVATIVE OF SECONDARY ELECTRON ENERGY DISTRIBUTION, dN/dE

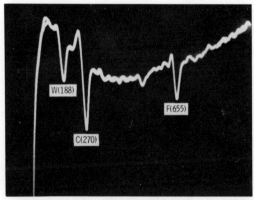

(a) DISK WITH PTFE FILM GENERATED BY 100-GRAM-LOAD SLIDING
AT VELOCITY OF 1 CENTIMETER PER SECOND AFTER 1 REVOLUTION
OF DISK

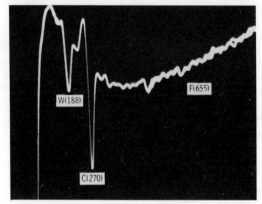

SECONDARY ELECTRON ENERGY, VOLTS ➡

(b) SAME SPOT DISK AS IN (A) AFTER 1-MINUTE TIME INTERVAL;
70-MICROAMPERE BEAM CURRENT.

Fig. 9. Auger spectrogram of tungsten disk.

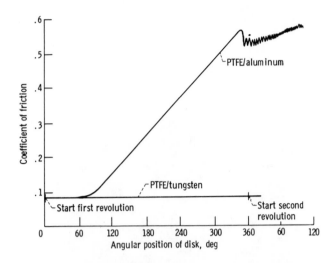

Fig. 10.   Coefficient of friction of PTFE on aluminum disk and
          tungsten disk in vacuum.   Siding velocity, 0.07
          centimeter per second; load, 250 grams.

DISCUSSION

The results of the study with PTFE and polyimide polymers in
contact with tungsten indicate that polymers will transfer to a
clean metal surface on simple touch contact. The transferred polymers
at the high field for helium ion imaging imply that the bond of the
polymer to the metal surface is chemical in nature. With PTFE it
is hypothesized that the bonding is that of carbon to the metal
surface because the carbon to carbon bond is the weakest bond in
the PTFE structure and the one most frequently seen broken on polymer
scission. Furthermore, the carbon could readily interact with the
clean tungsten to form bonds stable at the imaging and field
evaporation voltages applied in this study.

With the polyimide oxygen, nitrogen and carbon bonding to the
tungsten surface are possible. But again because of the imaging
and evaporation voltages employed the bond must be chemical in
nature.

The concept of chemical bonding of the polymers to the clean
metal surface necessitates the breaking of bonds in the organic
molecule and the subsequent formation of metal to carbon, nitrogen
or oxygen bonds to form organometallics. The breaking of organic
bonds by metal surfaces is observed with hydrocarbons contacting
metals in the field of catalysis. The tendency for such reactions

should be increased when the metal surface is atomically clean because of the enhanced surface activity of the metal.

The effect of loading in the transfer of polymers was examined. Larger amounts of polymers were observed with an increase in load. The polymer appears to remain on the tungsten surface in longer chain fragments indicating that fracture occurred deeper in the polymer body than was observed at light loads. When field evaporation was conducted the polymer chain length that adhered to the tungsten could be reduced to that observed with light loads.

It is of interest also to note that preferred orientation of the polymer chains toward the zone of contact has occurred. This resulted when the tungsten was pressed into the polymer under load. Adhesion of polymer occurred to the tungsten surface. Polymer bonds were broken in the bulk and the relative tangential motion of the polymer body along the radius of the tungsten tip resulted in texturing (preferred orientation of chain fragments).

CONCLUDING REMARKS

Based upon the results presented herein the following concluding remarks are made:

1. In adhesion experiments with PTFE and polyimide contacting tungsten results obtained in the field ion microscope indicate that the polymers bond chemically to the clean metal surface.

2. Polymer chain fragments which transfer to the surface of tungsten in field ion microscopy adhesion studies are highly oriented.

3. Auger emission spectroscopy analysis of PTFE transfer films to various metal surfaces indicate that the PTFE is bonded to the metal surface via the carbon atom.

REFERENCES

1. R. P. Steijn, ASM, Met. Eng., Quest, 7-9 (1967).
2. K. R. Makinson and D. Tabor, Proc. Roy. Soc. (London), A, <u>281</u>, 49 (1964).
3. D. H. Buckley and R. L. Johnson, SPE Trans., <u>4</u>, 306 (1964).
4. H. N. Hermance and T. T. Egan, Bell Sys. Tech. Jour. 37, 739 (1958).
5. S. M. Skinner, R. L. Savage and J. E. Rutzler, Jr., Jour. Appl. Phys., <u>24</u>, 438 (1953).

6. R. S. Fein and K. L. Kreuz, ASLE Trans., $\underline{8}$, 29 (1953).
7. R. S. Fein and K. L. Kreuz, ACS Div. of Pet. Chem., $\underline{13}$, 327 (1968).
8. E. W. Müller, Adv. Electronics Electron Phys., $\underline{13}$, 83 (1960).
9. E. W. Müller and O. Nishikawa, Adhesion on Cold Welding of Materials in Space Environments, ASTM Spec. Tech. Publ., No. 431, 67 (1968).
10. W. A. Brainard and D. H. Buckley, NASA TN D-6492 (1971).
11. W. A. Brainard and D. H. Buckley, NASA TN D-6524 (1971).
12. D. H. Buckley and S. V. Pepper, NASA TN D-6497 (1971).
13. S. V. Pepper and D. H. Buckley, NASA TN D-6983 (1972).

DISCUSSION OF PAPER BY D. H. BUCKLEY

M. O. W. Richardson (Loughborough University): In the light of the accepted concepts of the P.T.F.E. molecule being essentially a perfluorinated alkane in which the main chain undergoes a complete twist over a distance of about 26 carbon atoms and is surrounded by a tightly packed 'protective' sheath of fluorine atoms, I should like to enlarge my own understanding by asking the following: Is the P.T.F.E. main chain broken to form two separate free radicals <u>prior</u> to the apparent carbon/clean metal stable interaction?

If the answer is yes, then would you expect the P.T.F.E. molecule to go through a transition fluorine/metal state (as tentatively hypothesized in my paper yesterday for nC_5F_{12}/clean ion interactions) as a precursor to main chain breakage stimulated by metal catalytic activity?

If the answer is no, then how do the carbon main chain atoms approach the clean metal surface close enough for interaction without being sterically hindered by the fluorine atoms?

D. H. Buckley: We don't really know whether the polymer bonds directly to the metal on contact via the carbon atom or dissociates first. A metal to fluorine interaction appears unlikely because the bond should be more stable than the carbon to metal bond once formed. The fluorine would not be subject to electron induced desorption prior to desorption of the carbon.

D. T. Clark (University of Durham): Is it possible that the Auger analysis itself involving as it does an intense electron beam, modifies your results. It may well be that the initial interaction involves metal fluorine bonding and then specific cleavage under electron bombardment. In the case of your claimed PVC film, what evidence is there that the structure involves a simple polymerization rather than π alkene or σ vinyl complexes.

D. H. Buckley: Both elements of the polymer structure are present on the solid surface in the proper intensities ratios. It is only when the electron beam energy is appreciably increased that fluorine induced desorption occurs.

With respect to the PVC film, it may well be simple polymerization. This would appear most likely. It does not discount the possibility of the formation of either π alkene or vinyl complexes in addition thereto. When thick films are developed, they have all the properties of polyvinyl chloride.

R. F. Roberts (Bell Laboratories): Dr. Buckley, since Auger spectral peaks can also exhibit chemical shifts, I wonder if you observed any such shifts in the tungsten-tungsten/P.T.F.E. system which might substantiate your contention that the carbon and not the fluorine in P.T.F.E. is chemically bonded to tungsten?

D. H. Buckley: Chemical shifts, while observed were not of sufficient magnitude to draw any strong conclusions therefrom. We have observed significant chemical shifts with other systems, for example, the formation of aluminum oxide from the interaction of oxygen with aluminum on a copper-aluminum alloys where the shift is of sufficient magnitude to be really meaningful.

G. M. Robinson (3M Company):   Would you discuss any work you have done with studying the adhesion of a polymer containing a boundary lubricant on metal surfaces?

D. H. Buckley: We are currently examining molecular structures which contain surface active elements including sulfur, chlorine and phosphorus. Some of these compounds contain all three elements. The polymer acts as the source for the active species. Interestingly enough, frequently only one of the active species is found in the surface film.

G. Abowitz (Xerox Corporation):   Does the carbon bonding mechanism you discussed for clean W and Fe surfaces depend on the fact that they are BCC metals and have a high affinity for carbon? What about non BCC metal surfaces?

D. H. Buckley: Similar carbon bonding results have been observed on clean iridium and rhenium surfaces.

D. Tabor (Cambridge University):   Could the speaker comment on the mechanism by which polymer formation occurs? Does it arise from known processes such as the exposure of fresh surface, local hot-spots high local pressures, etc., or must one invoke a new process and coin another Greek word like Tribochemistry to cover our ignorance.

D. H. Buckley: It is not necessary for us to coin a new word to cover the concept of friction induced polymerization. There are many factors associated with the friction process which can supply the required activation for polymer film formation. The initial film on the iron surface is subjected to interfacial shearing, flash temperatures and clean strained metal, any one or all of which can initiate radical formation and subsequent polymerization.

H. Gisser (Frankford Arsenal):   I would like to comment on the polymer formation. The information coming from many sources not only do not contradict but support each other. From chemistry of the system, a free radical generated during the process polymerizes vinyl chloride; now we have seen evidence that such a polymer has been formed as demonstrated by Don Buckley with auger spectroscopy.

Yesterday, we heard that perfluoropentane came into contact with a clean metal surface and fragmented to form free radicals. In addition to that, there is some early work showing that aromatic compounds after rolling contact with metal surface readily formed free radicals or ion radicals. This resulted in a better load bearing capacity which is consistent with the fact that more polymer is formed when you have more aromatic compound present.

D. V. Keller (Syracuse University): I would like to add two statements. First, tribochemistry, we all recognize the fact that the metal surface is rough and it can deform magnificently. The metal atoms in that surface have an energy state quite abnormal in comparison to an equilibrium state. Secondly, it is well known that these are catalytic in nature, and it is possible that there are enough energy to cause many things happening on the surface. Therefore, loading conditions and high pressure on deformed sites might cause a lot of extra things to happen.

Experimental Investigation of the Effect of Electrical Potential on Adhesive Friction of Elastomers

Arvin R. Savkoor and T. J. Ruyter*

Laboratorium voor Voertuigtechniek

Technische Hogeschool, Mekelweg 2, Delft, The Netherlands

Using an electrically conductive, carbon black filled rubber slider, it is found that the frictional force may be increased when an electrical potential is applied across the interface. Under certain conditions a two to three fold increase in friction was observed for voltages up to 2000 V. The current flow is generally so small that the power consumption does not exceed a few watts. Such a remarkable influence on friction obtained at the expense of only a small power may be of interest in certain applications. It is therefore interesting to study phenomenological laws of friction under these conditions. It is found that Amonton's law of proportionality between normal load and frictional force is approximately valid for given speed, temperature and applied voltage.

Within the range of voltages used, the increase in friction was observed for rubber compounds with 50 p.p.hr

* Presently with Ballast - Nedam Dredging, Amstelveen, The Netherlands

carbon black content. No influence
was observed for compounds which contained
only 10 p.p.hr carbon black.

Some effort is made to understand
the mechanism which is responsible for
the observed increase in friction. It
appears that the increase results from
an additional electrical force which
supplements the normal force on the real
area of contact. An alternative explanation
based upon rise in the intrinsic coefficient
of friction is thought to be less plausible.
In order to elucidate the details of the
mechanism, certain experiments have been
performed. An electrical model of the
contacting bodies is considered.

INTRODUCTION

Adequate amount of carbon black in the compounding formulation
makes rubber conductive. The precise mechanism of conductivity is
a subject of a considerable discussion, although it appears that
the particle size and the polymer chain structure have a major
influence[1]. The conductivity of the carbon black loaded rubber is
strain sensitive[1]. During the course of experiments, Schallamach[2]
found that a thin layer of conductive rubber next to the contact
surface undergoes a severe deformation whereby its electrical
impedance is considerably increased. Further, he pointed out that
a D.C. potential applied between the sample and track during sliding
could give rise to electrostatic attraction between them; an
additional frictional force results. Using a new sample of carbon
black loaded compound, it is possible to measure the change of
conductivity due to deformation. The study reported[2] has been of
a cursory nature and no work was reported showing increased friction
with A.C. potential. The present paper is aimed at a more detailed
and systematic study of the laws of friction with applied electrical
potential and a consideration of the factors responsible for
increasing friction. Both D.C. and A.C. voltages have been used in
our experiments.

In the literature, the influence of electric current on contact
resistance and friction has been studied for sliding metal electric
contacts[3,4] and a mechanism of surface film breakdown is supposed
to account for the severe wear occurring with some brush materials.
Generally, the coefficient of friction is found to depend upon the
current in such a manner that above certain current value surface
film breakdown is supposed to occur. In our experiments, friction
is increased when an electrical potential is applied across the

contact surface of an electrically conductive compound which slides
on a conductive track. The increased force of friction for a given
applied normal load may be interpreted as an increase in the
coefficient of friction. Whether this increase is an apparent one
since the electrical potential increases the effective normal load
over and above the applied normal or whether the inherent coefficient
of friction is higher, is not obvious at first, in view of the two
ways of looking at the effects accompanying application of an
electrical potential.

EXPERIMENTAL

A simple constant speed tribometer has been used in our
experiments. The test arrangement is schematically shown in
Figure 1. The driving mechanism of the tribometer pulls a normally
loaded elastomer specimen at speeds ranging from 10^{-5} up to 1
centimeter per second. One of the surfaces of the specimen was
bonded onto a steel plate. The latter serves to transmit the
forces to the specimen and at the same time serves as one electrical
terminal. The other surface of the specimen is free to slide on a
countersurface of a stainless steel plate. The latter is fixed to
the tribometer body; it serves as the other electrical terminal.
The two electrical terminals may be connected to sources capable of
impressing D.C. or A.C. voltages.

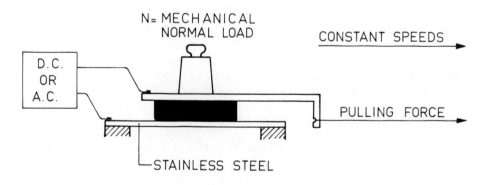

Fig. 1. Constant Speed Experiment

In order to obtain a low contact resistance across the bonded surface, a reasonably conductive adhesive was used. Some of the preliminary tests were carried out by making use of electrolytes interposed between these surfaces instead of the adhesive. It was found that the contact resistance using the adhesive was not excessive.

The test rubber compounds are composed mainly of SBR, reinforced with carbon black. The incorporation of carbon black makes the rubber more electrically conductive. In order to investigate the influence of carbon black, some of the specimens were prepared using different amounts of carbon black. The amount varied from zero to 50 p.p.hr. In the preliminary friction tests, it was observed that the electrical potential had practically no influence on friction of specimens containing less than 20 p.p.hr. carbon black. This observation is valid within the range of voltages from zero to 2000V which was available. The later tests have been performed with specimens of compounds containing 50 p.p.hr. carbon black. Using these compounds, it was observed that the frictional force was substantially increased when an electrical potential was applied across the two terminals.

An ammeter was interposed in the circuit to measure the current. The value of the current was usually amounted to a few milliamperes. This indicates that the power consumed in obtaining an increase in friction, is only of the order of a watt. The experiments have been carried out generally on dry surfaces. In order to understand the nature of the mechanism by which potential influences friction, a few tests have been conducted on surfaces which were wetted with water.

Both D.C. and A.C. voltages have been employed. The frequency of the A.C. voltage used was usually 50 cycles per second. In some tests an A.C. source with a variable frequency was used. The test frequencies were usually limited to a value lower than that at which an anomalous dielectric dispersion effects have been reported to occur. In some of the tests, frequencies up to 10 K Hz have been employed.

As a matter of convenience, the coefficient of friction is defined by the ratio of the tangential force to the mechanical normal load. The definition is only indicative; it may not be appropriate if the normal force over the contact is not equal to the mechanical normal load.

It is generally recognized that friction of elastomers depends upon both the speed and temperature. The latter two quantities are interrelated as far as their relation to friction is concerned. It is, therefore, convenient to describe the dependence by an isothermal friction-speed curve. The curve is obtained by measuring friction at various speeds which are low enough to maintain isothermal conditions at the sliding surfaces. The mechanical normal load is kept constant during a measurement.

When an electrical potential is applied, it is desirable to obtain the laws of friction. Since very little is known at present regarding the mechanism by which electrical potential causes friction to increase, the present investigation is primarily intended to examine the law of friction which relates frictional force to normal load.

In order to be able to interpret experimental results, it is interesting to know how the title effect depends upon voltage. With that view in mind, tests have been conducted at various values of voltages. It was observed that the effect is continuous in the range of voltages used. It is convenient to use a single voltage in order to study the relation between the force of friction and the various phenomenological quantities which describe friction. It is also interesting to know how the isothermal friction-speed curve as a whole, is affected by the application of an electrical potential. Experiments have been performed for various values of the mechanical normal loads ranging from 3 to 20 Kgf.

RESULTS OF FRICTION TESTS

Within the range of voltages, friction tests have been conducted mainly for compounds with 50 p.p.hr. carbon black.

The following observations have been made.

(a) Both D.C. and A.C. voltages were found to be effective in increasing friction. The increase would sometimes amount to 200 to 300%.

(b) In the constant speed tests, the electrical potential has been impressed at some instant during the uniform sliding motion of the specimen. It may be seen in Figure 2 that the tangential force builds up with time. The building up of force from its original value before the voltage is applied, to its steady value appears to be gradual. The reason for the gradual buildup does not appear to be related essentially to the electrical phenomenon.

A simple explanation is that the building up process depends upon the speed at which the specimen is being pulled. If it is assumed that the normal force is increased instantaneously when an electrical potential is applied, it may be conceived that the motion of the surface is suddenly checked. In order to recover its original sliding speed, an additional tangential

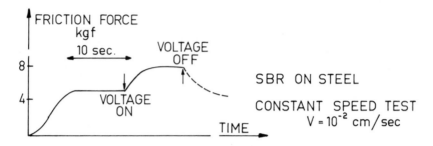

Fig. 2.   Increase in Friction on Application of Potential

force is needed. The additional force is supplied by the driving
mechanism of the tribometer, at a rate determined by the speed
of pulling. During this process, the speed at which the surface
slides, the deformation of rubber and consequently the tangential
force increase gradually. It may similarly be argued if, instead
of the normal force, the coefficient of friction should increase
instantaneously. It should be remembered that the force of
friction arises as a reaction to the applied tangential force.
A second reason is that when the normal force is suddenly
increased, some time is needed before there is a corresponding
increase in the real area of contact; the viscoelastic creep
process is time-dependent.

At present no measurements of surface displacements are available;
the possibility of some time-dependent electrical effects cannot
be ruled out.

(c) At a given speed of sliding, the observed increase in friction
    is, in general, not related linearly to the applied voltage.
    For lower values of voltages, the relation is approximately
    quadratic. As the voltage is increased, friction appears to
    increase less rapidly with voltage. A typical relation is
    shown in Figure 3.

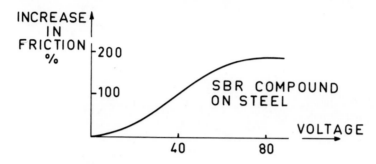

Fig. 3.   Increase in Friction with Voltage at Constant Sliding
          Speed (isothermal).

(d) It is interesting to know how the general shape of the friction-
    speed curve is affected by the application of voltage. Figure 4
    shows that the level of the isothermal friction-speed curve is
    raised when an electrical potential is applied. There is,
    however, no significant change in the shape of the curve. It
    appears, therefore, that electrical potential has no specific
    influence upon the properties which govern the shape of the
    friction-speed curve. It has been shown elsewhere[5] that these
    properties are none other than the viscoelastic properties of
    the elastomer.

(e) The most important relation as far as the law of friction is
    concerned is that concerning the influence of potential upon
    the load dependence of friction. When no potential is applied,
    it was observed that the coefficient of friction decreases as
    the normal load is increased. The friction-load relation
    obtained at a constant sliding speed is shown in Figure 5. In
    the same figure, it is seen that the friction-load relation
    obtained for a certain value of the impressed voltage, is
    similar to the original relation; only the absolute magnitude
    of friction is higher.

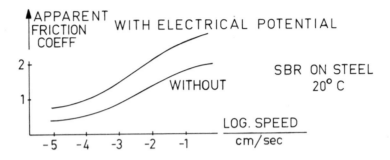

Fig. 4.   Isothermal Friction-Speed Curves with and without
           Electrical Potential.

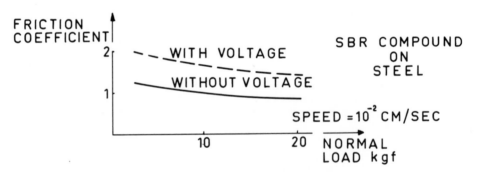

Fig. 5.   Load Dependence of Friction

An important point to note in these experiments is that the nominal area of the elastomer specimen is constant. The significance of this observation is discussed in the next section.

(f) The effects described above have been observed in both the cases in which D.C. and A.C. voltages are impressed so long as the surfaces are dry. The variation of A.C. frequency did not give rise to any additional effects as to the frictional behaviour.

On surfaces wetted with water, the electrical potential does not seem to have any effect on friction. This has been observed within the range of voltages available.

The values of current have been measured in the various tests. There is usually a small flow of current both under dry and wet condition of surfaces. On wet surfaces, the phase angle between the current and the applied A.C. voltage is small. On the other hand, a marked phase difference is observed between the A.C. voltage and current, on dry surfaces in static contact. The current has a leading phase with respect to voltage. It was observed that the value of the current increases as the frequency of the impressed A.C. voltage is increased. Under kinetic conditions, both the phase difference and the impedance increase. This aspect will be discussed in a later section.

PRELIMINARY DISCUSSION

It is desirable to conceive a theoretical model based upon the experimental observations such that the mathematical aspects of the observed relations are compatible with the physical picture of the model.

The observed increase in frictional force may be the result of a higher coefficient of friction for a given mechanical normal load. On the other hand, it may be caused by an increase in the normal force; the force of attraction results from the application of an electrical potential. In this case, the two surfaces act as the plates of a capacitor.

It is difficult to conceive how a large increase can occur in the actual coefficient of friction, since the coefficient characterizes the interactions which are intrinsic to the process of friction. It is possible that a breakdown of surface films; the value of the current is, however, too low to cause any appreciable heating at the contact surface. Moreover, if the electrical potential is reduced to zero, the specimen relaxes to attain the original value of friction. The distance traveled by the specimen during this period is much smaller than the linear dimension of the nominal

surface. It is probable, however, that the surface film breaks
down only locally at the areas of real contact. The friction tests
with the electrical potential were repeated several times over the
same nominal area. The probability of finding places where surface
film breakdown has occurred is thereby increased. In a subsequent
test without the electric potential, it was observed that the force
of friction did not rise above its original value before these tests.
It may be concluded that the surface properties are not noticably
modified by the application of an electrical potential. An increase
in the actual friction coefficient does not seem to be likely.

It is not unreasonable to expect that a normal force of
attraction arises when an electrical potential is applied across
the contacting surfaces which are separated by a dielectric. Some
of the aforementioned results suggest that the contact impedance
has a capacitive character. It is interesting to know the details
of the model which give rise to an increase in the normal force.

ELECTRICAL MODEL OF CONTACTING BODIES

In our experiments with an electrically conductive rubber
compound, a steady current flow has been observed when a D.C.
potential is applied across the contacting bodies. The rise time
of the current, if any, was not noticable. It may be concluded
that the real area of contact conducts the current. The same
applies to the bulk of the elastomer. The upper surface of the
conductive rubber specimen is uniformly bonded with a highly
conductive adhesive; the current density will, therefore, be
assumed to be uniform. The lower surface of the specimen which
serves as the sliding surface will give rise to a contact resistance.
A detailed study of contact resistance of conductive rubbers has
been reported by Norman[1]. The determination of the contact resistance is usually difficult; the reproducibility of measurements is
usually poor. Some of the data mentioned in the following discussion
may, therefore, be taken to be indicative of the order of magnitudes
involved. The contact resistance probably arises from the constric-
tions present in a dispersed contact, presence of surface contaminants like surface blooms and the non-uniform distribution of
conductive filler in rubber. Fortunately, the resistance measurements for low resistivity rubbers such as that used here, do not
depend sensitively upon the electrode material.

In order to illustrate the principal elements of the model, a
simple electrical scheme is shown in Figure 6. The bulk of the
specimen is taken to be purely resistive. The bulk resistance R_B
is in series with a contact impedance. The latter is depicted as
a resistance R_c parallel with a contact capacitance C. A further
elaboration is necessary before undertaking the task of identifying

the elements of the model with the various physical details of the contacting bodies.

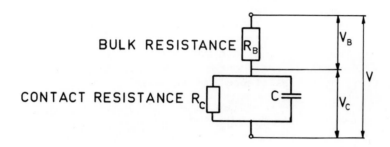

Fig. 6. Electrical Model of Sliding Rubber

Resistances

Bulk Resistance. Some measurements were made in order to estimate the bulk resistance of the conducting rubber specimen. Another specimen of the same rubber compound and having the same dimensions as the frictional test specimen was prepared. Both the opposite surfaces of the specimen were uniformly bonded to steel plates using the highly conductive adhesive. Neglecting any small contact resistance arising from bonding, the bulk resistance amounted to approximately 1 K Ω at an applied voltage of 30 V D.C. It was observed that the electrical behaviour is approximately ohmic in the range 1 to 30 V. The resistance decreases slightly as the voltage increases. The bulk rubber was also subjected to shear strains, comparable in severity to those experienced during the friction tests. The resistance increased with strain to about 1.5 to 2 K Ω. Some time-dependence of resistance was observed during this test; it is difficult to estimate the part played by the viscoelastic creep and the part played by the reformation of structural changes in the carbon black in relation to the rubber molecules.

Dry, Static Contact. In comparison to the magnitude of the
bulk resistance, the resistance of the frictional test specimen
which was only normally loaded against a dry steel plate, was found
to be considerably larger. The resistance varies with the time of
contact and with the normal pressure on the specimen. A few minutes
after loading with a nominal pressure of about 1/2 kgf/cm^2, the
resistance had a value of about 10 to 12 K Ω. The resistance
decreased when the normal pressure was increased. For a nominal
pressure of 1.3 kgf/cm^2, the resistance of 6 to 8 K Ω was measured.

The resistance R_c arises from the asperities which form the
area of real contact. The resistance comprises mainly of a constriction resistance and a resistance due to an alien surface film. If
the electrical resistivity of the material of the elastomer asperity
is ρ and if the area of real contact is assumed to be a circle with
radius a, the contact resistance for an asperity is[6]

$$R_c = \frac{\rho}{4a} + \frac{K}{\pi a^2} \ ,$$

where K is the resistance per unit area of the surface film. It is
assumed that the resistance owes nothing to the asperities of the
metallic countersurface. If the contact occurs over n identical
asperities, then the contact resistance R_c is given by

$$\left(R_c \right)_n = \frac{\rho}{4an} + \frac{K}{\pi a^2 n} \ .$$

This resistance is usually much larger than the resistance of the
bulk of the elastomer.

Considering that the contact resistance arises only from the
constriction resistance, Greenwood[7] has shown that the decrease in
resistance with an increase in normal pressure may be deduced from
the surface statistical data.

Dry, Kinetic Contact. Under kinetic conditions, the resistance
measured at 30 V D.C. was even larger than that for a static contact.
The resistance increased with the speed of sliding from a value of
approximately 25 K Ω at a speed of 10^{-4} cm/sec. to a value of over
100 K Ω at a speed of 10^{-1} cm/sec. The large increase in resistance
seems to occur at the transition from static to kinetic condition.
The intensity of shear strain by itself, does not appear to have a
dominant role to play in this observed increase. A test conducted
for a higher normal load such that the tangential force was less
than the force of static friction showed that a much smaller increase
in resistance occurred even when the intensity of shear strain was
comparable to the one in the previous test.

In the theory of friction of elastomers, it is customary to consider that the real area of contact varies with the speed of sliding on account of the viscoelastic properties of elastomers. It is, therefore, reasonable to expect that the variation of real area of contact with the speed of sliding if reflected in the variation of the kinetic contact resistance. It may be noted that the contact resistance is several times larger than the bulk resistance over the entire speed range. For a given applied voltage across the contacting bodies, the fraction of the voltage which acts across the contact resistance is, therefore, relatively insensitive to the variation of the contact resistance.

Wet Contact. In order to estimate the contact resistance under wet conditions, some water was introduced in between the surfaces of a statically loaded contact. The resistance was found to decrease to a value of about 5 K Ω. The load dependence of contact resistance almost disappeared under these conditions. The resistance in this state is still appreciably larger than the bulk resistance. This indicates that the layer of water offers a substantial resistance to the flow of current. The layer of water occupies the hollow spaces between the surfaces of the contacting bodies. Initially, the coefficient of friction under these conditions was not appreciably lower than that obtained when the surfaces were dry. Yet no change is observed in the force of friction when a moderate electrical potential is applied across the contacting bodies. Under wet condition, the kinetic contact resistance did not exceed the corresponding resistance obtained under static condition.

Some of these results are of significance in the following discussion to the purpose of locating the elements of the physical model which are responsible for the electrostatic attraction.

Contact Capacitance, Electrostatic Attraction

The source of the contact capacitance is relatively less tractable than that giving rise to the resistance. If the asperity material is considered as a real dielectric, there may be a capacitance associated with the aforementioned constriction resistance. Similarly, surface film, if any, may also contribute to the contact capacitance. It is also equally plausible that the nominal area as a whole may give rise to the capacitance with air serving as the dielectric medium. Some of the physical models which describe the sources of capacitance are now considered. The behaviour which follows from the assumption regarding each of these sources may be assessed in relation to that observed during the experiments.

If the nominal area of surface is represented as a plate of a parallel plate capacitor, it has to account for the various observed

relations which have been mentioned earlier. An important observation is that the frictional force increases when the normal load
is increased. The relation is linear to a first approximation.
Since the nominal area is kept constant, the only way in which the
capacitance can increase with the normal load is through the
reduction of the separation between the surfaces. It is now intended
to examine the model in which the nominal contact surfaces are
depicted as parallel plates of a capacitor.

The force of attraction N_1 between parallel plates of a
capacitor is given by Sears

$$N_1 = \frac{1}{2} \frac{Q^2}{\varepsilon A} \ .$$

In this Q is the charge; ε is the permittivity and A is the surface
area of a plate.

For an applied voltage V over a capacitor having a capacitance
C, the charge is given by

$$Q = C V$$

The capacitance C may be expressed in terms of the surface area and
the separation d,

$$C = \varepsilon \frac{A}{D}$$

Combining the equations,

$$N_1 = \frac{1}{2} V^2 \frac{\varepsilon A}{d^2}$$

It may be seen that the force increases with $\frac{1}{d^2}$; all other quantities
being kept constant.

The influence of normal load on the approach of surfaces has
been studied by Greenwood et al.[8]. They have described a relation
between the normal load and the separation. From that relation,
a doubling of normal load brings the surfaces near by a distance of
5 to 10% of the original separation. It appears, therefore, that
the above representation would not be able to account for the observed
increase in normal force.

Capacitance of the Void Area. The area bounded by the nominal
and the real areas may function as a capacitor. It is convenient
to refer to that area as the "void area". When the normal pressure
is small, the area of real contact is only a very small fraction of
the nominal area. Hence, a relatively substantial increase in the

area of real contact caused by a corresponding increase in the normal load may not alter the void area appreciably. The approximate proportionality between the electrostatic force of attraction N_1 and the mechanical normal load N which follows from the results shown in Figure 5, cannot thereby be explained.

Obviously, a more detailed model is needed for describing the capacitance of two rough surfaces. To that purpose, the surface statistical theory of Greenwood may be extended. Since the electrostatic force varies as $1/d^2$, it may be expected that only a small portion of the void area which lies in the vicinity of the real area of contact and surrounding it may make a significant contribution to the capacitance. Increasing the mechanical load increases both the real area of contact and an annular area surrounding it. In the annular area, the distances between the opposite points on the two contacting surfaces are very small. The contribution arising from the forces of attraction may be proportional to the mechanical normal load. The assumption that the annular region of the void area acts as a capacitor is, therefore, consistent with the experimental results (see Figure 7).

Capacitance in the Real Area of Contact. The dielectric properties of the elastomer material of the asperities in the immediate vicinity of the area of real contact may also be expected to contribute to the contact capacitance. Since the real area increases proportionately with increasing mechanical load, there is no conflict whatsoever with the experimental relation $N_1 \alpha N$.

DISCUSSION

In his work, Schallamach[2] has considered the capacitive contribution arising from the relative permittivity of elastomers. He did not, however, distinguish between the areas of real and nominal contact in his calculation for the thickness of the "top layer". He has invoked the strain sensitive resistance of a carbon black filled rubber, in order to explain the very high resistance near the contact surface. The highly resistive layer acts like a dielectric and, consequently, gives rise to the electrostatic force of attraction when an electrical potential is applied across it. He has attributed the high resistance to the increase in resistance of a surface layer by abrasion. In our experiments, no marked change in the contact resistance has been observed even when the surface of the specimen used was vigorously abraded with an emery paper.

It appears from the foregoing considerations of the contact resistance, that the constriction resistance may be very high. Holm[4] has shown that the main part of the constriction resistance is localized in the immediate neighborhood of the conducting region at the contact surface. The material in this layer can function as

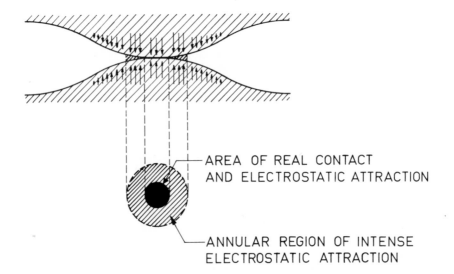

Fig. 7.   Physical Picture of the Mechanism

an effective dielectric medium. It appears, therefore, that the
influence of strain on the electrical impedance of the layer is
not essential to obtaining an electrostatic force of attraction.
It appears that the special property of the carbon-black filled
rubber merely enhances the effect by further increasing the constriction resistance. The imperfect nature of the physical contact
between bodies transmitting a normal load is probably the main
cause of this phenomenon. The high contact resistance enables a
fairly large share of the applied voltage to act across the contact
surfaces. The real areas of contact should be able to withstand a
sufficiently large voltage. In this context, the effect described
here is somewhat similar to the Johnsen-Rahbek effect described by
Holm[4].

The influence of increasing the voltage is to increase the
electrostatic force of attraction. From the simple formula of a
parallel plate capacitor, the relation has to be quadratic. It
appears from Figure 3, that this is indeed the case at low voltages.
At higher voltages, there is a gross deviation; the equilibrium
under the electrical and elastic forces is probably such that the
relation becomes more or less asymptotic.

PHASE ANGLES

A few tests have been conducted to evaluate the electrical and
the physical contact models using A.C. voltages. The experiments
have been conducted on the frictional test specimen which is in
contact with a stainless steel plate. The current flow, the applied
voltage and the phase angle between them were determined with the
help of an oscilloscope. A small series resistance was introduced
in the circuit in order to obtain a voltage signal proportional to
and in phase with the current. It was found that the impedance of
the circuit is resistive at low frequencies. The resistive character
is approximately maintained so long as the frequency does not exceed
a kilocycle/sec. At higher frequencies, the capacitive character
of the impedance begins to emerge. The phase angle has been determined from the elliptic trace on the oscilloscope screen. A distor-
tion in the elliptic form was noticable at higher voltages; the
behaviour at higher voltages appears to be non-linear.

The tests have been carried out on dry and water wetted surfaces,
both under static and kinetic contact conditions, a frequency of
about 5 KH_z. The phase lead for a dry static contact under a normal
pressure of 1/2 kgf/cm^2 amounted to 12°. Under similar conditions
the water wetted contact gave a phase angle $\emptyset = 6°$ irrespective of
static or kinetic conditions. The phase lead for a dry kinetic
contact was considerably larger, $\emptyset = 35°$; the speed of sliding was
0.04 cm/sec.

If the bulk impedance is ignored, the phase angle corresponding to the contact impedance is given by

$$\emptyset = \tan^{-1} R_c \omega C$$

Considering the aforementioned resistances and the observed phase angles, a contact capacitance of roughly 0.004 μF is obtained. For the dry contact, the contact resistances determined previously indicate that the contact capacitance of a static contact is larger than that of a kinetic contact. It also seems that wetting the contact by water does not appreciably alter the contact capacitance; only the contact resistance is changed. Noting the previously mentioned results, the contact resistance of a water-wetted contact is not insignificant when compared with the bulk resistance. It is, therefore, not clear why no influence of electrical potential on friction is observed. It is not unlikely that, although the force of attraction is present, the frictional force is limited by the shear strength of the water film. At present, no firm conclusions can be drawn. In view of the complex electrical effects which may arise when water films are present, it appears that a more elaborate model would be necessary. Considering the practical implications of the described effects, a study of the effect of water films would certainly be desirable.

CONCLUSIONS

1. The frictional force between conductive rubber and metal may be increased by applying an electrical potential across their interface. Both D.C. and A.C. potentials are effective. The power consumption is usually very small.

2. The aforementioned influence is not limited by the nominal area of the contacting surfaces. The frictional force increases linearly with the load to a first approximation. Sliding speed has no specific influence on the increase in friction; the friction-speed curve has roughly the same shape as that obtained without the applied potential.

3. The increase in frictional force appears to be a result of an increase in the normal force; the intrinsic coefficient of friction remaining constant.

4. The aforementioned results are valid for dry surfaces; when surfaces are wetted with ordinary tap water no influence of electrical potential is observed on the frictional behaviour.

5. The effects described here may be explained by the existence of constriction resistance at the contact of most surfaces in practice. Strain sensitive resistance of conductive rubbers

does not appear to be an essential requirement. Certain degree
of conductivity is, however, necessary for the effect to be
observed at moderately high voltages. The kinetic contact
resistance is considerably larger than the static contact
resistance. The former is found to be speed dependent. The
assumption of a speed dependent area of real contact in the
theory of friction of elastomers is sound.

ACKNOWLEDGEMENT

The authors appreciate the cooperation of the members of the
Vehicle Research Laboratory at Delft. Thanks are due to Professor
H.C.A. van Eldik Thieme and to Ir. E.G.M.J. de Vries for the many
valuable discussions.

REFERENCES

1. R. H. Norman, Conductive Rubbers and Plastics, Elsevier
    Publishing Co. (1970).
2. A. Schallamach, Proc. Phys. Soc. (London), B 66, 817 (1953),
    see also: Wear, 1, 384 (1958).
3. J. K. Lancaster and I. W. Stanley, Brit. J. Appl. Phys., 15,
    29 (1964).
4. R. Holm, Electrical Contacts, Springer Verlag, (1967).
5. A. R. Savkoor, Tech. Hogeschool Delft, Lab. voor Voertuigtechniek
    Report 093 (1968).
6. F. P. Bowden and D. Tabor, The Friction and Lubrication of
    Solids, Clarendon Press Oxford, Vol. 1, (1950).
7. J. A. Greenwood, A.S.M.E. Paper 65 - Lub 10, (1965).
8. J. A. Greenwood and J. B. P. Williamson, Proc. Roy. Soc.
    (London), A 295, 300 (1966).
9. P. J. Harrop, Dielectrics, London Buttersworth (1972).

DISCUSSION OF PAPER BY A. R. SAVKOOR AND T. RUYTER

J. J. Bikerman (Case Western Reserve University): Maybe I did not
hear it properly. Why do you need an electroconductive carbon?

A. R. Savkoor: We have used moderate voltages in our experiments.
In our preliminary tests, we did not find any noticable increase
in friction on applying the voltage, when a rubber compound with
as little as 10 p.p.hr. of carbon black was used. The effect was
observed when a compound having 50 p.p.hr. carbon black was used.
A certain amount of conductivity is essential.

According to our tentative hypothesis, it appears that the voltage
drop across the real area of contact and an annulus surrounding
each of the local area is responsible for an increase in the normal
force. It is obvious from the electrical model that the effect is
enhanced for a given voltage, if the bulk of the rubber compound
has a much lower resistance than that at the contact of the
asperities. This probably occurs in an enhanced form in the case
of black filled compounds; the large surface strains during sliding
may in effect, increase the resistance of the asperities considerably.
More experimental work needs to be done in order to examine the
validity of this argument.

Participant: Did you measure D.C. current? Was it ohmic?

A. R. Savkoor: Yes, for low values of voltages from 1 to 30 V.
As the voltage is increased, the current is found to increase more
rapidly; it also becomes time dependent. If we consider that the
normal force is increased on applying the voltage, the real area
of contact also increases. Moreover this increase is time-dependent
on account of the creep properties of the elastomer. The constriction
resistance is thereby dependent upon both voltage and time. The
rise time was, in relation to this, short.

D. Tabor (Cambridge University): In the contact between rubber and
a metal, there will be some electric-charge transfer. This may play
some role in the adhesion process. Is it possible that this was
influenced by the applied voltage? In particular, did you obtain
the same results if the polarity was reversed or if alternating
current was used? Again, do you obtain similar results for rubber
sliding on rubber?

A. R. Savkoor and T. J. Ruyter: We are not aware of the extent to
which electric-charge transfer affects adhesion. If the strength
of the adhesive forces is considered to be the determining factor
in the process of failure during the tangential separation caused
by the frictional forces, it would no doubt, have an important role
to play in friction. On the other hand, if we assume that adhesion
merely provides for the constraint of no relative motion between

points within the area of real contact; the strength properties of
the boundary layer determines the magnitude of the force of friction,
then, the cause of the adhesion may not be of significance so far
as the force of friction is concerned.

In reply to your following questions, it was found that the results
were not noticably affected (within the limits of reproducibility)
if the polarity of the D.C. voltage was reversed. An increase in
friction with electrical potential is also obtained when A.C.
potentials are used. The magnitudes of the potential drops across
the contact, in the two cases has not been estimated. The title
effect was also observed when rubber slides upon rubber (both
containing carbon black). It may also interest you to know that
we have obtained similar results using a steel slider on a counterface of a steel plate covered with an insulation paper such as the
one used in transformers.

The Direct Fluorination of Polymers and the Synthesis of Poly(carbon monofluoride) in Fluorine Plasma

R. J. Lagow, J. L. Margrave[*], L. A. Shimp, D. K. Lam, and R. F. Baddour

Department of Chemistry, Massachusetts Institute of Technology

Cambridge, Massachusetts 02139

The reactions of elemental fluorine with finely powdered hydrocarbon polymers and paraffins have been carefully controlled so that the products of the reactions are perfluorocarbon polymers. Potentially, this direct fluorination process is a new approach to the synthesis of fluorocarbon polymers. Polyethylene, polypropylene, polystyrene, polyacrylonitrile, polyacrylamide, "resol" phenol formaldehyde resin, and ethylene propylene copolymer have been fluorinated to produce perfluorocarbon polymers which are structurally similar to the hydrocarbon starting materials and have physical properties similar to known structurally related fluorocarbon polymers obtained by polymerization of fluorocarbon monomers.

Recently, there has been a large amount of interest, both academic and industrial in poly(carbon monofluoride) $(CF_x)_\eta$. This interest has been due primarily

[*]Department of Chemistry, Rice University, Houston, Texas 77001

to the development of some new synthetic
techniques at Rice University[1] and to
some degree to a process developed in
Japan for preparing a similar fluorinated
carbon material of lower molecular weight
from activated charcoals and carbon black[2].
The material, obtained from fluorination
of graphite is the most thermally stable
polymeric fluorocarbon material known.
It is indefinitely stable to 600° and
stable for short periods of time to 800°.
A new fluidized bed plasma synthesis for
poly(carbon monofluoride) is reported.

INTRODUCTION

In the patent literature, there have been hints that the
development of a technique for direct fluorination was possible
beginning with the work of Soll in 1938[1]. Notably, the surface of
polyethylene was first fluorinated successfully by Rudge in 1954[2]
and later repeated by Joffre[3] and others[4-7]. The fluorination of
polymer surfaces, of which polyethylene is the least difficult,
may in some cases be accomplished by simple dilution of fluorine
with nitrogen. Hydrocarbon polymer powders and delicate organic
compounds require increasingly more sophisticated control techniques
for smooth fluorination. The number of difficulties encountered
in this extrapolation is perhaps the principal reason why these
early efforts were regarded as isolated observations and direct
fluorination remained undeveloped until the sixties.

Potentially, this direct fluorination process is a new approach
to the synthesis of fluorocarbon polymers. Polyethylene, polypropylene, polystyrene, polyacrylonitrile, polyacrylamide, "resol"
phenol formaldehyde resin, and ethylene propylene copolymer have
been fluorinated to produce perfluorocarbon polymers which are
structurally similar to the hydrocarbon starting materials and
have physical properties similar to known structurally related
fluorocarbon polymers obtained by polymerization of fluorocarbon
monomers. High yield of fluorocarbon polymers approaching 100%
have been obtained. This direct technique used for fluorination
of hydrocarbons and polymers is called the LaMar process and has
been previously described in connection with the direct fluorination
of Lower molecular weight species[8-11].

Another fluorocarbon material, poly(carbon monofluoride) has
been reported to be a very useful lubricant by lubrication engineers
at the NASA-Lewis Research Center[12] and at the U.S. Army Frankford
Arsenal[13]. These studies have shown that carbon monofluoride as a

solid lubricant under extreme conditions such as high or low
temperatures, high pressures, or heavy loads is very much superior
to graphite or molybdenum disulfide. In addition, workers at the
U.S. Army Electronic Command at Ft. Monmouth, N. J.[14], and industrial
workers in Japan have concurrently demonstrated a high potential
for carbon monofluoride for use as a cathode material in high-energy
batteries.

Carbon monofluoride of a composition $CF_{0.92}$ was first prepared
by Ruff, Bretschneider, and Elert in 1934[15]. In 1959, Rudorff and
Rudorff reported a preparation of carbon monofluorides of compositions $CF_{0.68}$ to $CF_{0.98}$ which ranged from black to off-white in
color[16],[17]. At Rice University[18] techniques were developed which
employed both high-temperature and high-pressure syntheses to prepare the first perfluorinated carbon monofluoride which had a
stoichiometry of $CF_{1.12}$ and was snow white. These syntheses were
adaptable to the production of 40 g quantities of pure white carbon
monofluoride and also overcame the disadvantage of poor reproducibility which was characteristic of the previous syntheses.

A structure which was proposed by Rudorff for carbon monofluoride is shown in Figure 1. This structure has been shown to be
approximately correct except for the layer stacking[19] and the fact
that carbon atoms on the edges of each layer are divalent and therefore have both axial and equatorial fluorine. As one may determine
by examining the carbon/fluorine ratio of the most highly fluorinated
graphite, the $CF_{1.1}$ ratio is a result of these axial and equatorial
fluorines and the exact stoichiometry varies with the particle size.
Therefore only the snow white carbon monofluorides with fluorine/
carbon ratios greater than 1 lack the fluorine site deficiencies
characteristic of the earlier preparations.

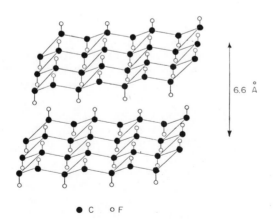

6.6 Å

● C   ○ F

Fig. 1. Rüdorff $(CF_x)_n$ structure.

Carbon monofluoride of lower stoichiometry has been normally prepared by the reaction of elemental fluorine with graphite at temperatures in the range 400-600°. It is interesting to observe that graphite does not begin to react with a fluorine atmosphere until 400°, while all carbon compounds containing double bonds will burn in fluorine under identical conditions. Examination of the thermodynamics of fluorine addition to normal double bonds reveals that this reaction is exothermic by approximately 68 kcal/mol of carbon fluoride bonds formed.

The π electrons in graphite are probably the most extreme case of carbon electron delocalization known since there is an equal probability of finding a given π electron in any of three adjacent rings around each carbon atom. It would be expected that the "stabilization energy" from this delocalization is very high since the "delocalization energy" of benzene is known to be 36 kcal/mol. It is probable that the reaction of fluorine molecules with graphite does not occur readily due to the extra activation energy required to attack the very stable π-electron system and that the dissociation energy of fluorine must be provided to facilitate the reaction. A Boltzman calculation reveals that at 400° there are only a marginal number of fluorine atoms available while at 600° the fluorine atomic population becomes kinetically significant. It is also possible that the activation energy for the reaction may be such that electronically excited fluorine atoms are required. It would appear that a glow discharge in elemental fluorine is ideally suited for preparing carbon monofluoride at low gas temperatures. A relatively high degree of dissociation of fluorine atoms may be achieved under appropriate plasma conditions and the likelihood of electronically excited fluorine atoms is also high. A fluidized bed reactor configuration was chosen to provide maximum surface contact of graphite particles with fluorine.

In the following sections, we review two of our major studies on the subject of new fluorinated polymers: (1) direct fluorination of polymers[22], and (2) synthesis of poly(carbon monofluoride) in a fluorine plasma[23]. Since both studies were published in non-engineering oriented journals, we would like to describe the experimental parts in full for those who are interested in the field of friction and wear.

EXPERIMENTAL

Direct Fluorination of Polymers[20]

Apparatus. The apparatus used was very straight forward and simple. A typical fluorine handling apparatus was used as shown in Figures 1 and 2. Items one and two are fluorine and helium sources respeceively. Item three is a Hastings-Raydist model LF50

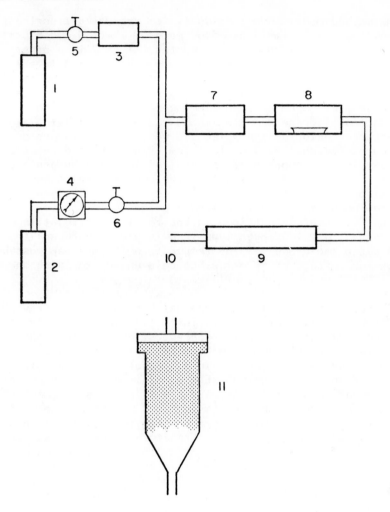

Fig. 2. Apparatus For Direct Fluorination

1. Fluorine tank
2. Helium tank
3. Mass flowmeter
4. Gas flowmeter
5. Needlevalve
6. Needlevalve
7. Brass mixing chamber
8. Reaction chamber
9. Fluorine absorber
10. Oxygen outlet
11. Alternate reactor for larger operation

mass flowmeter and model F50M transducer for measuring fluorine flow
rates. Item four is a simple gas flow meter. Items 5 and 6 are
needle valves to control the gas flow rates. Item 7 is a brass
mixing chamber packed with fine copper turnings. Item 8 is the
reaction chamber which consists of a 1" by 1.5' nickel tube containing a prefluorinated nickel boat, 7 1/2" long and 1/2" wide, used
to hold the sample. Item 9 is an alumina-packed cylinder to dispose
of the unreacted fluorine and Item 10 is the oxygen outlet. Item 11
is an alternate reactor design which may be used when larger quantities of product are desired. It is a semi-fluidized bed reactor in
which the sample is supported at the conical junction by a nickel
screen. The fluorine then passes into the conical region and
through the reactant on the screen. All connections between the
pieces of equipment are made with 1/4" O.D. heavy-wall copper tubing.
Whenever the fluorocarbon obtained in the reaction is volatile at
room temperature, a cold finger type trap is placed between Items
8 and 9 to catch the product. The temperature of the trap must be
high enough for passage of fluorine but cold enough to catch the
product.

Procedure. The general procedure has been described previously
[8-11] and its principal feature is infinite dilution initially followed by gradient changes of fluorine concentration. As the fluorination of a given material becomes more difficult to control, many
other factors must be considered. In all cases the direct fluorination resulted in the conversion of the polymeric material completely to a fluorocarbon polymer. If particle sizes greater than 150
mesh are used, a hydrocarbon core is retained in the polymer. A
fluorine flow of about 4cc/min into a reactor previously flushed
with helium was used in the 1" nickel reactor in most cases. When
larger quantities were desired, the 2 1/2 I.D. fluidized bed was
used. In this reactor the finely powdered hydrocarbon polymer or
paraffin sample was mixed in about a 1:1 volume ratio with potassium
fluoride. The initial flow was about 10 cc/min fluorine and 30 cc/
min He. After 8 hours, the flow was changed to 16 cc/min fluorine
and the helium flow was stopped. The potassium fluoride may act
as a catalyst but was used only in fluidized bed reactions and
absorbed some of the hydrogen fluoride byproduct. The fluorocarbon
polymer or perfluoroparaffin-potassium fluoride mixture was then
washed several times with water to remove all potassium fluoride
and hydrogen fluoride, filtered and dried at 100°C. In all cases
reaction times given are for purest products. The reactions were
essentially complete after 24 hours, and often after only 8 hours
exposure to fluorine. To remove all traces of hydrogen from the
polymer, however, a pure fluorine flow for about 48 hours was used.
It is important also that all oxygen and water vapor be eliminated
from the system or oxygen crosslinking will occur in a manner
detrimental to physical properties. This problem may be detected
in the infrared by features in the 1700 to 2100 cm^{-1} region.

The Direct Fluorination of Polyethylene. About 1 gram of finely powdered linear polyethylene (140 mesh) was placed in a nickel boat in the one inch nickel tube flow reactor system. The reaction chamber was flushed with helium at 20 cc/min for 3 minutes. The helium flow was then terminated and a 3 1/2 - 4 cc/min fluorine flow was passed over the sample at room temperature. The reaction was continued for 24 or 48 hours if a high purity sample was desired. A white perfluorocarbon polymer was produced which resembles its hydrocarbon precursor in appearance.

If the 2 1/2 O.D. fluidized bed reactor was used, the polyethylene was mixed with potassium fluoride in a 1:1 ratio and placed on the nickel screen. The initial flow rate was 40 cc/min helium plus 10 cc/min fluorine for 10 hours. Then the helium flow was terminated and the fluorine flow was 16 cc/min for 24 hours.

The product in both cases was a white powder which possessed similar physical properties to polytetrafluoroethylene. The powder did not have a sharp melting point and was thermally stable to 345°C in air, where it decomposed to fluorocarbon materials slowly if heated for long periods of time at 350-400°C. With conventional polyethylene starting material, a more crosslinked fluorocarbon polymer is obtained of higher molecular weight and slightly inferior thermal properties. The parent polyethylene melted at about 70°C in air.

The infrared spectra of the new compound was taken. The spectra between 600 cm^{-1} and 4000 cm^{-1} was devoid of any features except a strong broad carbon fluorine stretch from 1400 cm^{-1} to 100 cm^{-1} which was most intense at 1200 cm^{-1} and a small carbon-fluorine bending band indicating the complete replacement of hydrogen. The yield was over 99% of the weight calculated for conversion to a fluorocarbon polymer.

Analysis: Calculated for $[-CF_2-CF_2-]_n$: C, 24%; F, 76%

Found: C, 26.94%; F, 72.1%

The Direct Fluorination of Polypropylene. Exactly the same experimental procedure was followed with polypropylene as with polyethylene. Polypropylene was more easily decomposed and the flow rates were more crucial in this case.

The product was a white powder which was similar in appearance to polypropylene but was more particulate and harder. The physical properties of this material were similar to the fluorinated polyethylene. The powder was thermally stable to 350°C in air, softens at 360°C and 400°C to fluorocarbon fragments. The parent polypropylene powder melted at about 90°C and burned at 120°C in air.

The infrared spectra of the product were devoid of spectral features except for a strong carbon-fluorine stretch from 1400 to 1050 cm^{-1} which was strongest at 1235 cm^{-1} and a small carbon-fluorine bending band at 720 cm^{-1}. The 1235 band had side bands at 1350 cm^{-1} and 110 cm^{-1} due to the CF_3 stretches. The absences of absorption in the 3000 cm^{-1} region indicate the absence of hydrogen in the fluorinated product. The yield was ~100%.

<u>Analysis</u>: Calculated for $-CF_2-CF(CF_3)-$: C, 24%; F, 76%

Found: C, 27.64%; F, 72.05%

<u>The Direct Fluorination of Polystyrene</u>. Polystyrene foam was ground to a fine powder and placed in the reactor for fluorination. The experimental procedure was identical with that of the two previous examples including the flow rates.

The product was a very fine fluffy white powder. The product melted at 175°C to 250°C where it began to decompose. The parent polystyrene melted at about 65°C and burned at 120°C in air.

The infrared spectra were very different from the parent polystyrene and suggests perfluoroperhydropolystyrene. A strong and broad carbon-fluorine stretch was observed between 1410 cm^{-1} and 1050 cm^{-1} which was strongest at 1180 cm^{-1} with a possible side band at 1300 cm^{-1}. Also a weak carbon-carbon bending was seen at 100 cm^{-1}. The 3000 cm^{-1} region indicates that the sample contains no hydrogen by its absence of spectral features. This is much in contrast to the multibanded polystyrene spectra, often used to calibrate infrared spectrometers, which show strong carbon-hydrogen stretches at 3010 cm^{-1} and 2900 cm^{-1} and a multitude of aromatic double bond peaks. The yield was ~100%.

<u>Analysis</u>: Calculated for $[-CF_2-CF(C_6F_{11})-]_n$: C, 27.9%; F, 72.01%

Found: C, 27.87%; F, 72.08%

<u>The Direct Fluorination of Polyacrylonitrile</u>. The finely ground polyacrylonitrile powder was placed in the boat. The initial flow rate was 60 cc/min helium and 1 cc/min fluorine for 6 hours followed by termination of the helium flow and a 2 cc/min fluorine flow for 30 hours. Polyacrylonitrile was more difficult to fluorinate than the three previous hydrocarbon polymers but no visible decomposition occurred and the yield was approximately 95%. The product was a white powder which melted at 240°C and boiled at 250°C to 260°C.

The infrared spectra revealed a strong, broad carbon-fluorine stretch at 1350 cm^{-1} to 1050 cm^{-1} which was most intense at 1230 cm^{-1}. A sharp band of medium strength was observed at 935 cm^{-1}

corresponding to the nitrogen-fluorine stretch. Again the 3000 cm^{-1} region showed no sign of a carbon hydrogen stretch.

Analysis: Calculated for [-CF$_2$-CF(CF$_2$NF$_2$)- : C, 19.35%;

F, 72.67%; N, 7.65%

Found: 24.71%; F, 69.93%; N, 6.41%

The Direct Fluorination of Polyacrylamide. Identical experimental procedures and flow rates were used in this case as with the polyacrylonitrile. The product was a white powder which tended to have variable particle size. The product softened at 220°C and turned yellow. The melt began to decompose and turn brown at 230°C to 400°C. The product was also slightly hygroscopic and turned yellow in moist air, due to hydrolytic instability.

The infrared spectra had a broad strong carbon-fluorine stretch at 1300 cm^{-1} to 1000 cm^{-1} which was strongest at 1210 cm^{-1}. A small side band was observed at 1010 cm^{-} which was probably a nitrogen-fluorine stretch. Small bands at 1750 cm^{-1} and 1860 cm^{-1} were noticed which may have been due to the carbonyl stretch. In addition, a sharp band was observed at 740 cm^{-} which may be due to the carbon-carbon and/or carbon-nitrogen stretching.

Analysis: Calculated for [-CF$_2$-CF(CONF$_2$)-]$_n$ C,22.36%;

F, 59.0%; N, 8.695%; 0, 9.938%

Found: C, 29.56%; F, 50.2%; N, 5.56%; 0, 8.50%

The Direct Fluorination of Ethylene-Propylene Copolymer. About one gram of finely powdered ethylene-propylene copolymer (120 mesh) was placed in nickel boat in a one-inch nickel tube flow reactor. The reaction chamber was flushed with helium at 200 cc/min for 30 min. The helium flow was reduced to 4 cc/min and a 3 cc/min fluorine flow was initiated at room temperature. The helium flow was terminated after 10 hrs. and the 3 cc/min fluorine flow was maintained for 60 hrs. Care was taken to assure that the system was free of all oxygen and moisture during fluorination. A white perfluorocarbon polymer was removed after flushing with helium which resembles its hydrocarbon precursor in appearance.

The perfluorinated material softens slightly at 340° and decomposes slowly in air beginning at 360°C. The parent copolymer melted at about 65°C and burned at 110°C in air.

The infrared spectra consists of a strong, broad carbon fluorine stretch at 1210 cm^{-1}. The carbon-hydrogen stretch was completely absent.

Analysis: Calculated for $[-CF_2-CF_2-CF(CF_3)-CF_2-]_n$: C, 24%;

F, 76%

Found: C, 26.29%; F, 73.81%

The Direct Fluorination of Phenol-Formaldehyde Resol. A finely powdered, one-gram sample (120 mesh) was placed in a nickel boat. After placing the sample in the reactor and flushing with helium, a flow of 4 cc/min fluorine at room temperature was initiated. After 72 hrs. the reactor was flushed with helium and a white fluorocarbon material was removed. The white fluorocarbon polymer obtained stays white in air until about 350°C where it begins to rapidly decompose. It softens slightly at 330°C, suggesting that it may be moldable.

The infrared spectra consists only of a strong, broad carbon fluorine stretch at 1200 cm^{-1} and a small, broad band at 1010 cm^{-1} which may be due to trifluoromethyl groups. No carbon-hydrogen or double band activity were present.

Calculated percentages for the perfluorinated resol structure shown in Fig. 1: C, 27.7; F, 72.3. Found C, 28.21; F, 70.96.

Synthesis of Poly(carbon monofluoride) in a Fluorine Plasma[23]

Materials, Analyses, and Physical Measurement. The starting material was flake graphite (60 mesh 99%) which was selected to facilitate the particle fluidization and minimize particle losses. Fluorine was purchased from Allied Chemical Corp. (98%). The elemental analysis was done by Schwarzkopf Laboratories of Woodside, N. Y. Infrared spectra were obtained with a Beckman IR-20A spectrometer. KBr disk samples were used. Powder pattern X-rays were obtained on a Philips Electronics Model 12045B/3 X-ray generator using type 520 56/0 powder cameras. Copper (1.5405 Å) radiation was used in all X-ray spectra.

Apparatus. The apparatus is illustrated in Figure 3. Fluorine flow was monitored with a Hastings-Raydist Model LF 50M transducer. The fluorine flow was controlled with a Monel needle valve. A sodium fluoride trap (L), placed after the needle valve, was used to remove hydrogen fluoride from the fluorine and a -78° copper U-trap (K) was used to remove the last traces of hydrogen fluoride. The internal pressure was measured with a Hastings Model Sp-1 vacuum thermocouple gauge (T) which was isolated with a Monel valve which is not depicted. Cajon 1 in. Ultra Torr fittings were used as coupling at all connections labeled (C) in the diagram. Viton O-rings which were coated with Fluorolube stopcock grease were used as seals. The Cajon couplings above and below the reactor were

water cooled by soldering a piece of 1/4 in. copper tubing to them
and passing water through the tubing to keep the O-rings cool.
The items labeled (G) are Cajon stainless steel flexible bellows.
The glass items in the system were fabricated of Vycor. Nickel
screens (150 mesh) were used to support and contain the graphite
particles at the locations indicated in the diagram (F). These
screens were held in place with Teflon washers. The reactor used
was a 1 in. o.d. alumina (Al$_2$O$_3$) tube (A). Even quartz tubes will
not withstand a fluorine plasma for more than several hours. A
five-turn 1/4 in. copper coil was used in conjunction with a 2 kW
Raybond radiofrequency generator as a source for plasma generation.
A 20-M Hz operating frequency was selected. A particle return (D)
was fabricated from a Vycor erlenmeyer flask and a vibrator was
used (E) to prevent graphite particles from adhering to the walls.
This vibrator was connected to an electrical circuit which pulsed
the vibrations periodically. A particle trap was inserted after
the particle return to catch the fine white carbon monofluoride
powder produced. Unreacted fluorine was removed from the effluent
gases with an alumina trap (H) (8-14 mesh activated alumina, Al$_2$O$_3$
+6F$_2$ → 2 AlF$_3$ + $^3/_2$O$_2$). A liquid-nitrogen trap (I) was used to
protect the vacuum pump which was attached as indicated in the
diagram.

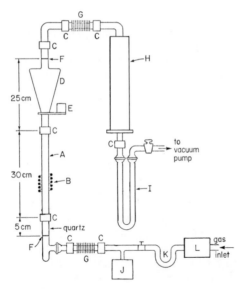

Fig. 3. Fluidized bed reactor for fluorine plasma synthesis.

    Experimental Procedure. About 1 g of flake graphite was
placed on a lower nickel screen (F). The system was then sealed
and evacuated to at least 0.1 mm. The alumina must be dry or such
vacuum is not possible. A fluorine flow of 30 cm^3/min is then

introduced into the system and the pumping speed is adjusted to
give an internal pressure of 2-5-mm. This fluorine flow fluidizes
the particles between the upper and lower nickel screens (F). The
fluidized particles are returned with the aid of the vibrator by
sliding down the walls of the particle return (D). Thus the particles pass continually through the plasma reactor (A) to ensure
maximum surface contact with the plasma. The water flow in the
Cajon joints (C) above and below the reactor (A) is then initiated.
The radiofrequency generator is then activated at a power output
of 480 W. This value was obtained by measuring the voltage across
the coil and the current between the blocking and tank capacitors.
A highly reactive fluorine plasma is produced as the power is
applied. The fluorine plasma is brilliant pink and turns to violet
if any air leaks are present. The particles are then fluidized
through the plasma for a 12-hr period. As the carbon monofluoride
formation occurs, breakup of the graphite flakes is observed. Some
of the snow white powder of stoichiometry $CF_{1.1}$ which is formed is
deposited on the upper part of the erlenmeyer particle return and
the majority of this carbon monofluoride passes through the upper
screen (F) into the particle trap. After 12 hr. the plasma is
terminated, the reactor is dismantled, and the white carbon monofluoride is removed from the particle trap and the uppermost walls
of the particle return.

The amount of snow white carbon monofluoride produced averaged
0.2 - 0.3 g per run. The material remaining on the lower nickel
screen (F) was found to be carbon monofluoride of lower stoichiometry
$\sim CF_{0.68}$ as determined by its infrared spectra and X-ray powder
pattern. If finely powdered graphite is substituted for the flake
graphite, more surface contact with the plasma occurs and as expected,
the reaction proceeds more rapidly. However, it is difficult to
keep the finer particles fluidized with the 30 cm^3/min fluorine
flow because they stick to the upper screen (F) out of the plasma
region. If this new synthesis is run on a larger industrial scale,
finer particles will be used and the apparatus will be appropriately
adapted. One other change would be necessary to perfect this system.
The Vycor particle return (D) and the Vycor inlet to the reactor
tube (A) are attacked by the fluorine plasma to such an extent that
they may only be used for two or three runs. This attack removes a
very significant amount of fluorine atoms from the plasma and this
efficiency loss results in longer reaction times. Fabrication of
these parts of alumina, although expensive, would completely
eliminate this problem.

Carbon Monofluoride. The total yield of carbon monofluoride
was 80% based on a mass balance study of the system. The major
by-product, carbon tetrafluoride, was recovered in the U-trap. A
10% yield of the snow white carbon monofluoride of composition
$CF_{1.19}$ was obtained. This white material was recovered in the

particle trap as previously indicated. Carbon tetrafluoride is the principal by-product and accounts for essentially all of the remaining 20%. Anal. Found: C, 34.2; F, 64.75. Calculated empirical formula: $CF_{1.19 \pm 0.04}$.

The infrared spectra of the samples consisted of a strong carbon-fluorine stretch at 1217 cm^{-1} and two medium bands at 1342 and 1072 cm^{-1} which are due to asymmetric and symmetric stretching vibrations of the peripheral CF_2 groups. A 332 cm^{-1} far-infrared band was observed and is due to bending in CF_2 groups.

The X-ray powder pattern contained nine lines. The observed "d" spacings (Å) were 5.80 vs, br; 3.4 w; 2.85 diffuse, m; 2.55 w; 2.2-2.25 vs, br; 1.83 w; 1.71 w; 1.66 w; and 1.289 s.

RESULTS AND DISCUSSION

Direct Fluorination of Polymers

Idealized structures representing the fluorocarbon polymer products are shown in Figure 4 which are in agreement with the elemental analysis. The analysis of the fluorinated acrylonitrile indicates that some decomposition of the nitrile groups occurred in our product. The nitrogen-fluorine stretch at 935 cm^{-1} and the elemental analysis indicate the presence of a large number of successfully fluorinated NF_2 groups.

The fluorination of polyacrylamide produced a fluorocarbon polymer with distinctly different physical properties and structure from the parent. There are significant number of molecular units of the proposed structure as evidenced by the carbonyl and nitrogen-fluorine stretches observed in the infrared spectra. It is likely that some of the amide groups have been destructively fluorinated to CF_3 groups. This phenomenon has been reported by Attaway, Groth and Bigelow[7]. However, no visible decomposition has occurred in the product which has a very white appearance.

The fluorination of ethylene-propylene copolymers yielded a perfluorocarbon polymer which only softens at 340°C in contrast to the physical properties of the commercially available tetrafluoro-ethylene-hexafluoropropylene copolymer. This well-characterized copolymer has a melting point of 290°C. It is probable that this difference in physical properties resulted from a greater degree of crosslinking or a higher molecular weight in the sample obtained by direct fluorination.

Resols are intermediates in the formation of highly crosslinked phenolformaldehyde polymers, and they usually have molecular weights in the 300-700 range. A typical resol structure is shown in Fig. 4,

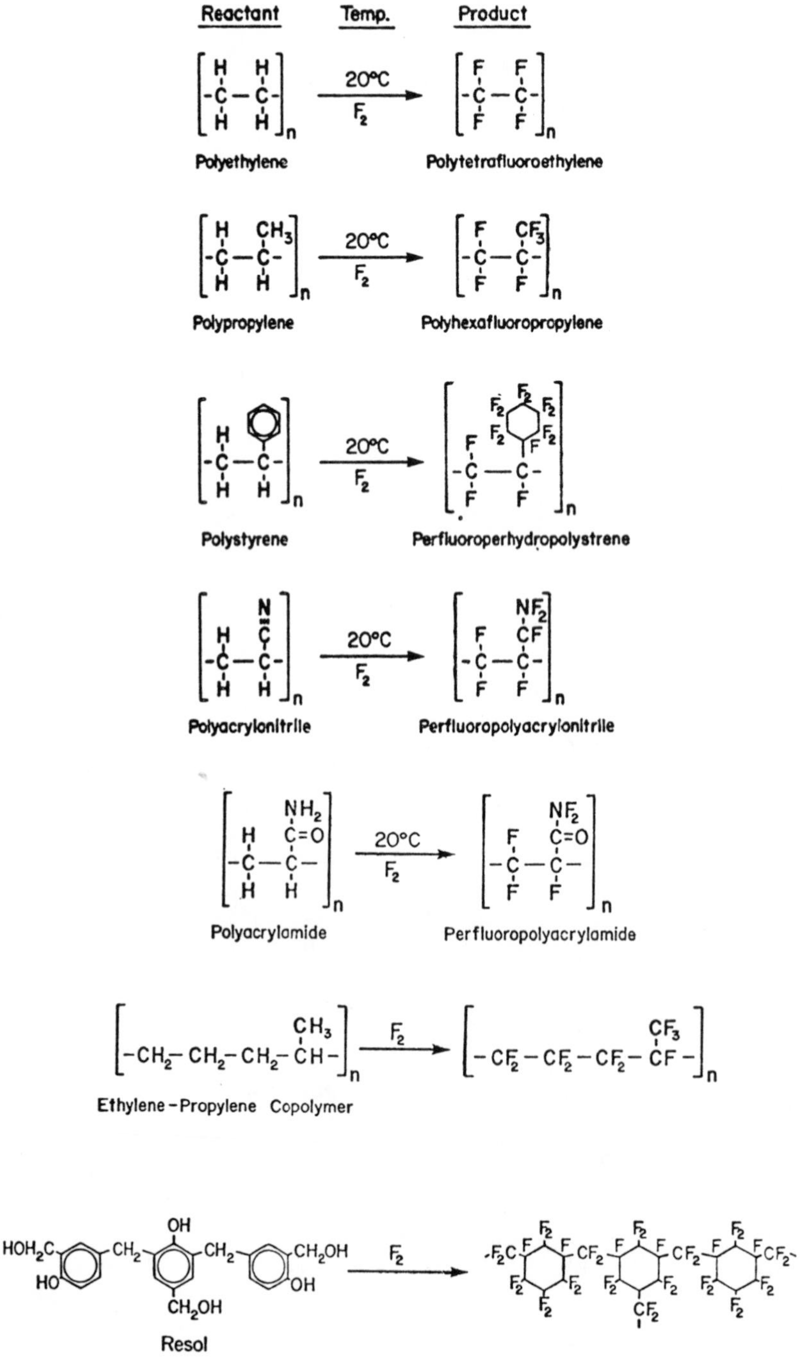

Fig. 4.   Fluorination of Polymers

although the actual structure may be more complex. The fluorocarbon structure postulated as a product in Fig. 1 is consistent with the elemental analysis obtained for the fluorinated polymer obtained. It appears that the crosslinking present in the new polymer may be similar to that indicated in the postulated structure.

The polymers produced with this process are different in several respects from the commercially available polytetrafluoroethylenes and polyhexafluoropropylenes which are made by polymerization of the monomers. In this new process. the cross-linkage, polymer length and particle size are similar to the hydrocarbon polymer precursor since the carbon-carbon bonds are not usually fractured. In general, the molecular weight of the polymer is usually increased due to a crosslinking process which is difficult to control. This crosslinking is apparent in the analytical data and insolubility of many of these materials encountered in the characterization of these materials in subsequent studies. A more dense product is usually obtained from the LaMar process than from the normal polymerization of tetrafluoroethylene which usually yields a loosely packed and flaky polymer, due to the steric factors of fluorine size and repulsion in the polymerization process. One of the obvious advantages of the LaMar process for fluorination of hydrocarbon polymers is that many different fluorocarbon polymer properties may be obtained from a single basic molecular unit by regulating the density and structure of the hydrocarbon polymer starting material.

New polymers may be produced by this new method whose fluorinated monomer is not easily produced or whose monomer will not polymerize easily as in the case of fluorinated polystyrene. This difference in properties is most striking in the case of the polymerization of hexafluoropropylene. Hexafluoropropylene is very difficult to polymerize due to the steric bulk and repulsion of the CF_3 group. Commercial polyhexafluoropropylene has a melting point (290°C) resulting from its low molecular weight, and has quite different properties than polytetrafluoroethylene. The fluorinated polypropylene, due to its higher molecular weight and cross-linkages, has a thermal stability similar to polytetrafluoroethylene.

One may then take advantage of the ease in obtaining various hydrocarbon polymer structures and produce the specific qualities desirable in a fluorocarbon for a given application. Also one may elect to fluorinate the polymer only partially to obtain intermediate properties.

The LaMar process also holds some physical advantages in certain instances over polymerization. Solid objects may be fluorinated to a depth of at least 0.2 mm (based on a crude weight gain calculation with a polyethylene object). Essentially, a fluorocarbon skin is

produced or a complete fluorination if the object is less than 0.2 mm
thick.   Polytetrafluoroethylene fibers, pellets, films, bottles and
solid objects; perfluoropolypropylene cloth, rope and pellets, and
perfluoroperhydropolystyrene film, lids and pellets have been prepared by the LaMar process. In the case of cloth and films, the
products are nearly completely fluorinated as evidenced by their
thermal decomposition to a brown solid at ~355°C, without melting.
A possible handicap here is that these products are somewhat brittle
since they have not been molded as fluorocarbon plastics, but this
may be overcome by selecting the proper hydrocarbon starting
material. The fluorination of solid objects made of polyethylene,
polypropylene, etc. is advantageous because polytetrafluoroethylene
powder is almost impossible to mold like other plastic materials.
It is usually extruded as rods or sheets and objects are prepared
by hand-machining methods.

Poly(carbon monofluoride)[21]

The plasma technique provides a new synthesis for poly(carbon
monofluoride) $(CF_x)_n$. The infrared spectrum is in substantial
agreement with that reported by Rudorff and Brodersen[10] and by
workers at Rice University[1]. The X-ray powder pattern "d" spacings
are in agreement with those of workers at Rice University[1] and in
rough agreement with the spacings of 6.0, 2.23, and 1.30 Å reported
by Palin and Wadsworth[11]. We have noted that the lines change
slightly with fluorine content. The elemental analysis is consistent
with the carbon monofluoride structure and corresponds to a stoichiometry of $CF_{1.19}$. This new synthesis for poly(carbon monofluoride)
has several advantages over previous high-temperature syntheses.
A much lower temperature was used. Measurement revealed that the
gas temperature of the plasma is less than 150°. The amount of
energy required to generate the plasma (480 W) is much less than
is required to maintain a furnace of corresponding size in the 600°
temperature range. Also it may prove easier to generate a larger
plasma than to keep a large reactor uniformly in the rather narrow
temperature range required for the thermal synthesis of the snow
white $CF_{1.1}$. The fluidized bed method also avoids the necessity
for batch processing since graphite may be continually added or
withdrawn from the system.

This synthesis provides an example of the use of a fluidized
bed plasma reactor design to prepare an industrially important
material. This synthesis also demonstrates some of the advantages
of a fluorine plasma in preparative fluorine chemistry.

REFERENCES

1. J. Soll, U. S. Pat. 2,129,289 (1938).
2. A. J. Rudge, Brit. Pat. 710,523 (1954).
3. S. P. Joffre, U. S. Pat. 2,811,468 (1957).
4. W. T. Miller, U. S. Pat. 2,700,661 (1955).
5. J. Pinsky, A. Adakonis, and A. Nielson, Modern Packaging, 6, 130 (1960).
6. W. R. Siegart, W. D. Blackley, H. Chafetz, and M. A. McMahon, U. S. Pat. 3,480,667 (1969).
7. J. A. Attaway, R. H. Groth and L. A. Bigelow, J. Am. Chem. Soc. 81, 3599 (1958).
8. R. J. Lagow and J. L. Margrave "The Reaction of Polynuclear Hydrocarbons with Elemental Fluorine", to be published.
9. R. J. Lagow and J. L. Margrave, Proc. Natl. Acad. Sci., 67, 4, 8A (1970).
10. N. J. Maraschin and R. J. Lagow, J. Am. Chem. Soc., 94, 8601 (1972).
11. N. J. Maraschin and R. J. Lagow, Inorg. Chem., 12, 1459 (1973).
12. R. L. Fusaro and H. E. Sliney, NASA Tech. Memo., NASA TMX-S262Y (1969).
13. H. Gisser, M. Petronio, and A. Shapiro, J. Amer. Soc. Lubric. Eng., 161 (May 1970).
14. K. Braeuer, Technical Report ECOM-3322; H. F. Hunger, Technical Report ECOM, U. S. Army Electronics Command, Fort Monmouth, N.J.
15. O. Ruff, D. Bretschneider, and F. Elert, Z. Anorg. Allg. Chem., 71, 1 (1934).
16. W. Rudorff and G. Rudorff, Z. Anorg. Chem., 217 (1947).
17. W. Rudorff and G. Rudorff, Chem. Ber., 80, 417 (1947).
18. R. J. Lagow, R. B. Badachhape, J. L. Wood and J. L. Margrave, "Some New Synthetic Approaches to Graphite-Fluorine Chemistry", J. Chem. Soc., May (1974), in press.
19. R. G. Bautista, D. W. Bonnell, and J. L. Margrave, "The Structure of Carbon Monofluoride", to be submitted for publication.
20. R. J. Lagow and J. L. Margrave, "The Controlled Reactions of Hydrocarbon Polymers with Elemental Fluorine", Polymer Letters, April (1974), in press.
21. R. J. Lagow, L. A. Shimp, D. K. Lam and R. F. Baddour, J. Inorg. Chem., 11, 2568 (1972).

DISCUSSION OF PAPER BY R. LAGOW

Participant (Duo Plastics): Can you carry out surface fluorination or chlorination under glow discharge?

R. Lagow: Yes, I have done some experiments like this. It might be feasible to gain a certain amount of control under the glow discharge. On the other hand, you might have guessed that our fluorination with organic or inorganic compounds would not work at all in the glow discharge. We used glow discharge only in the case of the fluorination of graphite for which we had difficulty in achieving the results otherwise. It (fluorination) has an energy control problem. Really, it does not need any extra energy from glow discharge. However, I won't be surprised that someone will develop a smooth glow discharge process in the future.

H. Gisser (Frankford Arsenal): The direct fluorination of poly-ethylene film leaves a solid structure equivalent to the original state of low crystallinity unless recrystallization or molecular motion has taken place. These may be time and temperature dependent. Have any friction or other measurements been run on directly fluorinated polyethylene to explore the nature of the surface?

R. Lagow: Yes, but most of these studies have been done by indust-rial researchers whose firms may be reluctant to release information on promising results. There have been studies which have indicated promise of definite applications for surface fluorination to change a number of surface properties including some unknown lubrication. These have been done on a number of classes of polymers.

Structure, Bonding and Dynamics of Surface and Sub-surface Polymer Films Produced by Direct Fluorination as Revealed by ESCA

D. T. Clark[†], W. J. Feast, W. K. R. Musgrave and I. Ritchie

Chemistry Department

University of Durham, Durham City, U. K.

The nature of the surface (i.e. ~50Å) of a polymer is of crucial importance in determining many physical, chemical and mechanical properties and this arises since solids communicate with the rest of the universe by way of the surface. A detailed ESCA examination has been made of the nature of the surface fluoropolymer produced by interaction of high density polyethylene films with fluorine diluted in nitrogen which demonstrate the great potential of ESCA in this area. The topics discussed will include <u>inter alia</u>:

(i)   The preparation and surface characterization of polymer films.
(ii)   Direct measurement of fluorination depth (in the 0-50Å range) as a function of reaction conditions.
(iii)   Stoichiometry and structure and bonding of surface and subsurface as a function of reaction conditions.
(iv)   Kinetics and mechanisms of the processes involved.

INTRODUCTION

General

The technological importance of polymers, polymer coatings
and films has resulted in a considerable interest in the chemistry
and physics of their surfaces, and in particular in the changes
which occur during contact with the various environments experienced
in manufacture, fabrication and use. Not surprisingly therefore,
a great number and variety of techniques have been employed in
investigations of surfaces, including inter alia, gloss and gonio-
photometric measurements[1], abrading the surface and examining the
removed material, reflectance and transmission spectroscopy, and
studies of diffraction and scattering of light, electrons and
X-rays[2,3]. In recent years, the spin-off from the intensive interest
in the solid state physics of thin films has benefited surface
studies generally[4]. However, despite the considerable effort and
originality put into many of these investigations, it appears that
surprisingly little detailed information has been acquired concerning
molecular structure (and in particular changes in molecular structure
during exposure to chemical or physical agents) in the surface* and
immediate subsurface regions. It is, however, clear that the
immediate surface must be of prime importance in determining many
of the chemical and physical properties of a solid.

The reasons for the paucity of detailed information about
changes in molecular structure particularly for polymeric systems
in this region stem from limitations inherent in most of the techniques which have been applied. These limitations can be classified
under two major headings. Firstly, many techniques are limited by
the depth of the surface sampled, in that small changes in, say, the
outer 50Å of the surface are not detectable because the sampling
method records information about a layer, several hundreds or
thousands of Angstroms thick, and even some of the techniques which
do deal with the immediate surface layers, depend on the periodicity
of the structure (e.g. LEED) and are hence not applicable to the
analysis of less regular structures. Secondly, many techniques
provide information at too coarse a level or of a kind that cannot
be directly related to changes in molecular structure.

In the previous paper[5], we have outlined in some detail the
application of ESCA to the study of structure and bonding in polymer
systems. It should be evident from this that ESCA has particular
advantages as a technique for studying surface and subsurface

---

* 'Surface' has several meanings of which only the geometrical
   definition is precise. Throughout this paper the word 'surface'
   and immediate subsurface is used to denote the outermost few tens
   of Angstroms of a solid.

phenomena. In this work we describe the application of ESCA to
studies of structure, bonding and dynamics of surface and subsurface
polymer films produced by direct fluorination of polyethylene.

The direct fluorination of surfaces is of course of some
considerable current interest,and an ESCA study of this process
for polyethylene is therefore a logical sequel to our earlier
studies which provided us with fairly detailed correlations between
molecular structure and shifts in observed binding energies for
fluoropolymers and some confidence in the theoretical model used
for the interpretation of spectra[6]. As will become evident this
background is of vital importance in unravelling the complexities
of surface fluorination by ESCA.

It should perhaps be emphasized that ESCA adds a new dimension
to many aspects of polymer physics and chemistry,and this has
considerably broadened the initial objectives of our investigations
to include inter alia, surface oxidation, surface degradation by
thermal and photochemical means and the study of diffusional
problems. The work described here is, therefore, in the nature of
a preliminary sketch of a small part of a larger canvas that we
envisage taking many man years to complete.

Surface Fluorination of Polyethylene

The possibility of submitting the cheap and readily fabricated
polyethylene to a surface treatment wnich would improve its properties in one way or another has inspired many investigations.
Treatments such as surface oxidation by corona discharge, bombardment
polymer films would be particularly apposite at this time.

Detailed ESCA examinations have, therefore, been made of the
fluorination of pressed films of high density polyethylene, and
have been directed at the following points of interest.

(i)   Preparation of polymer films.

(ii)   Establishment of fluorination depth as a function
        of fluorinating conditions.

(iii)   Overall stoichiometries as a function of depth for
        various fluorinating conditions.

(iv)   Molecular structure of the surface and immediate
        subsurface.

(v)   Kinetics and mechanism of the fluorination process.

EXPERIMENTAL

Preparation of Samples and ESCA Examination
of the Surface Prior to Fluorination

In the previous paper[5], we have outlined the general means of preparation of samples convenient for ESCA studies. Of the methods discussed, it is clear that the pressing of films to some extent offers many advantages both in terms of convenience and of limiting the possibility of contamination (e.g. from solvent). Our previous studies of PTFE films produced by this process reinforces the conclusion that under carefully controlled conditions it should be possible to produce uncontaminated films of polyethylene suitable for direct ESCA examination and for the fluorination experiments. For comparison purposes, a variety of commercially available poly-ethylene films have also been studied. To complement the ESCA investigations, the more traditional techniques of Transmission Infra Red (TIR) and Multiple Attenuated Total Reflectance (MATR) have also been employed. Whilst TIR samples the bulk, under appropriate conditions MATR can provide information concerning the outermost few hundred Å of a sample. The three techniques are, therefore, complementary in nature. Fig. 1 shows the C_{1s} and O_{1s} spectra for commercially produced low density polyethylene films both before (a) and after (b) desiccation. Two features are clearly evident. Firstly, the substantial signals corresponding to the O_{1s} levels exhibit an unresolved structure, one component of which disappears on storing the samples over P_2O_5. Comparison with previously measured binding energies indicates that the higher binding energy component of the O_{1s} levels corresponds to H_2O whilst the lower binding energy peak corresponds to $>C=0$. Secondly, the C_{1s} levels show a small shoulder to the high binding energy of the main peak, the shift of ~3 eV again being consistent with carbon in $>C=0$ environments. Comparison with MATR measurements employing single crystal Germanium is instructive in this respect since strong absorptions are present in the carbonyl region (but not in the OH region). Careful double beam TIR experiments on the ~200μ films also shows the presence of carbonyl groups. The three techniques taken together, therefore, emphasize that whilst the carbonyl groups are distributed throughout the bulk of the sample, hydrogen bonding involving extraneous water is localized at the surface. The carbonyl groups may be attributable in part to various additives such an anti-oxidants, anti-blocking agents, etc.

In preparing films of high density polyethylene for the fluorination experiments, three methods of preparation have been investigated. In all of these, films were pressed between sheets of clean aluminum foil at the minimum temperature necessary for plastic flow, with a hand press (~200°C).

(c) The samples were pressed in air.

(d)   The samples were pressed in a nitrogen atmosphere.

(e)   The polyethylene powder was pumped down and let up
      to a pure nitrogen or argon atmosphere for several
      cycles before being transferred in an inert
      atmosphere into the press.

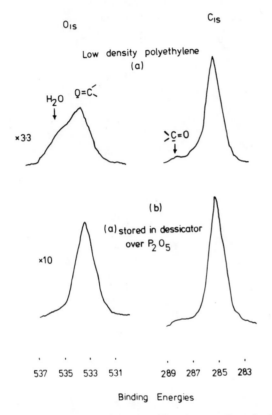

Fig. 1. C_{1s} and O_{1s} for low density polyethylene samples.

The O_{1s} spectra corresponding to samples from these three
modes of preparation are shown in Fig. 2 and are quite striking.*
ATR and TIR experiments did not reveal the presence of any oxygen
function (-OH, $>$C=O, C-O-C etc.) and in fact the spectra were
virtually identical. This in itself demonstrates the great power
of ESCA in distinguishing minute differences between samples when

* It should perhaps be emphasized that direct study of the powdered
  sample (prior to pressing) by ESCA reveals no trace of O_{1s} signal.

such differences are localized at or near the surface. Comparison
with the data for low density polyethylene and with model monomer
systems shows that the O_{1s} signal arises from $\underline{>C=0}$ type environments. The three methods of preparation clearly indicate that
'unoxidized' surfaces may most readily be prepared by excluding
all traces of oxygen during the pressing stage. On leaving samples
exposed to the atmosphere for some time, hydrogen bonding to the
surface $>C=0$ groups from extraneous water in the atmosphere occurs
and the O_{1s} peak then acquires the characteristic doublet nature.
Again MATR and TIR do not reveal any changes since the hydrogen
bonding is localized at the surface. Clearly ESCA adds a new
dimension to the study of surface properties foremost amongst which
might be listed, friction, wear, aging, and triboelectric phenomena
which all depend to a greater or lesser extent on detailed consideration of the surface and immediate subsurface of polymer films.

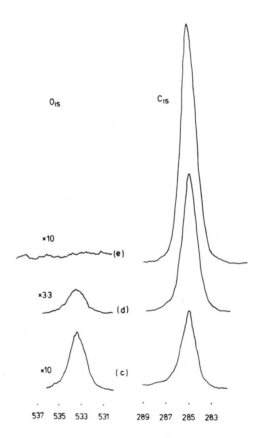

Fig. 2. C_{1s} and O_{1s} for pressed high density films of
         polyethylene.

It is perhaps of some interest to obtain some information on the relative frequency of $>C=O$ features in the oxidized films and this may be obtained as follows. From extensive studies of compounds containing both carbon and oxygen, infinity values (I_α) may be obtained for the C_{1s} and O_{1s} core levels in homogeneous solids[11]. If we make the reasonable assumption that because of temperature gradients, diffusional phenomena, and the relatively low concentration of oxygen (the only available oxygen being that entrained in the powdered polyethylene sample), that in the sample prepared by method (d) oxidation is likely to be limited to the first monolayer (i.e. ~5Å), we may then with a knowledge of the likely escape depth dependences (Λ_1 and Λ_2 for the O_{1s} and C_{1s} core levels) calculate the surface concentration of $>C=O$ structural features. Taking Λ_1 and Λ_2 as 8Å and 10Å (corresponding to MgK$\alpha_{1,2}$ exciting radiation), the result obtained is that if all the O_{1s}'signal corresponds to $>C=\underline{O}$ located in the first monolayer then there is ~9% monolayer coverage of this structural feature. Since such films are routinely easy to prepare in a reproducible manner, the initial fluorination experiments have been with this material rather than the more difficultly prepared samples produced by method (e). As will become apparent, detailed analysis of structure and bonding of the surface fluorinated films provides strong support for this crude estimate of the extent of surface oxidation.

One question which immediately arises is that of the structure at the surface of polyethylene films produced in this manner. The question of polymer crystallography and morphology has been extensively discussed and documented by Wunderlich[12]. For the most part, discussion has centered around samples crystallized from solution or produced from the melt under pressure. There are two considerations which make it difficult to extrapolate from this work. Firstly, our prime consideration is the structure at the surface of samples whereas the published data refer rather to the bulk structure. Secondly, our mode of sample preparation differs considerably from those for which information concerning the bulk structure are available.

Of particular relevance to our studies however, are the extensive investigations of Schonhorn[13] on nucleation and transcrystalline growth at polyethylene-metal interfaces. The major points of interest arising from this work are that heterogeneous nucleation and crystallization of polyethylene melts against high energy surfaces such as aluminum, results in marked changes in the surface region morphology; electron microscopy indicating a region of transcrystalline growth extending to a depth of 25-50μ. Schonhorn also commented that the mechanics of peeling the polymer film from the aluminum determines the thickness of polymer remaining on the metal surface and that for thin aluminum, failure should occur closer to the metal polymer interface.

In the preparation of the polymer films used in this work, the procedure entailed pressing the high density polyethylene powder between clean sheets of aluminum foil. The aluminum foil was then stripped from the surface. In previous ESCA studies, we have investigated in detail the surface of aluminum foil[14]. The main points which emerge from these studies are that the oxide layer in commercially produced annealed (domestic) foil is typically ~20Å thick and that a tenaciously held hydrocarbon type layer is present at the surface which is not readily removed by either degreasing treatment or heating under very high vacuum conditions.

ESCA provides a convenient tool, therefore, for investigating the nature of the peeled surfaces. Fig. 3 shows the O_{1s}, C_{1s} and Al_{2p} levels for the surface of the aluminum foil used for pressing and of the peeled foil appropriate to the polyethylene sample (d) of Fig. 2. (It should be stated at this stage that no trace of Al_{2p} core levels could be detected for the sample 2d.) The most significant feature is that both the aluminum and oxygen core levels are of appreciable intensity in the peeled foil and this can only be interpreted on the basis that failure occurs very close to the surface. From the relative increase in intensity of the peak due to the C_{1s} levels (taken in conjunction with an escape depth of 10Å for electrons of kinetic energy ~968 eV) a reasonable estimate for the thickness of polymer adhering to the peeled foil would be ~10Å.

In an investigation of the wettability of melt crystallized polyethylene films formed by nucleation at an aluminum (aluminum oxide) surface, Schonhorn measured[13b] a contact angle of 65 degrees and derived a percentage crystallinity for the surface region of 63.2%. We have investigated therefore the contact angles for both of the peeled surfaces using glycerol as the wetting liquid to correspond to the measurements of Schonhorn. For both surfaces a contact angle of ~66° was obtained which would strongly suggest that our samples are of comparable surface structure to Schonhorn's.

Experimental Arrangement for Studying Surface Fluorination

For these preliminary experiments, a very simple apparatus was assembled and a schematic of the experimental set up for fluorination films is shown in Fig. 4. Fluorine was obtained from an electrolytic cell containing an electrolyte of hydrogen fluoride and potassium fluoride (approximate composition KF,2HF), thermostatically maintained at 82 ± 5°C. The maximum load was 10 amps at 10 volts giving a maximum fluorine output of approximately 6.5 gms/hour.

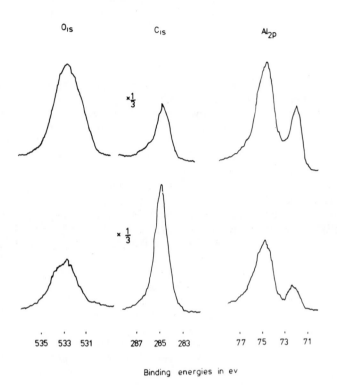

Binding energies in ev

Fig. 3. C_{1s}, O_{1s}, Al_{2p} for peeled foil before and after
pressing.

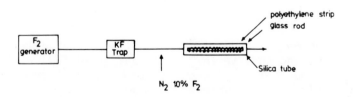

Fig. 4.   Schematic of apparatus for surface fluorination

Prior to carrying out a fluorination, the cell was run at low current for ~4 hrs followed by a further hour at high current to ensure fluorine of adequate purity. No attempt has been made in these preliminary investigations to quantitatively determine the purity of the fluorine used; however from previous investigations, a reasonable estimate would be >98%, with the major impurities being hydrogen fluoride, oxygen and traces of water.

During a run, the fluorine cell was typically run at 5 amps, the output being taken through a KF trap to remove HF and then mixed with a metered flow of oxygen free nitrogen such that the dilution corresponded to 10% fluorine in nitrogen. Strips of poly-ethylene produced by the hot pressing process were wrapped around a glass rod and inserted into a silica tube. A graded seal was used to connect a ground glass socket to the silica tube at the end remote from the fluorine generator and through which the poly-ethylene samples could be mounted. The glass rod, on which the polyethylene strips were wound, was provided with 'legs' at either end to hold the samples clear of the inside walls of the silica tube during the experiment. Prefluorinated copper pipework and brass connectors were used to connect the fluorine cell and KF trap via glass tubing, a teflon coated isolation valve and graded seals to the silica tube. Previous to any experiment, the apparatus was swept with oxygen free nitrogen stream for ~1 hr before connecting the fluorine generator. During an experiment, a given sample was exposed to the 10% fluorine in nitrogen for a given period of time. Initial experiments indicated that concentration gradient effects were absent and that identical ESCA spectra were obtained from fluorinated samples taken from opposite ends of the glass rod. Further, since the samples were loosely wound around the glass rod, identical results were obtained from both the surface facing toward and away from the glass. In the initial experiments, the apparatus was made light tight and in a later paper we will discuss the effect of irradiating the reaction cell during the fluorinating process.

DISCUSSION OF RESULTS

Introduction

As a preliminary study, films were fluorinated for ~1 hr. The C_{1s} spectra revealed that the original peak at 285 eV arising from -CH_2-CH_2- units had been replaced by a complex structure extending to ~292 eV. This together with an intense fluorine 1s peak indicate that extensive fluorination had taken place. This was readily confirmed by MATR experiments with single crystal Ge which indicated the presence of C-F groups, and indeed careful double beam TIR studies also indicated this. These results overall indicated, therefore, that fluorination had proceeded well into the bulk and on the basis of the MATR and TIR data a reasonable estimate would be

>100Å. The initial stages of the process were therefore investigated
and a convenient range of reaction times established of 0.5 sec -
300 sec.

To illustrate the great sensitivity of the technique for
studying the initial stages of reaction, Fig. 5 shows the C_{1s}, O_{1s}

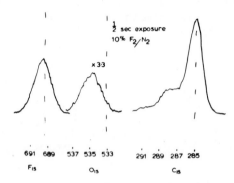

Fig. 5. C_{1s}, O_{1s} and F_{1s} core levels for polyethylene
sample fluorinated for 0.5 sec.

and F_{1s} core levels for a sample fluorinated for ~0.5 sec. There
are several features of major interest. Firstly, the appearance
of a strong F_{1s} signal; the absolute binding energy being characteristic of a C̲F type system[6]. Secondly, there is an increase in
overall intensity of the O_{1s} signal; however, it is clear from the
characteristic binding energy that this is due to water and most
probably arises from hydrogen bonding to the surface carbonyl
features from the trace of water impurity in the system. Finally,
the C_{1s} spectrum shows in addition to the main peak of polyethylene
at 285 eV, a broad shoulder centred ~288 eV characteristic of C̲F
type environments, and a low intensity tail at higher binding energy
centred ~291 eV attributable to C̲F$_2$ type environments. It would
clearly be feasible to detect fluorination occurring on a much
shorter time scale and a reasonable lower limit with present
instrumentation would be ~10^{-3} sec although of course it would be
technically difficult to fluorinate a sample on this time scale.
It does indicate, however, that experiments to detect fluorination
employing much more dilute fluorinating conditions would be perfectly
feasible.

Fig. 6 shows the F_{1s}, O_{1s} and C_{1s} core levels for films
fluorinated for 0.5 sec and 30 secs. The differences, although
quite striking, go undetected by either MATR or TIR emphasizing
the surface nature of the initial reaction. After 30 secs exposure,
the original C_{1s} peak has almost completely disappeared, appearing
only as a shoulder on a broad peak centred ~286 eV. Further broad
peaks are centred ~288.5 eV and 291 eV with a tail extending to

~292 eV. Comparison with the binding energies as a function of
structural environment discussed in the previous paper readily
allows an assignment of these major features[5]. Firstly, the most
intense peak at ~288.5 eV corresponds to $\underline{C}F$ type environments whilst
those at ~286 eV and 291 eV corresponding to carbons subjected to
secondary substituent effect from fluorine and from $-\underline{C}F_2-$ type
environments respectively. This data alone, therefore, demonstrates
that extensive fluorination has taken place at the surface and
immediate subsurface of the polyethylene films. Repeat experiments
were completely reproducible.

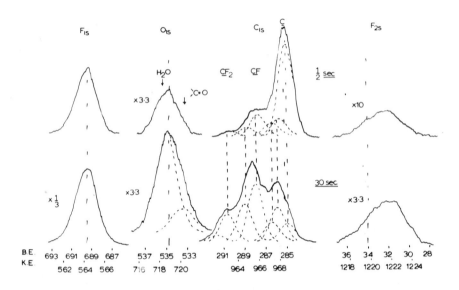

Fig. 6. C_{1s}, O_{1s}, F_{1s} and F_{2s} levels for fluorinated
samples of polyethylene (0.5 sec and 30 sec).

Quantitative Analysis of Results

Depth of Fluorination as a Function of Time. Having established
that ESCA is capable of detecting surface fluorination at a very
early stage a more refined analysis of the data may be attempted.
Firstly, we may investigate the application of ESCA to the investigation of film thickness.

In discussing the valence bands of polymers in the previous
paper[5], we have emphasized that a peak appearing at ~35 eV binding
energy in fluoropolymers may be assigned to F_{2s} levels which are
essentially core like. It is clear from Fig. 6 that the F_{1s} and
F_{2s} levels with kinetic energies of ~564 eV and 1222 eV should
correspond to a considerable difference in escape depth dependence.
Extrapolation from the generalized curve of mean free path versus

kinetic energy (Fig. 13 of the review paper[5]),for example,would
suggest ~8Å and ~14Å respectively for the F_{1s} and F_{2s} levels. We
can now proceed to use this as a basis for estimating the thickness
of the fluorinated polymeric films.

Before outlining the background theory however, it is worth
digressing on the likely kinetics and mechanism of the fluorination
process. A more detailed discussion will be presented in a later
section,and at this stage all we need to note is that fluorination
will proceed via a chain reaction as illustrated in Fig. 7. The
low bond dissociation energy of fluorine (37 Kcals/mole) coupled
with the large negative enthalpy changes in reactions 2 and 3 ensures
an efficient chain reaction. From studies of abstraction reactions
of fluorine atoms with simple alkanes, it may be inferred that the
activation energies for both steps 2 and 3 are extremely low and
that hydrogen abstraction by fluorine is a relatively unselective
process[15]. We will return to this point later on, but we may note
that comparison of the likely rates for reactions 2 and 3 with the
likely rate of diffusion of fluorine through the polymer film strongly
suggests that fluorination of the bulk will be diffusion controlled.

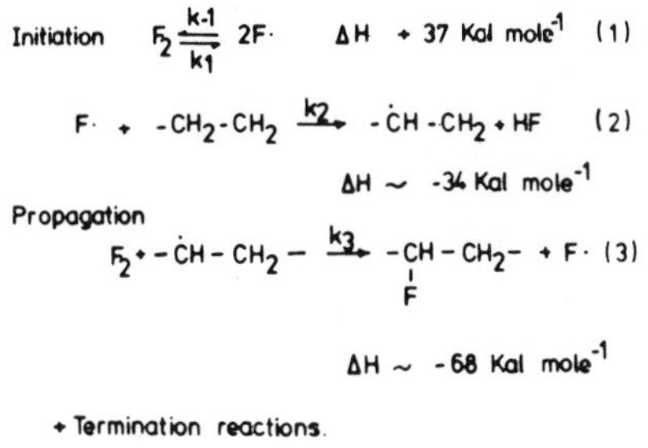

Initiation $\quad F_2 \underset{k_1}{\overset{k_{-1}}{\rightleftharpoons}} 2F \cdot \qquad \Delta H + 37 \text{ Kal mole}^{-1} \quad (1)$

$F \cdot + -CH_2-CH_2 \xrightarrow{k_2} -\dot{C}H-CH_2 + HF \quad (2)$

$\Delta H \sim -34 \text{ Kal mole}^{-1}$

Propagation

$F_2 + -\dot{C}H-CH_2- \xrightarrow{k_3} -CH-CH_2- + F \cdot \quad (3)$
$\qquad\qquad\qquad\qquad |$
$\qquad\qquad\qquad\qquad F$

$\Delta H \sim -68 \text{ Kal mole}^{-1}$

+ Termination reactions.

Fig. 7. Mechanism of free radical fluorination of alkanes.

As will become apparent, ESCA provides evidence that this must indeed
be the case. Irrespective of the detailed structure of the polymer
therefore, we might expect that since the activation energies for
(2) and (3) are so low that fluorination would tend to proceed in
a relatively uniform manner. We, therefore, start with this as a
working assumption which may be subjected to detailed scrutiny as
the analysis proceeds.

Consider now the situation (Fig. 8) of a fluoropolymer film
(which need not of course be homogeneous in composition throughout
its depth) of thickness d formed at the surface of a polyethylene
film. (Since in general the films which were fluorinated were in
excess of 50μ (i.e. 5×10^5Å) thick,we may safely assume that the
overall thickness of the polymer is infinite compared with typical
mean free paths of electrons).

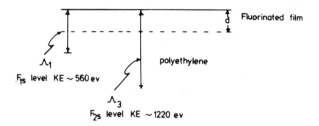

Fig. 8. Electron mean free paths in relation to fluorinated
         film thickness for F_{1s} and F_{2s} levels (Schematic).

We may write the intensity of the F_{1s} and F_{2s} levels as
Equations (4) and (5).

$$I_{F_{1s}} = I_{\alpha F_{1s}}(1-e^{-d/\Lambda_1}) \tag{4}$$

$$I_{F_{2s}} = I_{\alpha F_{2s}}(1-e^{-d/\Lambda_3}) \tag{5}$$

where the symbols have the usual meanings and Λ_1 and Λ_3 are the
electron mean free paths appropriate to the F_{1s} and F_{2s} levels
respectively. Now the ratio of infinity values for the F_{1s} and
F_{2s} levels is directly proportional to the measured area ratios
for the two peaks (Equation 6). From studies of simple homopolymers
as discussed previously[5],

$$\frac{I_{\alpha F_{1s}}}{I_{\alpha F_{2s}}} = \frac{LA_{\alpha F_{1s}}}{A_{F_{2s}}} \tag{6}$$

a value for the area ratios may readily be obtained and under our
experimental conditions is equal to 10.12.

Equations (4), (5) and (6) may now be combined to give (7), the
proportionality constant L cancelling.

$$\frac{I_{F_{1s}}}{I_{F_{2s}}} = \frac{\cancel{L}A_{F_{1s}}}{A_{F_{2s}}} = \left(\frac{\cancel{L}A_{\alpha F_{1s}}}{A_{\alpha F_{2s}}}\right)\frac{(1-e^{-d/\Lambda_1})}{(1-e^{-d/\Lambda_3})} \tag{7}$$

This equation relates the measurable properties of the system (viz.
the area ratios of the F_{1s} and F_{2s} peaks), and the film thickness
and electron mean free paths. Denoting the measured ratios for the
fluorinated films by y and the infinity values measured for simple

homopolymers by K, Equation (7) may be recast in the form (8).

$$(y_{/K}-1) = \left(y_{/K}e^{-d/\Lambda_3}e^{-d/\Lambda_1}\right)$$

$$(8)$$

At this stage, we obtain $y_{/K}$ from experiment and the generalized curve of mean free path as a function of kinetic energy gives some idea of the likely range of values for Λ_1 and Λ_3.

A computer analysis may now be used in a self consistent process. Taking the measured intensity ratios, the right hand side of Equation (8) may be evaluated for a range of values of d and of Λ_1 and Λ_3. Tables were constructed corresponding to values of d in the range 2-41Å at 1Å intervals and for ratios of Λ_3/Λ_1 from 1-2 in increments of 0.1 with values of Λ_1 varying in the range 5-15Å in 1Å steps. The data corresponding to 24,200 separate computations for five separate fluorination times is too lengthy to reproduce fully here but sufficient data has been included to illustrate the basic philosophy behind our analysis. This may be done by reference to the two fluorinations so far discussed viz. 0.5 sec and 30 secs. The corresponding measured values of y were 15.64 and 10.83 respectively yielding values of the LHS of Equation (8) of 0.545 and 0.07. Inspection of the Tables revealed that for the 0.5 sec run no fit was possible within the range of parameters taken if the ratio of Λ_3 to Λ_1 was <1.5. The fits within the range of 1.6 - 2.0 for this ratio bracketed the value of Λ_1 in the range 4-12Å and film thickness d in the range 3-11Å. Considering the data for the 30 sec run and limiting the ratio of escape depths to be compatible with those for the 0.5 sec run, considerably narrows the range of values for Λ_1 for which a fit to the experimental data may be obtained to 5-9Å with a corresponding range for the film thickness of 22-42Å. The 30 secs run, bracketing as it does the escape depth, now further limits the range of film thickness for the 0.5 sec run to 3-9Å.

It is interesting to note that the ratio of escape depths, Λ_3 to Λ_1 suggested from the generalized curve, is ~1.7-in the middle of the range bracketed by the analysis for just two runs.

Table 1 presents the measured F_{1s}/F_{2s} ratios for fluorination experiments corresponding to reaction times of 0.5 sec, 2 secs, 15 secs, 30 secs and 300 secs. Computer fits to this data further reinforces the analysis presented above. Film thicknesses have therefore been computed for values of $\Lambda_1 = 7$Å with $\Lambda_3/\Lambda_1 - 1.7$Å. This gives a value of 11.9Å for Λ_3, both points fitting very well on the generalized curve previously discussed. The computed data corresponding to these parameters are shown in Fig. 9. The derived thickness of the fluorinated films are 3.5Å, 6Å, 16Å, 30Å and 36Å respectively for reaction times of 0.5, 2, 15, 30 and 300 secs respectively. This preliminary analysis is most encouraging.

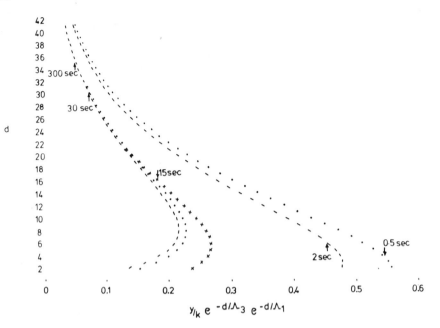

Fig. 9.   Estimation of fluorinated film thickness from the F_{1s} and F_{2s} levels.

For direct comparison of the validity of this data, an independent analysis may be made from the C_{1s} spectra. Consider the fluoropolymer film of thickness d formed by reaction of fluorine with polyethylene. Taking standard lineshapes and linewidths and employing the basic philosophy of lineshape analysis previously discussed in detail elsewhere[16], the C_{1s} spectra may be unambiguously deconvoluted into component peaks and this is shown for the 0.5 sec and 30 sec runs in Fig. 10. Evidence will be presented in the next section that the C/F stoichiometry of the fluoropolymer is ≤ 2 for the films produced in this study. This being the case, the peak at binding energy 285.0 eV may be assigned to unreacted polyethylene. The area ratio of this portion of the C_{1s} peak to the remaining higher binding energy part arising from various C_{1s} environments of the fluoropolymer film, (whose assignment will be discussed in detail later) forms the basis for estimating the film thickness d.

The analysis proceeds as follows. The intensities of the C_{1s} peaks arising from the fluoropolymer film and polyethylene may be expressed as (9) and (10).

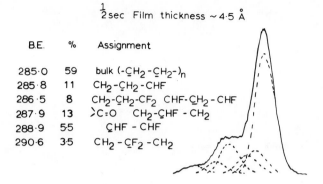

$\frac{1}{2}$sec Film thickness ~ 4.5 Å

B.E.	%	Assignment
285.0	59	bulk $(-CH_2-CH_2-)_n$
285.8	11	CH_2-CH_2-CHF
286.5	8	$CH_2-CH_2-CF_2$ $CHF-CH_2-CHF$
287.9	13	$>C=O$ $CH_2-CHF-CH_2$
288.9	5.5	$CHF-CHF$
290.6	3.5	$CH_2-CF_2-CH_2$

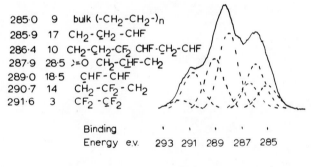

30 sec Film thickness ~ 27 Å

285.0	9	bulk $(-CH_2-CH_2-)_n$
285.9	17	CH_2-CH_2-CHF
286.4	10	$CH_2-CH_2-CF_2$ $CHF-CH_2-CHF$
287.9	28.5	$>=O$ $CH_2-CHF-CH_2$
289.0	18.5	$CHF-CHF$
290.7	14	$CH_2-CF_2-CH_2$
291.6	3	CF_2-CF_2

Binding
Energy e.v.   293 291 289 287 285

C_{1s} levels

Fig. 10. Deconvolution of the C_{1s} spectra for fluorinated
polyethylene samples (0.5 sec and 30 sec)

Fluoropolymer $\qquad I_{C_{1s}}^{F_{pol}} = I_{\alpha C_{1s}}^{F_{pol}} (1-e^{-d/\Lambda_2})$ $\qquad$ (9)

Polyethylene $\qquad I_{C_{1s}}^{Pet} = I_{\alpha C_{1s}}^{Pet} e^{-d/\Lambda_2}$ $\qquad$ (10)

The measured relative area ratios may therefore be written as
(11).

$$\frac{A^{F_{pol}}}{A^{Pet}} = \left(\frac{A_{\alpha}^{F_{pol}}}{A^{Pet}} \right) (e^{d/\Lambda_2}-1) \qquad (11)$$

TABLE 1

ANALYSIS OF COMPONENT PEAKS FOR FLUORINATED POLYETHYLENE SAMPLES
(ALL BINDING ENERGIES IN eV)

1. 0-5 sec

C_{1s} Spectrum

Peak No.	Binding Energy	Area %	C_{1s}/F_{1s}	Area Ratios F_{1s}/F_{2s}	C_{1s}/O_{1s}
1	–	–			
2	290·6	3·5			
3	288·9	5·5			
4	287·9	13·5	1·81	15·64	8·15
5	286·5	8			
6	285·8	11			
7	285·0	58·5			
F_{1s}	689·4				
O_{1s}	534·7	89			
	532·8	11			

2. 2 sec

C_{1s} Spectrum

Peak No.	Binding Energy	Area %	C_{1s}/F_{1s}	F_{1s}/F_{2s}	C_{1s}/O_{1s}
1	–	–			
2	290·9	7			
3	289·0	11·5			
4	287·9	17·5	0·94	14·69	6·76
5	286·5	8			
6	285·8	16			
7	285·0	40			
F_{1s}	689·1				
O_{1s}	534·4	85			
	532·7	15			

Table 1 (continued)

3. <u>15 sec</u>

C_{1s} Spectrum

Peak No.	Binding Energy	Area %	Area Ratios C_{1s}/F_{1s}	F_{1s}/F_{2s}	C_{1s}/O_{1s}
1	291·5	1·5			
2	290·9	9·5			
3	289·1	18			
4	287·9	22·5	0·76	11·97	4·07
5	286·5	10			
6	285·8	18·5			
7	285·0	21			
F_{1s}	689·2				
O_{1s}	534·6	80			
	532·7	20			

4. <u>30 sec</u>

C_{1s} Spectrum

Peak No.					
1	291·6	3			
2	290·7	14			
3	289·0	18·5			
4	287·9	28·5	0·59	10·83	3·98
5	286·4	10			
6	285·9	17			
7	285·0	9			
F_{1s}	689·2				
O_{1s}	534·6	79			
	532·7	21			

Table 1 (continued)

5. <u>300 sec</u>

C_{1s} Spectrum

Peak No.	Binding Energy	Area %	C_{1s}/F_{1s}	Area Ratios F_{1s}/F_{2s}	C_{1s}/O_{1s}
1	291·9	6			
2	290·8	15·5			
3	289·1	24			
4	288·0	23	0·59	10·61	3·98
5	286·5	16			
6	285·7	9·5			
7	285·0	6			
F_{1s}	689·5				
O_{1s}	534·7	78			
	532·8	22			

TABLE 2

CALCULATED FILM THICKNESS d FOR SURFACE FLUORINATED POLYETHYLENE FILMS

Experiment	$\dfrac{A^{F_{pol}}_{C_{1s}}}{A^{C_{1s\ Pet}}}$	From C_{1s} Spectra	d(Å) From F_{1s} and F_{2s} Spectra	Av.
0.5 sec	0.71	5.5	3.5	4.5
2 sec	1.50	9.0	6.0	7.5
15 sec	3.76	15.0	16.0	15.5
30 sec	10.10	24.0	30.0	27.0
300 sec	15.67	28.0(46)*	36.0	32(41)*

* Revised estimate see text section 3(a)

From measurements on simple homopolymers, it may be shown that

$$\left(\frac{A_\alpha^{F_{pol}}}{A_\alpha^{Pet}}\right) \approx 1,$$ (11) therefore reduces to a particularly simple form (12)

$$\frac{\left(A_{C_{1s}}^{F_{pol}}\right)}{\left(A_{C_{1s}}^{Pet}\right)} = (e^{d/\Lambda_2}-1)$$ (12)

The mean free paths of ~7Å and ~12Å for F_{1s} and F_{2s} levels respectively (corresponding to kinetic energies of ~560 eV and ~1220 eV), given by the previous analysis and the generalized curve of escape depth dependence on kinetic energies suggests a value of ~10A for Λ_2 where the kinetic energy corresponding to photoionization from C_{1s} core levels is ~960 eV. The measured area ratios of C_{1s} peaks are given in Table 1. As a first approximation taking $\Lambda_2 =$ 10Å, the derived values of d (from Equation (12)) are 5.5Å and 24Å for the 0.5 sec and 30 sec runs respectively. These are in good overall agreement with those derived from the F_{1s} and F_{2s} levels. A comparison for all five runs with the two methods of determination are given in Table 2. The close correspondence between these two sets of data suggests that the logic of the overall analyses is correct. The deviation between the two methods of analysis is greatest for large film thicknesses. This is not unexpected, since it is clear from Fig. 9 that the analysis based on the F_{1s} and F_{2s} levels is less sensitive for large d whilst that based on the C_{1s} spectra will be subject to greatest error for a relatively high degree of fluorination since any slight surface contaminant hydrocarbon (B.E. 285.0 eV) arising during sample manipulation after fluorination will tend to lead to d being underestimated.

Stoichiometries of Fluorinated Films. In previous papers, we have shown how in the particular case of homogeneous fluoropolymer films, two independent means are available for establishing overall stoichiometries viz. from F_{1s} to C_{1s} area ratios and from the component peaks of the C_{1s} region[17].

Considering firstly the F_{1s} and C_{1s} levels, the measured area ratios of F_{1s} peaks to C_{1s} peaks (arising from the fluorinated film) are given in Table 1. We may express ratios as Equation (13).

$$\frac{A_{F_{1s}}}{A_{C_{1s}}} = \left(\frac{A_{\alpha F_{1s}}}{A_{\alpha C_{1s}}}\right) \frac{(1-e^{-d/\Lambda_1})}{(1-d^{-d/\Lambda_2})}$$ (13)

In studying homopolymers of fluoroethylenes, an infinity ratio for

$$\frac{A_{\alpha F_{1s}}}{A_{\alpha C_{1s}}} = 2.05 \text{ was obtained}[6].$$

Equation (13), of course, applies to equal number of F_{1s} and C_{1s} core levels and therefore dividing the measured area ratios by those calculated for a unit stoichiometry gives the actual stoichiometry. As a starting point, d, has been taken as the average of those obtained by the two methods indicated earlier. The computed stoichiometries are given in Table 3.

TABLE 3

C/F STOICHIOMETRIES

Expt.	I C/F From C_{1s}/F_{1s} Spectra	II C/F From C_{1s} Spectra	III C/F From C_{1s} Spectra Taking Surface Oxidation into Account	Av. I and III
0.5 sec	2.01	1.60	2.05	2.03
2 sec	1.43	1.40	1.54	1.48
15 sec	1.37	1.27	1.40	1.38
30 sec	1.16	1.12	1.23	1.20
300 sec	1.17	1.04	1.12 (1.16)*	1.16

* see section 3(a)

Deconvolution of the C_{1s} spectra as previously indicated in Fig. 10 allows an independent means of calculating the stoichiometries if assignments can be made of the component peaks. In this objective, our previous work on both homopolymers and copolymers has proved invaluable[6]. Fig. 10 also illustrates the detailed assignment of peaks for the 0.5 sec and 30 sec fluorination experiments. If we ignore for the time being the contribution from the peak at 287.9 eV arising from surface carbonyl features previously discussed, the stoichiometries may be worked out quite straightforwardly as shown in Table 4 for the 0.5 sec and 30 sec runs. The preliminary data for the two methods of calculating the stoichiometries are collected in Table 3. Overall there is seen to be a remarkable measure of agreement between the two methods with the relatively minor discrepancies arising at the two extremes.

TABLE 4

STOICHIOMETRIES FOR FLUORINATED POLYETHYLENE FILMS
FROM THE C_{1s} SPECTRA

0.5 sec. run

Peak	Binding Energy	Structural Feature	Area%	Contribution to overall stoichiometry C	F
1	-	-	-	-	-
2	290.6	$-CF_2-$	3.5	0.035	0.070
3	288.9	$-\overline{CFH}-$	5.5	0.055	0.055
4	287.9	$-\overline{CFH}-$	13.5	0.135	0.135
5	286.5	$-CH_2-$	8	0.08	-
6	285.8	$-CH_2-$	11	0.11	-

C/F 1·60

30 sec. run

Peak	Binding Energy	Structural Feature	Area%	Contribution to overall stoichiometry C	F
1	291.6	$-CF_2-$	3	0.030	0.060
2	290.7	$-CF_2-$	14	0.14	0.28
3	289.0	$-CFH-$	18.5	0.185	0.185
4	287.9	$-CFH-$	28.5	0.285	0.285
5	286.4	$-CH_2-$	10	0.100	-
6	285.9	$-CH_2-$	17	0.170	-

C/F 1

We may now proceed to a more refined analysis. As we have previously indicated the polyethylene films used in these experiments contained surface carbonyl features. Consider now the data in Table 1 pertaining to the 0.5 sec run. From the mode of preparation of the film, it seems highly likely that the oxidation will be localized within the first monolayer, which taking appropriate bond lengths would correspond to the first ~5Å of the sample, very close in fact to the estimated thickness of the fluorinated film for the 0.5 sec fluorination experiment. For unit stoichiometry, the area ratio of the C_{1s} to O_{1s} peaks for the fluorinated film is given by

$$\frac{A_{C_{1s}}}{A_{O_{1s}}} = \left(\frac{A_{\alpha C_{1s}}}{A_{\alpha O_{1s}}}\right)\frac{(1-e^{-d/\Lambda_2})}{(1-e^{-d/\Lambda_4})} \tag{14}$$

From extensive studies of thick homogeneous films of simple organic molecules, a value of $\dfrac{A_{\alpha C_{1s}}}{A_{\alpha O_{1s}}} = 0.5$ has been established[11].

From the generalized curve for escape depth as a function of kinetic energy, a reasonable estimate appropriate to the O_{1s} levels (KE ~ 720 eV) would be $\Lambda_4 = 8\text{Å}$. The sensitivity factor $\dfrac{A_{C_{1s}}}{A_{O_{1s}}}$

for a film of thickness 4.5Å is then found from Equation (14) to be 0.42 for unit stoichiometry. From the overall data pertaining to the C_{1s} and O_{1s} levels, the ratio of

$$C_{1s}^{\text{Fluoropolymer}} : O_{1s} \text{ from C=O } = 26$$

may readily be calculated.

Multiplication by the derived sensitivity factor then gives a stoichiometry of C/O of 11.0 for the surface film viz. a surface coverage of ~9%.

If we consider now the original unfluorinated film, (of thickness large compared with d), the area ratios for unit stoichiometry corresponding to the C_{1s} and O_{1s} levels are given by Equation (15) where d corresponds to the film thickness at the surface within

$$\frac{A_{C_{1s}}}{A_{O_{1s}}} = \frac{A_{\alpha C_{1s}}}{A_{\alpha O_{1s}} (1-e^{-d/\Lambda_4})} \tag{15}$$

which the oxidation is confined. This is readily evaluated giving an area ratio for unit stoichiometry of 1.16. The measured area ratio is 52.5 which gives an overall stoichiometry of 61 for $C:O$. As will become apparent, there is no evidence for oxidation accompanying fluorination so that the stoichiometry of the first monolayer should correspond to that established for the 0.5 sec run viz. 11.0. To give the required stoichiometry, therefore, for the unfluorinated sample, the C_{1s} signal would have to arise from 5-6 monolayers. If the average depth of a monolayer, viz. average separation between adjacent chains (since the surface structure should largely have the chains oriented parallel to the surface), is taken as ~5Å, >94% of the signal intensity derives from the first 27.7Å.

We may now turn to the discrepancy between the stoichiometries for the 0.5 sec fluorination experiment computed by the two methods discussed previously. The overall stoichiometry of C to O for the fluorinated film of 11.0 allows a correction to be computed for the

C to F stoichiometry from the C_{1s} spectra since a percentage of the peak at 287.9 eV assigned to $\underline{C}HF$ type environments arises from >C=O structural features. With this taken into account, the refined $\overline{C}$ to F stoichiometry is 2.05 in excellent agreement with the figure derived from the C_{1s}/F_{1s} area ratios. A similar analysis for the experimental data leads to the revised figures shown in Table 3. The two methods are seen to be in excellent agreement.

Discussion

Depth of Fluorination and Stoichiometry as a Function of Time.
From a plot of the depth of fluorination (as an average of the two methods discussed in 2(a) versus log t, it is evident that the points for the first four experiments in the range 0.5-30 secs fit on a smooth curve whilst that for the longest run of 300 sec would appear to indicate too low a fluorinated film thickness than the initial analysis. The most likely cause of this would be a sub-monolayer contaminant hydrocarbon film arising after the fluorination experiment. In each experiment after exposure to the stream of fluorine diluted in nitrogen for the requisite time, the sample chamber was flushed with a nitrogen stream to remove any traces of fluorine before the samples were removed and stored in a desiccator. This flushing process was lengthened for the longer runs to ensure complete removal of any fluorine. Samples were transferred from a storage desiccator to the sample probe immediately before insertion into the spectrometer and in this respect all samples were treated in the same manner. The only likely differences in the possibility for surface contamination therefore arose in the flushing process. In considering the form of the analyses for sample depths previously presented, it is clear that the experiments corresponding to more extensive fluorinations would be more critically dependent on even trace amounts of sample contamination.

For the 300 sec experiment in particular, in the analysis based on C_{1s} levels, if a few percent of the peak at 285.0 eV (characteristic of not only polyethylene but also hydrocarbon type contamination) arose from contaminant, then,since ratios are involved,the results would alter appreciably if this were not accounted for. By contrast for the shorter experiments, the results will be relatively insensitive to small traces of contamination. We have addressed ourselves to this problem in two ways. Firstly, blank experiments indicated that if fluorinated samples were placed in the reactor and flushed with nitrogen for extended periods the peak at 285.0 eV increased in intensity. (Although of course in absolute terms the intensity increase was small). Secondly, we have developed a self-consistent analysis of the experimental data for the 300 sec run based on the premise that part of the intensity of the peak at 285.0 eV derives from surface hydrocarbon contaminant (arising after the fluorination) rather than solely from bulk unreacted polyethylene. This is not

unreasonable as is clearly evident from Fig. 11 where a plot has
been made of the intensity of the peak at 285.0 eV versus log t.

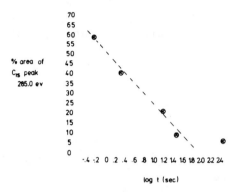

% area of
C_{1s} peak
285.0 ev

log t (sec)

Fig. 11.   Plot of relative area of C_{1s} peak at 285.0 eV vs.
log t for fluorinated polyethylene.

For the first four runs, a reasonable linear correlation is evident
the data point corresponding to the longest run being such that the
peak is much more intense than expected. Consider not the area
ratios corresponding to complete monolayer coverage of the fluorinated film, Equation (16), where we have assumed that d the monolayer

$$\left(\frac{A_{C_{1s}}^{Contaminant}}{A_{C_{1s}}^{F_{pol}}} \right) = \frac{(1-e^{-d/\Lambda_3})}{e^{-d/\Lambda_3}} \tag{16}$$

thickness is small compared with the thickness of the fluoropolymer
film for which our previous estimate of ~32Å may be taken as a lower
limit. Taking a typical hydrocarbon type contaminant monolayer
thickness to be ~5Å, complete monolayer coverage would result in an
area ratio of contaminant C_{1s} signal to fluoropolymer signal of
0.649 (viz. the C_{1s} signal at 285.0 eV would be ~39% of the total
C_{1s} signal). The observed 6% could, therefore, arise from only ~12%
of monolayer coverage emphasizing the extreme sensitivity of the
technique. If we take an error limit of ±1% for the area ratios
then at the lower limit, if 1% of the signal intensity at 285.0 eV
in the 300 sec experiment arises from unreacted polyethylene, the
calculated film thickness would be ~46Å. It may readily be shown
that the remaining 5% signal intensity at 285.0 eV attributable in
this case to ~10% monolayer coverage by contaminant, does not affect
in any significant way the analyses of film thickness based on F_{1s}/F_{2s} area ratios. However, the C to F stoichiometry is now in even
better agreement with that calculated from the C_{1s}/F_{1s} spectra viz.
1.16 vs. 1.12 compared with 1.17. This analysis would, therefore,

suggest that for the 300 sec run the likely film thickness would be
in the range 36-46Å. Fig. 12 shows now a log plot of average film
thickness vs. duration of fluorination.

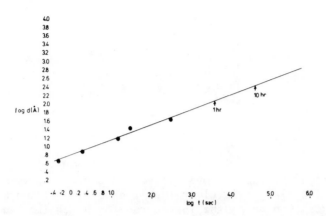

Fig. 12.   Log plot of average film thickness vs. duration
           of fluorination.

     Considering the large number of steps in the analysis the
correlation is surprisingly good and indicates that overall the
analysis must be essentially correct. For such short fluorinations,
the reaction is localized close to the surface and this explains why
both TIR and MATR fail to detect any reaction. Since most previous
experiments reported in the literature have referred to extended
fluorinations where reaction has proceeded into the bulk, it is of
interest to take the correlation in Fig. 12 to estimate the depth
of fluorination for a typical experiment lasting for say 10 hours.
The extrapolation suggests ~270Å into the bulk, whilst for a 1 hour
fluorination experiment the calculated film thickness is ~120Å.
This accords with the fact that fluorination in this case can be
detected by MATR and careful TIR double beam experiments. It should
be emphasized, however, that in general our reaction conditions
(10% fluorine in nitrogen) are considerably more moderate than most
other investigations have employed.

     In setting up the experiments, our initial program involved
fluorinations for 0.5, 2, 30 and 300 secs which were carried out in
duplicate in a sequential manner. Fortuitously, preliminary analysis
of the data derived from these studies suggested carrying out a
further experiment corresponding to a 15 sec fluorination and this
particular experiment was carried out after the fluorine cell had
stood unused over a weekend period. Although relatively efficiently
capped, the design of the fluorine cell, particularly when the nature
of the electrolyte is taken into account, is such that traces of
moisture are taken up and that on initial start up, oxygen as well
as fluorine is liberated at the anode. This explains the otherwise

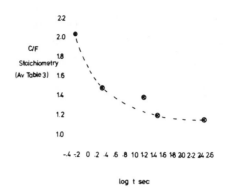

Fig. 13. C/F stoichiometries vs. log t

anamolous carbon to fluorine overall stoichiometries shown in Fig. 13 (plotted vs. log t). It is clear that the points fit onto a smooth curve, the exception being the 15 sec run for which the carbon to fluorine stoichiometry is too high indicating that there has been less fluorination than expected. By contrast, the depth dependence on time indicates that the film thickness for this experiment is normal. This in many senses is a key observation, since it is clearly most readily explicable on the basis that the fluorination is diffusion controlled. The explanation of the data is then that in the 15 sec run diffusion into the bulk had proceeded to the requisite depth but the presence of oxygen (or fluorine oxide) in increased amounts inhibited the chain reaction thus giving a higher than expected carbon to fluorine stoichiometry. This is not unreasonable in terms of the dynamics of the overall process which will be discussed in the next section.

The increasing fluorine content coupled with a consideration of the likely nature of the fluorination process strongly suggests a variation in composition of the fluorinated polyethylene film as a function of depth into the sample. A crude estimate of this may be obtained from the available data. Considering first the results for the 0.5 sec experiment, the derived film thickness suggests that it is reasonable to consider the analysis in terms of reaction of the polyethylene chains in the transcrystalline region at the very surface of the polymer. We will return to this point in the next section; however, it is interesting to note that the overall stoichiometry of C/F ~2 for the experiment corresponds in composition to polyvinylfluoride,and in fact the largest component peaks at 287.9 eV and 285.8 eV are in excellent agreement with those determined for the homopolymer[16]. As a starting point, therefore, we take the stoichiometry of the first 'monolayer' to be C/F = 2.03.

Consider now the normalized data pertaining to the C_{1s} levels of the fluorinated portion of the polyethylene given in Table 5.

TABLE 5

C_{1s} LEVELS FOR FLUORINATED FILMS OF POLYETHYLENE
(NORMALIZED CONTRIBUTIONS)

Sample	1		2		3		4		5		6	
	BE	%	BE	%	BE	%	BE	%	BE	%	BE	%
0.5 sec	-	-	290.6	9.6	288.9	14.9	287.9	25.5	286.5	21.3	285.8	28.7
2 sec	-	-	290.9	12.2	289.0	21.1	287.9	24.4	286.5	14.4	285.3	27.8
30 sec	291.6	3.3	290.7	17.4	289.0	21.7	287.9	25.0	286.4	12.0	285.9	20.7
300 sec	291.9	6.5	290.8	1717	289.1	27.4	288.0	19.4	286.5	18.3	285.9	10.8

(These results are derived from those in Table 1 with due allowance being made for the contribution at 287.9 eV arising from $>C=0$ features previously discussed.)

The progressive increase in proportion of the higher binding energy components is readily apparent although it is clear that for the 300 sec run, for example, this represents the average over surface and subsurface features. For the 2 sec fluorination experiment for which a film thickness of ~7.5Å is suggested by the previous analysis, we may consider what proportion of the signal intensity derives from the topmost 4.5Å corresponding to the depth of fluorination for the 0.5 sec run. This is readily computed to be 68% viz. in a film of thickness 7.5Å, 68% of the signal intensity for the C_{1s} levels derives from the topmost 4.5Å. The activation energy for abstraction of hydrogen (in a polyethylene chain) by fluorine is very small and progressively increases as the degree of fluorination increases. As a reasonable working hypothesis, therefore, we may use the initial stoichiometry established for the 0.5 sec experiment (in terms of a given intensity distribution amongst the component C_{1s} peaks) as the stoichiometry for the 32% signal intensity not arising from the first monolayer for the 2 sec run. This then allows a new stoichiometry to be computed for the first monolayer and this is indicated in the Scheme A. The derived stoichiometry of C/F 1.34 would appear to be entirely reasonable. An extension of this analysis may therefore be carried out on the data pertaining to the 30 sec and 300 sec runs. For both these cases, the topmost 4.5Å is computed to contribute ~39% and ~37% respectively of the total signal intensity and again taking the composition of the remainder of the film as being that appropriate to the initial fluorination, where diffusion to the surface of the polymer (from the gas phase) rather than through the bulk is involved, the derived stoichiometries for the first monolayer are 0.78 and 0.61 respectively. Fig. 14 shows a plot of the derived stoichiometries for the first monolayer as a function of time. The analysis although crude, would seem to be entirely reasonable and

SCHEME A

Derivation of Approximate Composition of First 4.5Å of Polyethylene
Film (Designated as First Monolayer) as a Function of Exposure Time
to Reactant Stream

C_{1s} Levels % Composition

Sample	2	3	4	5	6	
	CF_2	CF	CF	CH	CH	
0.5 sec	9.6	14.9	25.5	21.3	28.7	
2 sec	12.2	21.1	24.4	14.4	27.8	
32% of 0.5 sec experiment	3.1	4.8	8.2	6.8	9.2	
Signal intensity from first 4.5Å of 2 sec sample	8.9	16.3	16.2	7.6	18.6	Derived C/F stoichiometry
						1.34

would indicate that in the 300 sec run the surface composition is
between that for polytrifluoroethylene and PTFE in terms of carbon
to fluorine stoichiometry.

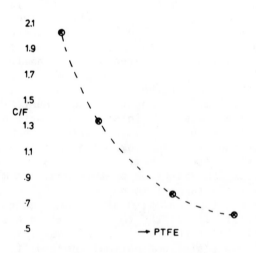

Fig. 14.   C/F Stoichiometries as a function of time for
first monolayer.

   Dynamics of the Fluorination Process. We have previously outlined in Fig. 7 the basic mechanism for the fluorination of polyethylene. The activation energy for the abstraction reaction[2] is
known to be very small (~0.2 Kcal/mole for straight chain hydrocarbons[15]), and this is at least an order of magnitude smaller than
the typical activation energies quoted for diffusion in polyethylene
and simple fluoropolymers in general (cf. ref. 18). It is entirely
reasonable, therefore, that reaction other than at the very surface
of the polymer will be diffusion controlled.

   Although the initial abstraction reaction[2] has a low activation
energy,successive replacement of hydrogen by fluorine will be expected
to increase the activation energy both from the concomitant strengthening of C-H bonds and adverse polar effects in the transition state.
Of some relevance in this connection are the relative rates of
fluorination at different sites reported for gas phase studies of
1-fluorobutane[15]. These are shown in Fig. 15. It is clear that
introduction of fluorine deactivates both α and β sites. We may use
this data as a basis for discussing the fluorination of polyethylene.

   For the initial surface fluorination corresponding to the 0.5
sec experiment, an overall C/F stoichiometry of the surface film of
~2 is observed. Scheme B shows the calculated relative rates of
fluorination for a four carbon fragment of the polymer chain based
on the data for 1-fluorobutane. For a stoichiometry C/F of exactly
2, the possibilities on this simple model would be one CF_2 group or
two CHF groups in the four carbon chain. It is clear that the
preference for the latter should be ~3 x that for the former. The
data in Table 5 provides striking confirmation of this. The two
major peaks of approximately equal intensity of 287.9 eV and 285.8 eV
corresponds extremely well in binding energy with those previously
reported for polyvinyl fluoride (288.0 eV, 285.9 eV). By contrast,
the peak at 290.6 eV corresponding in binding energy to CF_2 groups,
(cf. 290.8 eV in polyvinylidene fluoride) with adjacent $-CH_2-$ groups
represents approximately 0.3 x the intensity of these peaks. It
might also be inferred from Scheme B that progressive fluorination
would tend to proceed relatively uniformly so that for a stoichiometry
C/F of 1.0 a $(CHF-CHF)_n$ backbone would predominate. The relatively
large increases in intensity of the peaks at ~289.0 eV (Table 5)
would tend to support this. Clearly however, a great deal more
experimental work is required before a more detailed discussion is
possible.

   Thus far, the discussion has been in terms of a straightforward
chain reaction in which hydrogen abstraction (2), Fig. 7, is followed
by the chain propagating reaction (3). The overall analysis would
suggest that this is reasonable and represents the major reaction
pathway. It should be emphasized, however, that reaction (3) is
exothermic to the extent of ~68 Kcals/mole. If this were relatively
localized the possibility of molecular elimination of HF cannot be

Relative rates of fluorination F_2/N_2 at 20°C

$$F—CH_2—CH_2—CH_2—CH_3$$

$$0.3 \quad 0.8 \quad 1 \quad 1$$

Fig. 15.   Relative rates of fluorination at different sites
in 1-fluorobutane

SCHEME B

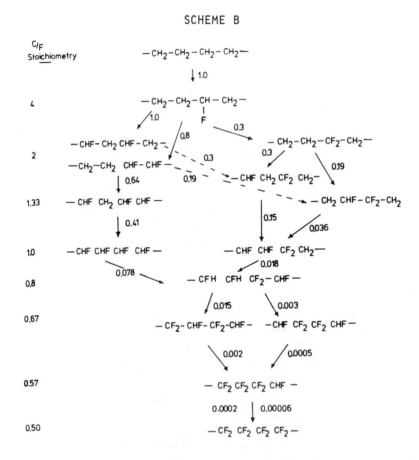

This scheme is illustrative rather than exhaustive.

discounted, although in gas phase fluorinations of alkanes this
possible side reaction is unimportant.   The introduction of a double
bond coupled with the production of free radicals from abstraction
reactions of the type (2) could provide a mechanism for cross linking.
The available data, however, would make this seem unlikely as a major
route for the overall reaction.

CONCLUSIONS

In this investigation, we have attempted to use ESCA to provide information on the initial stages of the fluorination of polyethylene. The results, together with the detailed analysis, illustrates the great potential of ESCA in studying structure and bonding of the surface and immediate subsurface of polymer films which complements our previous studies on the bulk properties of polymers. The preliminary study described here has opened up a whole new range of investigations which are currently in progress and will be reported on in due course.

ACKNOWLEDGEMENTS

Thanks are due to S.R.C. for provision of equipment and a Research Fellowship to one of us (I.R.).

REFERENCES

1. P. S. Quinney and B. J. Tighe, Br. Polym. J., 3, 274 (1971).
2. CF. B. G. Brand, L. J. Nowacki, W. Mirick and E. R. Mueller, J. Paint Technol., 40, 396 (1968).
3. Cf. H. Gregengack and D. Hinze, Phys. Stat. Sol., 8, 513 (1971).
4. C. B. Duke, 'Electron Scattering by Solids', Chapter 1, 'Electron Emission Spectroscopy', Ed. E. Dekeyser, D. Reidel, Boston, U.S.A. (1973).
5. D. T. Clark, 'The Application of ESCA to Studies of Structure and Bonding in Polymers', This Symposium.
6. Cf. Ref. 5 and references therein.
7. (a) R. H. Hansen and H. Schonhorn, Polym. Lett., 4, 203 (1966);
    (b) - J. Polymer Sci., B, 4, 203 (1966);
    (c) - J. Appl. Polymer Sci., 11, 1461 (1967);
    and references therein.
8. A. J. Rudge, British Patent, 710,523 (June 1954).
9. J. L. Margrave and R. J. Lagow, Chem. Eng. News, 48, 40 (1970).
10. Cf. Ref. 9 and (a) K. Tanner, Chimica, 22, 176 (1970);
    (b) H. Schonhorn, P. K. Gallagher, J. P. Luongo and F. J. Padden Jnr., Macromolecules, 3, 800 (1970);
    (c) H. Shinohara, M. Twasaki, S. Tsujimura, K. Watanabe and S. Okazaki, J. Pol. Sci. A1, 10, 2129 (1972);
    and references therein.
11. D. M. J. Lilley, Ph.D. Thesis, University of Durham (1973).
12. B. Wunderlich, 'Macromolecular Physics', Vol. 1, Academic Press, New York, (1973).
13. (a) H. Schonhorn, Polymer Lett., 467 (1964);
    (b) - Macromolecules, 1, 145 (1968).
14. (a) D. T. Clark and K. C. Tripathi, Nature Phy. Sci., 241, 162 (1973);

(b) - Nature, Phy. Sci., 244, 77 (1973).
15. Cf. J. M. Tedder, Quart. Rev. Chem. Soc., XIV, 336 (1960).
16. D. T. Clark, 'Chemical Applications of ESCA', Chapter 6,
     Electron Emission Spectroscopy, Ed. W. Dekeyser, D. Reidel,
     Boston, U.S.A. (1973).
17. Cf. D. T. Clark, W. J. Feast, D. Kilcast and W. K. R. Musgrave,
     J. Pol. Sci. Polymer Chem. Edn., 11, 389 (1973).
18. Cf. (a) R. A. Pasternak, M. V. Christensen and J. Heller,
               Macromol., 3, 366 (1970);
          (b) R. A. Pasternak, G. L. Burns and J. Heller, Macromol.,
               4, 470 (1972);
          (c) R. G. Gerritse, J. Chromatography, 77, 406 (1973);
     and references therein.

DISCUSSION OF PAPER BY D. T. CLARK

R. W. Phillips (Aerospace Corporation): On what basis did you resolve your carbon 1s peak into six or so component peaks?

D. T. Clark: From studies of simple homopolymers, line widths and line shapes may be established for individual core levels. Our line shape analysis then follows a logical procedure starting from the simplest (0.5 sec) to the most complex (300 sec) fluorinated samples making use of calculations on model systems and experimental data from homopolymers to assign binding energies to the possible structural features. The procedure has been discussed elsewhere.*

* D. T. Clark, "Chemical Application of ESCA", pp. 373-507, in Electron Emission Spectroscopy, Ed. W. Dekeyser, D. Reidel Publishing Co., (1973), Dordrecht, Holland.

S. R. Aggarwal (General Tire and Rubber): 1. Have you studied the direct fluorination of polypropylene? Polypropylene will be more susceptible for fluorination. 2. Can you clearly distinguish by ESCA measurements between C-F bond, from adsorption of fluorine?

D. T. Clark: We have not as yet studied the direct fluorination of polypropylene but it does feature in our future plans now that we have sorted out the complexities of the surface fluorination of polyethylene. I agree that abstraction of hydrogen should be some-what more facile than for polyethylene leading perhaps to more selective fluorination - the relative rates of abstraction being in the order

$$
\begin{array}{ccccc}
CH_3 & & CH_3 & & \\
| & & | & & CH_3 \\
-C- & > & -CH_2-C & > & | \\
T & & | & & -CH- \\
H & & H & &
\end{array}
$$

ESCA can readily distinguish between F_{1s} levels appropriate to covalently bond C-E type environments and adsorbed F_2 of HF the former being some 4 eV and the latter ~2 eV higher in binding energy. F^- on the other hand would be ~2 eV to lower binding energy.

Author Index

K

L

Subject Index